2E
ENERGY TECHNOLOGY
SOURCES OF POWER

Anthony E. Schwaller, PhD
St. Cloud State University

Anthony F. Gilberti, PhD
St. Cloud State University

Thomson Learning
TOOLS

Publisher: Brian Taylor
Marketing Manager: Dan Dale
Project Manager: Suzanne F. Knapic
Editor: Christine M. Kunz

Cover Photo © Andrew Child Photography
Cover Design: Sandy Weinstein

Library of Congress Cataloging-in-Publication Data

Schwaller, Anthony E.
 Energy technology: sources of power / Anthony E. Schwaller, Anthony F. Gilberti.—2nd ed.
      p.   cm.
 Includes index.
 ISBN 0-538-64469-9
 1. Power resources. 2. Power (Mechanics) I. Gilberti, Anthony F., 1956– . II. Title.
 TJ163.2.S347   1996   94-48282
 333.79—dc20   CIP

Contents

Preface

INTRODUCTION

In the past twenty years, increased emphasis has been placed on how energy is used within society. Energy use and consumption is part of the political, environmental, economic, technological, and social aspects of our society. Today's energy awareness is a result of the energy crisis of the early 1970s as well as the events that led the United Nations into the Gulf War of the early 1990s. Energy awareness has resulted from increased environmental concerns, continued depletion of our natural resources, and increased energy consumption.

SCOPE OF THE TEXTBOOK

Because of increased energy awareness, a number of scientists believe that solutions to many of the energy problems lie in the field of education. The overall purpose of *Energy Technology: Sources of Power* is to provide the reader with a comprehensive survey of energy and its technology—including energy supply and demand; depletable, nondepletable, and renewable energy resources; and, energy utilization. These topics and their advances in the future cannot be discussed without integrating them into the basic study of society. Throughout this textbook, all topics are consistently interrelated with economic constraints, political issues, environmental concerns, technological advances, and sociocultural values of the United States. In addition, many comparisons are made to other societies and how they use energy.

SECTIONS AND CHAPTER LAYOUT

The textbook is divided into four major sections with a total of ten chapters. Section I, **Introduction to Energy Technology**, includes three chapters. Chapter 1 deals with the energy supplies and sources available. Chapter 2 deals with the energy demand. Chapter 3 is concerned specifically with many of the terms used to study energy and power technology.

Section II, entitled **Nonrenewable Energy Resources**, includes four chapters. Chapters 4 through 7 deal with the fossil fuel resources of coal, petroleum, and natural gas, and with nuclear energy. All four of these resources are considered nonrenewable or depletable resources.

Section III, entitled **Nondepletable and Renewable Resources**, includes two chapters. Chapter 8 addresses the renewable energy resource of solar energy. Chapter 9 deals specifically with many of the additional renewable resources including wind, bioconversion, ocean thermal energy conversion, hydroelectric energy, tidal power, and geothermal energy.

Section IV, entitled **Energy Utilization**, addresses a variety of technologies needed to convert energy resources into usable power. Chapter 10 deals with energy conversion, energy storage, and energy conservation technology.

COURSE/PROGRAM DESCRIPTION

This textbook is designed to be used primarily for technology education classes, both at the secondary and postsecondary levels, in which energy and power technology is considered part of the overall curriculum. In addition, many science programs may benefit from the use of this textbook.

The textbook is applicable to both secondary upper level energy technology programs as well as courses provided during the first two years of college curriculum in technology education. Many power technology instructors may also find the information concerning supply and demand, solar energy, alternative resources, and energy utilization very stimulating.

SPECIAL FEATURES

There are many special features of this textbook.

Math Interface—Throughout each chapter there are various mathematical problems that are presented. These problems reinforce many of the energy concepts and much of the terminology studied in each chapter.

Energy Issues—Throughout most chapters there are various energy issues presented. These issues serve as excellent group discussion activities so one can see issues from political, environmental, social, economic, and technical points of view.

Point/Counterpoint—At the end of each chapter a major issue is presented. The issue is often debatable because of the various viewpoints presented.

Appendices—Extensive appendices have been included at the end of the textbook. These appendices contain various formulas, energy data, conversions, and additional technical information to help better understand concepts of energy technology. In addition, many of the appendices are used to help you easily complete the mathematical problems.

Alternative Futures—The study of energy is an ever-changing body of knowledge. Therefore, a description of various future energy technologies has been included at the end of each chapter.

Summary/Review Statements—To aid you in studying energy technology, each chapter includes various summary and review statements. These statements represent a review of the major concepts and principles within each chapter.

Discussion Questions—At the end of each chapter a series of discussion questions has been included. These discussion questions help you to answer many of the value-related issues dealing with energy technology.

Objectives—Each chapter begins with a set of outcome-based statements to aid you in identifying the content of the chapter.

Terms to Know—A list of energy terms has been included at the beginning of each chapter. Each term listed is highlighted in the text, and many can be found in the glossary at the end of the textbook.

Glossary—A glossary has been provided at the end of the textbook. The glossary includes definitions from each of the terms listed in the Terms to Know section of each chapter.

I

Introduction to

Energy Technology

Section 1 deals with energy supplies, resources, consumption, and energy terminology. Specifically, energy supply and demand are discussed in relation to various sectors or consuming parts of society. These include the residential/commercial, transportation, industrial, energy, and petrochemical users or sectors. Section 1 addresses three major questions in the study of energy technology:

1. What are the usable energy resources and supplies available to humankind?
2. Where and how is energy being consumed in society?
3. What basic terminology is currently being used to help explain the study of energy technology?

Chapter 1 introduces energy technology and the various resources and supplies now being consumed. Energy problems within society and basic laws of energy are also discussed. In addition, recent patterns of energy demand are introduced.

Chapter 2 introduces the concepts of energy demand and consumption. The manner and rate of energy consumption, and the amount of energy being consumed within the sectors are also discussed.

Chapter 3 introduces basic energy terminology. Terms studied in this chapter reflect all facets of the study of energy systems. These terms serve as a reference for the remaining chapters in the textbook.

chapter 1

Energy Supplies and Sources

INTRODUCTION

A Glance Backward

The earth is about 5 billion years old, and life on earth is estimated to have begun about 3 billion years ago. Historical records date back only about 6,000 years. In terms of human understanding, it has only been four centuries since we came to know that the earth is not flat, about three centuries that we have known of electricity, a little over one century since the introduction of the "horseless carriage," and about a half of a century since we began to understand nuclear fission. In the period of human history, then, it is only in the most recent instance that humans have ascertained how to change the course of life on the earth in acute ways. This is especially true regarding the importance of energy use.

The Importance of Energy

Humans' social and cultural development, their distinction and detachment from other animals, have been made possible by the exploitation of energy sources. This is not to suggest that energy use has made humanity what it is. Rather, it is the fact that, without energy use, humans would not be at the state of development they are today. Four abilities further distinguish humans from other species: the ability to reason abstractly, the ability to communicate through the spoken and written word, the ability to create and use complex tools and machines, and the ability to make and control fire. With the ability to convert an energy source, humans were able to gain more control of their living conditions. Fire allowed them to live in cooler habitats. It expanded their food base and provided some protection against other animals.

Human use of energy dates from the discovery and creation of fire, some 8,000 years ago. This command of energy became basic for survival in many parts of the world when the last ice age forced an adjustment to colder climates and to foods that required cooking. The development of settlements into cities increased the use of fuel reserves, and the first "energy crisis" quickened the collapse of early civilizations as fuel supplies diminished. The valley areas of the Tigris and Euphrates Rivers, Greece, North Africa, and Italy were all in turn gradually deforested by 350 B.C. Often, large land tracts within these areas became permanently barren and unproductive for agricultural and human development.

While it is true that humankind will always experience resource shortages, usually it is possible to find acceptable substitutes. With energy, however, humans have found no substitute—there is no method of creating it. To say that "technology will find an answer to this dilemma" merely reduces fears over the impending shortage and overshadows the real issue of using the energy available in a safe, efficient, and prudent manner. Society requires energy for the production of goods and services. Individual consumers depend upon it to heat homes and cook food. People use it to transport themselves and products to various destinations. The list is endless, but the truth remains that energy in its various forms is essential for human life. Yet today, there are greater discrepancies than at any time in history among societies in their access to energy, in their material standard of living, and in their potential for survival. This is perhaps one of the greatest paradoxes of contemporary society. Despite the ability to find energy, to predict its potential, and use it, society could not forecast accurately its social consequences. Energy is understood scientifically. It is controlled technically, but society has not mastered it as a major social factor in contributing to a more just and humane society.

PURPOSES

Goals of Chapter One

Energy can be studied from several viewpoints. This textbook reviews energy from a technical point of view. In addition, economic, political, environmental, and social aspects of energy are examined. The general goals of chapter 1 are to study energy supplies, energy forms, basic energy laws, known resources, growth patterns, and the importance of developing an energy-conscious society. At the completion of this chapter, you will be able to

1. Compare and contrast the various forms of energy.
2. Explore how two basic laws of physics relate to the study of energy technology.
3. Classify the five resources currently being used to supply energy.
4. Describe the amount that each energy resource contributes to the aggregate energy supply.
5. Analyze the growth patterns of the five basic resources used to supply energy.
6. Identify the historic events that led the United States into an energy crisis, and thus to greater levels of energy consciousness.
7. Evaluate the status of energy supplies in the United States and in the world.
8. Identify a model used to make sound energy decisions.

Terms to Know

By studying this chapter, the following terms can be identified, defined, and used:

Btu
Energy
Energy Conversion
Energy Utilization
Entropy
GDP and GNP
Growth Patterns
Growth Rates
Hydrocarbons
Power
Thermodynamics

INTRODUCTION TO ENERGY

Energy Use

Throughout history and particularly over the past 15 to 20 years, much attention has been focused on the study of energy and its technology within our society. Energy use within the United States, as well as other countries, has risen. Greater consumption means there must be an increased energy supply. Much of this increase in demand is based upon an enlarging need for various goods and services within a society. In the past, the *Gross National Product (GNP)* has been used as a means of measuring the affluence of a nation. The GNP was a measure of a nation's market value, determined from national income and product accounts, of all goods and services produced in a year. Thus, the GNP measured all the goods and services produced by workers and capital supplied by United States residents and corporations. More recently, however, the federal government is using the *Gross Domestic Product (GDP)*. The GDP measures all the goods and services produced by workers and capital located in the United States. Either measure provides an indication of goods and services produced.

The GDP per capita is also considered to be a measure of the quality of life within a country. Thus, if the GDP per capita in the United States were twice that of another country, supposedly citizens would have twice as many material things; hence, twice as much "happiness" as the inhabitants of that other country. It is not astounding that the United States has adopted the GDP as a measure of the quality of life since its economy is driven by the concept of infinite resources and unlimited growth. Both the GDP and GNP have been criticized as a measure of quality for many reasons. As a measure, all production is treated equally; $2 million of armaments is considered the same as $2 million worth of food. Environmental demise, loss of species, or resource depletion are not considered when determining the GDP. Other items are also not included, such as the value or care spent on family or creating a more just and humane society. Finally, there are questions concerning advertising, leisure and entertainment products, military expenses, and transportation. These areas and their associated goods and services are not considered a measure of happiness or productivity, and there is a growing consensus that they should be reduced since they contribute to a reduction of natural resources.

Figure 1-1 illustrates changes in energy consumption over 40 years in relation to the Gross Domestic Product (GDP). The horizontal axis shows years. The vertical axis shows energy consumption in quadrillion Btu's and GDP in 1987 dollars.

The GDP reflects consumer spending and has a relationship to energy consumption. Increases or decreases in the GNP/GDP generally lead to similar changes in consumer energy usage. This relationship can be seen in Figure 1-1. Essentially, as the Gross Domestic Product increases, the energy use (United States Energy Consumption) also increases.

GDP and Energy Consumption— Energy Issue 1-1

If the GDP declined over several years, what effect would it likely have on the amount of energy consumption within society? How would it affect living patterns and the production of goods and services? What can you do or what have you done as an individual to help reduce the amount of energy consumption in your daily life?

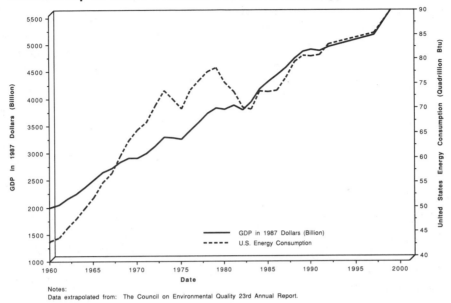

Relationship of Gross Domestic Product and Energy Use

Notes:
Data extrapolated from: The Council on Environmental Quality 23rd Annual Report.

FIGURE 1-1 There is a nearly direct relationship between the Gross Domestic Product and the amount of energy consumed within a society. As the GDP increases, energy use also increases.

Increased population in the past has also contributed to more productivity. Although many technologies are more efficient today, consumer spending has continued to increase. More consumer goods continue to be purchased each year, thus causing the energy demand to increase. Without considerable availability of energy, the United States could not produce such a large proportion of the world's goods and services. Such energy-intensive production contributes to an increased GDP. Clearly, the energy use within the United States is a concern to citizens and government. However, the question remains: *"How can the United States reduce its energy consumption while maintaining an appropriate standard of living for its citizens?"* Issues surely to be reflected upon eventually may focus on reducing the GDP; reducing population growth; developing more efficient energy systems; creat-

ing user taxes on energy; and redefining of concepts related to quality of life, progress, and an appropriate standard of living.

Each year the GDP varies with the percentage of increase of energy available. During the oil embargo of 1973, our society was increasing energy consumption at approximately 3 percent each year. By reviewing past energy usage, it can be seen that since the oil embargo of 1973, energy consumption has increased between 2 and 4 percent per year. The **growth rate** of energy consumption (or anything else that grows) can be determined for any time period by the formula shown below:

$$\text{Growth Rate in \%} = \frac{\text{(New Value minus Original Value)}}{\text{Original Value}} \times 100$$

For example, if a home energy bill were $120 one month and the next month the bill increased to $128, the growth rate or percentage of increase is calculated as follows:

$$\text{Growth Rate in \%} = \frac{(128 - 120)}{120} \times 100$$

$$= \frac{8}{120} \times 100 = .066 \times 100 = 6.6\%$$

Figuring Growth Rates—Math Interface 1-1

Growth rates of energy can help consumers determine how much energy they are using each year. Knowing growth rates can provide consumers with a good reason to save more energy. Solve the following growth-rate problems:

1. If the United States used 50 MB/DOE in one year and the next year consumed 55 MB/DOE, what is the percentage of increase in energy use during the year? Growth rate for this year equals

   _____.

2. If a consumer's electrical demand increased from 11.1 kilowatt-hours one month to 14.6 kilowatt-hours the next month, what percentage increase of electrical growth did the consumer have? Percentage of increase equals

   _____.

3. Based on the answer obtained in question 2, what would be the increase if this rate of growth continued for an entire year (assume that this rate of growth is being compounded each month)? Annual percentage increase would equal _____.

Electrical Demand

The generation of electricity and electrical demand in the United States is also increasing. Electric utilities are both major producers and consumers of energy. By 2010, approximately one-fifth of the energy consumed will be in the form of electricity. Electric utilities, already one of the larger energy-consuming sectors in the nation, will use approximately 40 percent of all the primary energy consumed in 2010 [End-use energy consumption, also referred to as primary end-use energy consumption, is the sum of fossil fuel consumption by the four end-use sectors (i.e., residential, commercial, industrial, and transportation) and generation of hydroelectric power. These sectors are presented in detail in chapter 2]. Electrical energy demand has roughly doubled in the last half century.

The generation of electricity and electrical demand in the United States is also increasing.

Besides the weather, variations in economic growth account for the largest influence on the demand for electricity in the United States (economic factors, world energy prices, and government stability cause variations in the amount of electricity being used). Since electricity is used to produce and consume most goods and services, the demand for electricity increases during periods of economic prosperity. During the 1970s, the GNP (now GDP)grew at an average annual rate of 2.8 percent while sales of electricity increased by an average of 4.2 percent. Over the next twenty years, estimates show that the trend for electrical growth will have approximately a one-to-one relationship with the GDP. The GDP is expected to increase through 2010 an average of between 1.8 to 2.7 percent per year, while sales of electricity will also grow by 1.8 to 2.7 percent per year. Electrical demand will be further discussed in chapters 2 and 10.

U.S. consumers use a great deal of energy to maintain their present standard of living. For example, each person uses the equivalent of about 7 gallons of oil per day to maintain his or her present standard of living. By contrast, in Africa each person uses only about 0.7 gallons of oil per day. Some estimates indicate that, at current United States consumption rates, all present petroleum and natural gas reserves will be expended by the end of the twenty-first century. These estimates do not include reserves that may be discovered as price increases encourage exploration. In any case, rapid and increased consumption will cause more rapid fossil fuel depletion.

> ***Energy Per Person—Energy Issue 1-2***
> *Provide examples of how each person in the United States can consume 7 gallons of oil (equivalent) each day. In other words, give examples of how a person uses energy to make up a total of several gallons. Compare and contrast this use to that of an individual in a developing country. Provide a listing of what this energy is being used for in both countries.*

Energy Worldwide

The United States far surpasses other countries of the world in energy usage. The United States is the world's largest energy consumer and the second largest energy producer. Should the decline of economic activity in the former Soviet Union continue, the United States may also become the leading producer of energy. Figure 1-2 presents graphically a comparison of population and energy consumption for three selected countries. Note that the United States uses considerably more energy per capita than industrialized Japan or the developing country India.

The United States, with about 5 percent of the world's population, consumes between 25 and 28 percent of the world's total energy. Figure 1-3 shows a comparison of energy consumption of several geographical areas in the world. Note that North America consumes more energy than any other region.

Energy Efficiency and Conservation

Does the achievement of economic and social gains always mean increased energy use? Using past energy statistics, one could conclude that economic and social achievements were a result of increased energy availability and usage. However, it is not sufficient to accept the

Proportion of Energy Consumption and World Population

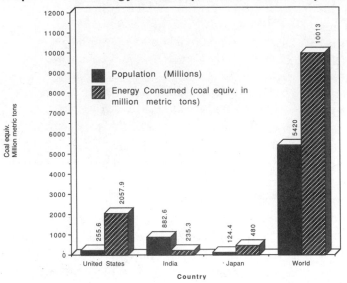

FIGURE 1-2 The United States contains about 5 percent of the world's population but consumes between 25 and 28 percent of the world's total energy.

Notes:
Energy consumed is based on the consumption of coal; lignite; petroleum products; natural gas; and hydro-, nuclear, and geothermal electricity. Data extrapolated from the *Statistical Abstract of the United States, 1991.*

Approximate World Energy Consumption

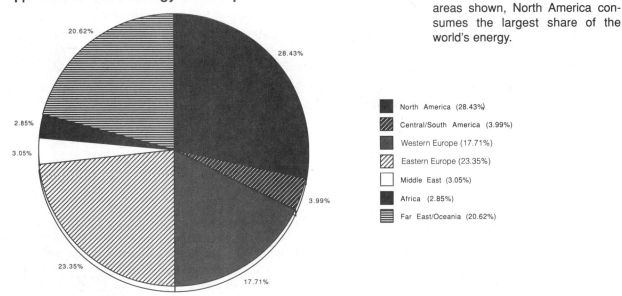

FIGURE 1-3 Of the geographical areas shown, North America consumes the largest share of the world's energy.

Notes:
Eastern Europe includes the former Soviet Union.
The United States consumes about 87% of the total energy for North America.
Data extrapolated from *United States Department of Energy, 1990 Energy Facts.*

increase in energy usage as a socioeconomic necessity without looking at a related question: *Does the United States utilize energy efficiently?*

Efficient use of energy has become important in the past few years. Efficiency is related to the amount of energy used and the amount wasted. The goal of many manufacturers has been to conserve energy in their manufacturing processes and to produce more energy-efficient products. Energy efficiency, conservation, and the use of renewable energy resources must play an important role today and in future uses of energy. Energy efficiency will be addressed more closely in chapter 3. Energy conservation is further detailed in chapter 10.

Factors that Control Energy Use

Several factors affect energy consumption in society:

Population growth. Although population growth has generally slowed in the United States, total world population is still increasing. The greater the population, the more energy required by the society. This is true of most developing countries where population is increasing faster than in the United States.

Laborsaving and convenience equipment. People are buying more products that contribute to domestic convenience. More products purchased means energy usage in total will increase—from the energy required to manufacture the products, to the energy expended in transportating the product, to the energy consumed from the use of the product.

The ability of electricity to reach a wider segment of the population. Greater electrification increases the number of appliances in use. This further increases the overall demand for electrical energy.

Inefficient use of energy. Energy continues to be wasted through operational inefficiency. Many homes, for example, still lack sufficient insulation and, therefore, have unnecessary heat losses. People still practice inefficient driving habits and poor car maintenance. These practices, of course, increase fuel consumption. In addition, many products still need technical updating to improve energy efficiency.

The lack of knowledge about conservation of energy. Even though energy conservation measures have been encouraged, many people still waste energy needlessly. The United States wastes an estimated 45 to 50 percent of all the energy it consumes.

Growth in business and industry. Growth in business and industry tends to increase energy usage. This is caused by more products being manufactured, effective communication (i.e., advertising), and transportation of products to diverse markets. All of which, again, calls for more energy.

Obviously, energy technology needs to be carefully studied because of these factors. As a society we must use energy more effectively.

ENERGY DEFINED

Energy and Power

Energy is defined as the ability to do work. Coal, oil, and natural gas are forms of energy called fossil fuels or hydrocarbons. Energy is often confused with power. The difference between the two terms is as follows:

Energy - the ability to do work.

Power - the measure of work done over time or rate of work.

For example, consider the energy in gasoline. Gasoline provides the energy needed to operate an engine. The engine is able to yield or produce output power usually measured as horsepower, which will be further discussed in chapter 3.

Forms of Energy

The majority of the energy used by humans comes from the sun, either directly or indirectly. The atmosphere is warmed by solar energy and provides a temperature range in which organisms can survive. Photosynthetic conversion of solar energy provides humans with plant and animal food. Humans benefit from solar radiation in the form of winds and rain. In addition, the energy sources of coal, natural gas, and petroleum represent stored solar energy. To help examine how energy is used within a society, six different forms have been categorized—radiant, chemical, mechanical (gravity, motion, sound), thermal, nuclear, and electrical energy. For purposes of clarity, the forms are further defined.

Photosynthetic conversion of solar energy provides humans with plant and animal food.

Radiant Energy

Radiant energy is that energy that comes from a source in the form of electric or magnetic fields. The source may be natural, as in the sun, or artificial, such as the light from a camera flash. Radiant energy, sometimes called light energy, has wavelike characteristics—similar to a water wave cresting along the shore of a beach. The wavelength is the distance in space the wave must move before the wave repeats itself. An example of the wavelength of radiation can be seen in Figure 1-4.

Wavelength of Radiation

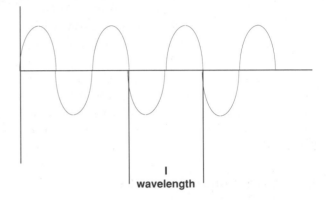

FIGURE 1-4 Illustrated is an example of a radiation wave. The wavelength is the distance in space the wave must move before the wave repeats itself.

Radiant energy can be found in numerous forms, such as gamma rays or X rays. Each form of radiant energy produces a band (or spectrum) of wavelengths that vary. For example, ultraviolet radiation has a wavelength of about 10^{-5} cm; whereas, infrared waves can be 100 times longer. Figure 1-5 illustrates the relationship of various forms of radiant energy to visible light. Because many waves are short, they are normally measured in millimicrons (mμ) (1 millimicron is equal to 1 billionth of a meter, or 25 millionths of an inch) or in Angstrom units (Å or A. U.) (1 Angstrom is equal to one-tenth of a millimicron).

Radiant energy also has properties that are particlelike. In certain forms, radiant energy may arrive at a destination in discrete blocks called photons. Photons can be described as flowing in groups or wavelengths. A photon is

Wavelengths of Various Forms of Radiant Energy

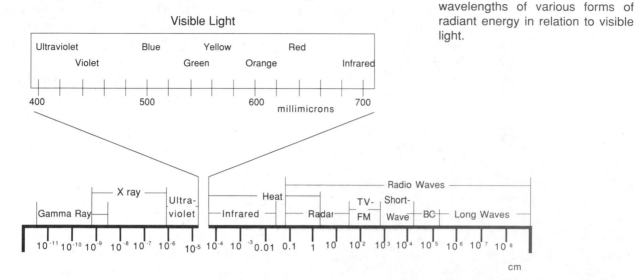

FIGURE 1-5 Illustrated are the wavelengths of various forms of radiant energy in relation to visible light.

related to the size of an object that can absorb it. For example, the photons of ultraviolet and infrared radiation can be absorbed by atoms and molecules. However, it would take an antenna to absorb the photons found in the radio waves of radar or shortwave, since their wavelengths may be hundreds of meters in length.

Radiant energy can also be classified in terms of its frequency. The frequency of a wave is a measure of its speed of vibrations over a certain time period. For example, radiant energy can also be measured in cycles per second. This is a determining characteristic of its energy value. High-frequency radiant energy has large amounts of energy. Gamma and X-ray forms of radiant energy are examples of high-frequency radiant energy; whereas, radio waves are examples of low-frequency radiant energy.

Radiant energy is generally given off by anything that has heat in it. In theory, any object that is above *absolute zero* gives off radiation. (The coldest temperature attainable registers −273° C or −459° F. This temperature is

referred to as absolute zero.) Light bulbs, fire, and other glowing objects are examples of sources of radiant energy. The hotter the object, the higher the frequency. The colder the object, the lower the frequency. Microwaves, visible light waves, infrared waves, ultraviolet waves, gamma waves, and X rays are common forms of radiant energy having distinct frequencies. The radiant energy from

Radio waves are examples of low-frequency radiant energy. (Photo by Richard Younker)

the sun has many frequencies and can easily be collected to produce thermal or electrical energy. In Figure 1-6 radiant energy from the sun is being collected by solar collectors.

These collectors are called solar photovoltaic cells. They are designed to convert the radiant energy from the sun directly into electricity.

FIGURE 1-6 These solar collectors at the Kirk Patrick Center in Oklahoma City, Oklahoma, use radiant energy for input. (Courtesy of the American Petroleum Institute)

Note that radiant energy can interact with matter in four distinct ways: 1) It can bounce off or be reflected; 2) it can be absorbed by matter, thus setting the molecules or atoms of the matter into vibratory motion (this may cause the matter to reradiate its absorbed energy); 3) after being absorbed it can be dissipated into the substance or into space as heat; or, 4) it can be absorbed into the matter to produce a chemical change. One good example of the last would be the process of photosynthesis, which occurs when radiant energy from the sun strikes a living plant. The radiant energy striking the plant is converted to chemical energy inside of the plant. Figure 1-7 provides a graphic illustration of the first three ways listed that energy and matter can interact.

Chemical Energy

Chemical energy is produced from the rearrangement of atoms in a molecule or by combining free atoms. As previously indicated, one example of chemical energy comes from

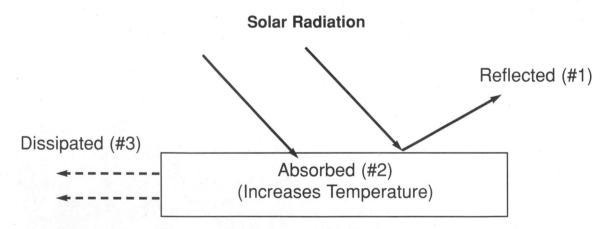

FIGURE 1-7 Radiant energy can interact with matter in four distinct ways. The energy can be: 1) reflected from the object, 2) absorbed as heat, 3) reradiated, or 4) absorbed to produce a chemical change. The fourth example is not illustrated.

the sun's radiation through photosynthesis. Photosynthesis is the formation of carbohydrates in the chlorophyll-containing tissue of plants that have been exposed to the sun's radiation. In this process the energy of sunlight is used by plants to separate molecules of carbon dioxide and water and to rearrange their atoms into carbohydrates (starches and sugars). Oxygen is a by-product of photosynthesis. This process is responsible for the energy in foods that we eat, which are either of plant origin or derived from plant-eating animals.

Another example of chemical energy is that found in fossil fuels. Chemical energy is in coal, oil, and natural gas. Fossil fuel energy is derived from dead or decaying organic matter. An example of a fossil fuel that is derived from decaying organic matter is peat. As a form of chemical energy, peat consists of partially decomposed plants, vegetable matter, and inorganic minerals.

Energy found in batteries and fuel cells can also be categorized as chemical. This stored energy is the product of various chemical reactions, which eventually yield electrical energy.

Today, chemical energy in the form of fossil fuels is our most important commercial energy source. Virtually every society is dependent upon this form of chemical energy for the production of goods and services. Yet, this energy source is very small when compared to nuclear energy stored on the earth or radiant energy available from the sun. As fossil fuels continue to be exhausted and become more costly, humans will have to utilize other energy forms to meet their needs and wants.

Mechanical Energy

Mechanical energy is found in objects that are in motion. Objects in motion have a tendency to remain in motion, while objects at rest have a tendency to remain at rest. This notion is based upon *Newton's First Law of Motion* which states: A body at rest remains at rest and a body in motion remains in uniform motion in a straight line, unless acted on by an external force. For example, a child's ball that is rolled on the ground will tend to roll until stopped by an object or by friction. The same ball would roll much farther on smooth ice before stopping. This is because the horizontal interaction, called friction or frictional resistance, between the ball and the ground is greater than the friction between the ball and the ice.

A body at rest remains at rest and a body in motion remains in uniform motion in a straight line, unless acted on by an external force. For example, a child's ball that is rolled on the ground will be slowed down by friction.

There are two types of mechanical energy, kinetic and potential. *Potential energy* is often defined as a subset of mechanical energy. Potential energy exists because of the relative position of one object to another. Some examples of potential energy are springs that are compressed and air that has been compressed within a confined space. Each of these examples have the potential for doing work. The force, generated by an object being pulled downward by gravity, can also be a form of mechanical energy. For example, the water being held back by a dam is considered mechanical energy. The water is potential energy before moving and kinetic energy when moving or falling. Potential energy can also be found in chemical and other forms of energy as well. For example, there is potential energy in a gallon of gasoline. *Kinetic energy* can be defined as the energy possessed by a body by virtue of its motion. The water falling from a waterfall or dam possesses kinetic energy since it is a body in motion.

The energy derived from pressure waves is known as sound energy, another form of mechanical energy. Vibrating objects, such as a tuning fork or human vocal cords, produce sound energy that travels through the air. When one object hits another, sound energy is produced. Other examples of common

The energy derived from pressure waves is known as sound energy, another form of mechanical energy. Vibrating objects, such as human vocal cords, produce sound energy that travels through the air.

mechanical energy sources include that of wind, fluids, inertia, the result of magnetic forces, gravity, and waves.

Thermal Energy

The total amount of thermal energy within a given object or substance can be accounted for by the sum of the molecular kinetic and potential energies of all its molecules. According to the molecular theory of matter, all substances are constituted of millions of submicroscopic particles called molecules. The term *molecule* describes the smallest possible particle of a given substance that still retains the properties of the substance. Molecules are made up of smaller particles called atoms. Atoms themselves are composed of yet smaller elemental particles that include—but are not limited to—protons, electrons, and neutrons. Therefore, thermal energy would include the internal energies that result from rotation or vibration of atoms and the potential energy of combining forces within atoms.

Thermal energy can be further defined with the following example. When two objects of different thermal energy are placed in contact, the energy (kinetic and potential) contained within them is transferred from one to the other. For example, suppose a bucketful of hot rocks is dropped into a container of cool water. The thermal energy will be transferred from the rocks to the water until the system reaches a stable condition, called *thermal equilibrium.* In thermal equilibrium, the rocks and the water would produce the same sensation of hot or cold when touched. Thus, there is no more transfer of thermal energy once equilibrium is reached.

A common example of thermal energy is the burning of fuel. In Figure 1-8 natural gas (chemical energy) is being burned to produce

FIGURE 1-8 When natural gas (chemical energy) is burned, it produces thermal energy. (Courtesy of the American Petroleum Institute)

thermal energy. In the burning process, a portion of the potential energies found in the gas are converted through burning to thermal energy. Fuels such as coal, oil, and natural gas have chemically stored the sun's radiation through the photosynthesis of the plant forms from which they are derived. Technically, there is thermal energy in any object whose temperature exceeds –273° C (–459° F). Thermal energy and its relationship to heat will be explored further in chapter 3.

Nuclear Energy

Nuclear energy is usually derived from the fission (or splitting) of certain heavy, unstable atoms. Nuclear energy can also be produced by the fusion (or joining) of hydrogen or certain other atoms. When atoms are fused or fissioned, the energy is released mainly as thermal energy. Sometimes the energy within the atoms is considered chemical energy. However, when nuclei of atoms have been split or combined to produce thermal energy, it is then categorized as nuclear energy. Figure 1-9 illustrates a view of a nuclear power plant. In this power plant, the chemical energy of uranium is changed into thermal energy and stored in

water. The thermal energy is then converted to mechanical energy by the creation of steam. The mechanical energy (steam) then turns an electrical generator to produce electrical energy for consumer use.

Electrical Energy

Electrical energy is the movement of electrons from molecule to molecule within conductors. Electrical energy may power numerous machines—such as home appliances, industrial motors, golf carts, and so on. Electrical energy is typically stored in chemical form in batteries. Most electrical energy today is produced by large coal, oil, or nuclear power plants.

ENERGY CONVERSION

Energy Utilization

One important property of energy is its ability to change from one of the six forms into any of the others. For example, the sun's energy is radiated to the plant and animal life on the earth. This energy then is converted into chemical energy, in the form of hydrocarbons. The chemical energy in hydrocarbon fuels (oil, natural gas, and coal) is converted to thermal energy when burned. Thermal energy can be transformed into electrical energy in an electrical generating plant with a dynamo or generator. The electrical energy may then be converted to radiant energy (such as light bulbs) or mechanical energy (such as motors) by home appliances. Thus, *energy utilization* is technology that converts energy from one form to another to produce work. For example, the sun's energy can be changed into a variety of forms depending on the energy utilization equipment applied and the end use intended. Therefore, it is possible

FIGURE 1-9 Nuclear energy is considered one of the six basic forms of energy. This nuclear power plant changes uranium into thermal and, finally, into electrical energy. (Courtesy of Becthel Group Incorporated)

to use photovoltaics to produce electricity, or a solar flat-plate collector to produce thermal energy for space heating.

Law of Energy Conservation

To help understand energy, *thermodynamics* (heat and motion) is used. Two important *thermodynamic laws* help to explain energy and energy utilization. The first law is called the *Law of Energy Conservation.* This law states: Energy cannot be created or destroyed; it can only be transformed from one form to another. The total energy is constant within a closed system. This is also called the *First Law of Thermodynamics.* The law states that there will always be energy within a particular system, if there is energy in the system. For example, the energy within a gallon of gasoline (chemical energy) is transformed, by combustion, into thermal energy. The thermal energy is converted into

mechanical energy by a piston-type engine. The remainder of the thermal energy is unusable because of inefficiencies of the engine. The total energy is still constant within the system, in this example. However, some energy (i.e., exhaust heat) is in such a state that it is no longer usable by the existing technology.

The Law of Energy Conservation might suggest to some that a large-scale energy shortage is not a possibility since energy is never used up. However, this is not the case. Although some of the energy is not being used, it is downgraded to a point that it can no longer be converted into a useful form by existing technology.

Second Law of Thermodynamics

Energy and energy technology can be better understood by analyzing another law of physics, called the *Second Law of Thermodynamics.* This law states: A natural process always takes place in such a direction as to cause an increase in the entropy (randomness) of the universe.

Entropy

Entropy is a measure of the unavailable energy in a closed system and is considered a measure of the degree of disorder of a particular energy form. According to the laws of thermodynamics, energy must always flow in such a direction as to increase the entropy. In simpler terms, entropy means that naturally things tend to become more random. This process applies to *energy conversion*—converting energy from one form to another. The direction of the process is always from a lower to a higher entropy or greater randomness. For example, in an automobile the chemical energy converted by the combustion process is converted to heat. The cooling system absorbs some of this heat with the remainder being transferred through the engine components and exhaust system. Therefore, this heat and the energy contained within it has become more random (i.e., its level of entropy has increased).

The Second Law of Thermodynamics suggests that although energy is never lost, it tends to be transformed into other less useful or random forms, such as thermal energy. Every time energy is converted from one form to another, some energy moves to a more random or less concentrated state. The amount will be a function of the efficiency of the energy conversion technology.

In order for energy to be commercially and economically useful, it must be obtained in adequate quantities and in a form that allows easy conversion. Energy must also be readily stored and transported. Currently only five resources of energy meet these criteria; namely: coal, oil, and natural gas (chemical); hydroelectricity (mechanical); and nuclear energy. However, in the past 10 to 20 years, other sources have been studied extensively—such as solar, bioconversion, and ocean thermal energy. These and other renewable energy resources are further defined in chapters 8 and 9 of this textbook.

**Energy Laws and Shortages—
Energy Issue 1-3**

In light of the law of energy conservation, which asserts that energy cannot be created or destroyed, explain how an energy shortage can occur.

ENERGY SOURCES IN USE— RENEWABLE, NONRENEWABLE, AND NONDEPLETABLE

Renewable Energy Sources

In order to conceptualize better the various forms of energy, the sources of energy may be divided into three broad categories. These categories are renewable, nonrenewable, and nondepletable energy sources. A renewable resource reflects a perception of time. A resource is renewable if it can be replenished at a rate that allows it to be used in a significant or commercial manner. While there is no limit on the total quantity of energy humans can obtain from a renewable resource, there is a limit on the rate at which energy can be extracted. Thus, renewable energy sources are effectively exhaustible if they are not managed to assure replenishment.

Renewable energy sources have had a prominent role in the social development of the human species. Prehistoric human beings, the hunter-gatherers, used firewood to cook wild food to sustain their existence. These early humans lived at the mercy of nature, as it was not until many years later that they learned how to cultivate plants and to domesticate animals. The renewable energy from the cultivated plants and domesticated animals supplied milk, meat and other foods, transportation, and hides for clothing and shelter—all of which aided human survival. In terms of human use, renewable energy sources have been used the longest. Examples of renewable energy sources include: animal power; human power; and the chemical energy stored in plants, animals, and organic wastes. Plants and organic wastes as energy sources are often referred to as biofuels or biomass.

An example of a renewable energy source is human power.

Biomass

Hydrocarbons in *biomass* are vegetable materials produced from carbon dioxide, sunlight, and water. These sources of energy have an advantage over fossil fuels in that they are renewable. Like fossil fuels, they can be converted from a solid fuel by burning or converted into a liquid or gaseous fuel. There are two kinds of biofuels. These are plants grown for their specific energy content and fuels from various waste products.

An example of a plant that can be grown for its specific energy content is corn. Using corn, one can produce ethyl alcohol—a hydrocarbon fuel. In one simple production method of ethyl alcohol, various grains (corn stalks, cheese whey, or vegetable garbage) are placed in a still and vaporized by heat. The vapor is then condensed by a cooling device

to obtain the fuel. More sophisticated processes of fermentation may also be used for fuel collection.

An example of a biomass fuel produced from a waste product is the use of spent cooking liquor and lignin from the production of paper. In the papermaking process, the fibrous raw material of wood or cellulose must be collected. This cellulose may be collected by a chemical pulping process that separates the wood from lignin. (Wood is made up of multilayers of cellulose fibers bound together by lignin. The lignin is a natural compound that acts as a binder and support for the wood fibers.) In this chemical pulping process the cellulose is obtained using heat, pressure, and a chemical mixture known as cooking liquor. Waste products of this process are the spent cooking liquor and lignin. These waste products can be burned in a recovery boiler contributing more than 50 percent of the energy needed by the mill for the production of paper.

While the technologies to convert biomass into commercially available fuels are similar, the methods of collection, conversion, delivery, and end use differ significantly. Hydrocarbon resources from biomass will be presented in more detail in chapters 9 and 10.

Nonrenewable Energy Sources

Energy resources are nonrenewable if the pace of formation is so slow as to be meaningless in terms of replacement in the human life span. Nonrenewable energy sources are exhausted or depleted once they are used. Coal, natural gas, and petroleum were formed over many thousands of years. While these resources are still being formed today, for all practical purposes, they are not renewable. Likewise, the fissionable elements of uranium and thorium are also not renewable.

Only since the mid-nineteenth century have humans made a significant shift away from renewable energy to nonrenewable sources for space and water heating and industrial processing. Prior to 1850, most space and process heating in the world was accomplished with wood. In Western Europe, the use of wood for shipbuilding and for making charcoal (used for space heating and industrial processes) had severely reduced wood supplies. Thus, Britain began to use the nonrenewable fuel of coal for heating and industrial processing on a large scale by the early nineteenth century. In the United States, the renewable energy source of wood continued as the predominant fuel until about 1875. In fact, most locomotives in North America used this fuel source until being converted to coal in the latter part of the nineteenth century. Nonrenewable energy sources include: coal, geothermal heat traps (i.e., heat contained in steam and water trapped in pockets within the upper section of the earth's crust), natural gas, oil shale, petroleum, tar sands, thorium, and uranium. For purposes of clarity, the nonrenewable energy sources can be further subdivided into hydrocarbon resources and nuclear energy.

Hydrocarbon Resources—Fossil Fuels

One major source of stored energy is a group of fuels known as **hydrocarbons.** Hydrocarbons can be found in two distinct forms—fossil fuels and biomass. Hydrocarbons in fossil fuels represent the remains of material (plant and animal) that lived long ago. They consist mostly of hydrogen and carbon atoms. Because of the different pressures, temperatures, and time factors involved in their natural formation, hydrocarbons exist in different forms. The most common hydrocarbons are coal, oil, and natural gas.

Coal is a solid fuel, while petroleum is a liquid fuel. Natural gas is very light in weight, invisible, and very difficult to smell (unless odorous chemicals are infused with it). Coal, as with other fossil fuels, stores energy in chemical form.

Coal is mined and transported to electric utility companies, where it is converted into thermal energy. Once converted, it is used to make steam in a boiler. The steam is then used to obtain mechanical energy from turbines, and finally to create electrical energy from a generator for home and industrial use.

Oil or petroleum is transported from the well to a refinery where it is broken down chemically to yield a variety of products. Some common products include lubricating oils, gasolines, tars, asphalts, diesel fuels, home heating fuels, and jet fuels.

Natural gas is also transported directly from the well to a refinery for further processing. Figure 1-10 shows a part of a refinery for extracting various products from natural gas. The task of a natural gas refinery is to refine or separate out methane, propane, liquefied petroleum gas (LPG), butane, ethylene, and

FIGURE 1-10 This refinery breaks down natural gas into various products such as propane, butane, ethylene, and methane. (Courtesy of the American Petroleum Institute)

hydrogen from the natural gas. Once separated, an odor is added to the gases before being used by individual consumers or industry. This odor helps consumers identify gas leaks within a home or business.

Large amounts of natural gas are also transported directly through a pipeline to fuel heating, processing, and production systems in commercial, industrial, and residential applications.

Fossil fuels are *finite* resources. Finite resources are limited in quantity. These could eventually be depleted by extensive overuse.

Nuclear Energy

Simply stated, all matter is made up of atoms. When atoms are forced to split (i. e., the fission process), thermal energy is released. Nuclear energy is produced by the fission or splitting of certain radioactive atoms. These atoms are usually in specially prepared radioactive pellets. Note that nuclear energy can also be generated by fusion or the combining of specific radioactive atoms. Both fission and fusion are discussed in detail in chapter 7.

Nondepletable Energy Sources

A nondepletable energy source is one that cannot be exhausted through collection or extraction. These energy sources do not significantly decrease in their availability as they are being consumed. Thus, these energy sources are limitless and offer the greatest potential for achieving future economic and social growth.

Like renewable energy sources, nondepletable energy supplies have been used for thousands of years. Among the first uses included the drying of food, salt, and animal hides by the sun. Additional uses throughout history included harnessing the wind by sails for travel and using windmills and waterwheels for irriga-

tion, grinding grain, cutting lumber or stone, or crushing ores. Clearly, civilization received a vital impetus from these technological advancements. In the future, the large scale use and conversion to nondepletable energy resources may help to reduce the exhaustion of nonrenewable fuels. Nondepletable energy resources include solar radiation, water and wind power, tidal energy, geothermal energy contained within the mantle and deep crust of the earth, and ocean thermal energy conversion (i.e., the extraction of heat energy from oceans to operate a turbine-generator). While nondepletable and renewable energy sources form a small percentage of the total energy consumed, they will play an important role in the future as nonrenewable forms of energy are depleted. A more in-depth examination of nondepletable and renewable energy sources is presented in chapters 8 and 9.

Energy for the Future—Energy Issue 1-4
After comparing and contrasting the benefits and disadvantages of renewable, nonrenewable, and nondepletable energy sources, which appears to be the most available and benign energy resource (socially and environmentally) for the future? Describe the specific energy source that holds the greatest potential for alleviating the world's energy crisis.

ENERGY COMPARISONS

Energy Mix

Knowing the relative mix of energy resources can aid one in making future decisions about energy supplies. To get an idea of the relative mix of energy within the United States, refer to Figure 1-11. This figure represents the per-

United States Energy Mix

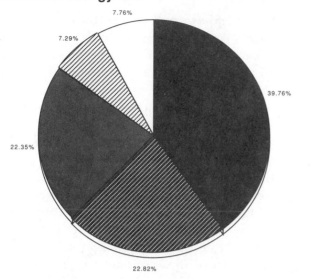

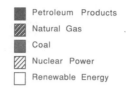

Notes:
Total may not sum due to rounding.
Data extrapolated from the *Energy Information Administration, 1992 Annual Energy Outlook with Projections to 2010.*

FIGURE 1-11 The United States gets most of its energy from these five resources. Although percentages may change each year, the chart illustrates that the United States is extremely reliant on hydrocarbon energy resources.

centages of energy being produced from each of the five major resources.

Oil or petroleum supplies the greatest percentage of the energy consumed in the United States. Natural gas is second in importance; coal, nuclear power, and hydroelectricity constitute the balance. Although these percentages are fairly accurate, yearly and monthly fluctuations occur. In any case, the United States clearly is a significant consumer of hydrocarbons. Although the percentages may vary, most other countries also rely heavily on hydrocarbons for their energy needs.

Oil Shortage—Energy Issue 1-5

Why is oil used more in society than other energy resources? If oil were in severe shortage, which resources would be used to make up the difference? What would be some immediate social effects on consumers if this were to happen?

Energy Units

In the United States the most common energy unit is the *British Thermal Unit* or ***Btu***, which is used to measure thermal energy. One Btu represents the thermal energy needed to raise one pound of water one degree Fahrenheit. However, when studying total energy consumption, the quad (quadrillion) is used. One quad is equal to 1,000,000,000,000,000 Btu's (1×10^{15}). For example, Figure 1-12 shows the amount of energy in quads per year consumed in the United States over a period of years and projected into the future.

Another unit used is the *MB/DOE*. The MB/DOE represents *m*illion *b*arrels *p*er *d*ay *o*il *e*quivalent. This unit measures total daily energy use or consumption equivalent to millions of barrels of oil. To determine the MB/DOE, it is assumed that all forms of energy used have been converted to barrels of oil as the base unit. Using energy units such as quads and

MB/DOE helps energy companies determine how much total energy is being used by a society per day or per year.

Converting Different Energy Units—Math Interface 1-2

Energy can be expressed in many units of measure. Often these units are converted from one to another. Be sure to review Appendix C to understand all of the metric decimal prefixes. Solve the following conversion problems:

1. If one quad equals 1,000,000,000,000,000 Btu's, how many barrels of oil would constitute a quad? (Hint: Review Appendix L, "Energy Content of Fuels.") One quad is equal to _____ barrels of oil.

2. The energy content of one barrel of crude oil (42 gallons) is equal to how many cubic feet of natural gas and how many kWh's (kilowatt-hours) of electricity? (Hint: Review Appendix D, "Fuel and Energy Equivalents.") One barrel of crude oil equals _____ cubic feet of natural gas. One barrel of crude oil equals _____ kWh's of electricity.

3. Five short tons (2000 lbs each) of coal is equal to how many Btu's? (Hint: Review Appendix D, "Fuel and Energy Equivalents" and Appendix L, "Energy Content of Fuels.") Five tons of coal equal _____ Btu's.

Energy Growth Patterns

Energy *growth patterns* are used to show how society is using energy. Growth patterns help to illustrate which energy resources are being used over a period of years. Figure 1-13 shows a chart of the United States' use of energy over several years (with projections into the future). This chart shows that coal usage is increasing; nuclear energy production is increasing; and new resources, such as solar energy and other alternatives, are expected to increase. These factors will affect the total energy mix. Additionally, this chart shows how energy usage dropped because of the 1973 oil embargo. Note that the embargo caused a downward turn of the economy. Charts such as this help to predict future energy patterns based upon past energy consumption rates.

When one examines the charted energy growth patterns, certain conclusions can be drawn.

1. There seems to be a steady increase in the total amount of energy being consumed within the United States.
2. Between 1973 and 1975, there was a decrease in total energy usage. This was caused by several factors, including (a) an economic recession, (b) the Arab oil embargo of 1973, and (c) increased energy conservation techniques.
3. Natural gas usage appears to remain stable in the future because of the available supply and economic and environmental advantages.
4. The use of nuclear energy will remain stable or increase slightly. This, of course, will be affected by federal restrictions placed on the use of reactors and by environmental concerns. Fuel prices will also be a significant factor affecting the development of nuclear energy.

United States Energy Consumption since 1950 (Quadrillion Btu) with Projections to 2010

FIGURE 1-12 This graph illustrates energy consumption measured in quadrillion Btu's. Notice the significant increase over the past 40 years.

Notes:
Data extrapolated from: The Council on Environmental Quality, Environmental Quality 23rd Annual Report and Energy Information Administration, 1992 Annual Energy Outlook with Projections to 2010.

Energy Growth Patterns

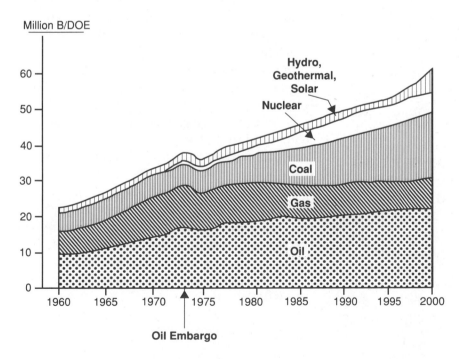

FIGURE 1-13 This chart illustrates, for the U.S., the relative amounts of energy generated from each of the major resources, as well as expected growth patterns.

5. Coal consumption shows a notable increase because substantial deposits continue to be available within the United States. This availability has produced more mining and exploration of coal. Efforts are proceeding to continue to reduce environmental pollution and acid rain that are consequences of coal use.

6. Oil supplies will remain relatively stable without large increases in the future. Oil supplies increased slightly in the early 1980s because of the Trans-Alaska Pipeline Act of 1973. However, total oil supplies will definitely decline as we move into the future. In fact, many experts are predicting declines in oil supplies in the early twenty-first century. A review of oil usage of the past 20 years reveals two major trends. The first is that energy use in the United States has grown and an increasing portion of that energy is supplied by imports. Of the 17 million barrels per day of petroleum consumed in the United States in 1990, approximately 42 percent was imported. This figure is expected to rise to between 53 and 69 percent by 2010. This is surely a concern for the United States as securing energy becomes an even greater economic reality. On a more positive note, the second major trend is that the United States is using energy more efficiently. In the past 20 years, the energy efficiency in the United States has improved by approximately 25 percent.

7. Hydroelectric, geothermal, and solar energy production show small increases. Hydroelectric energy resources are currently a small portion of the total United States' energy mix. Very little technological advancement can take place without significantly increasing dams on rivers, lakes, or reservoirs. Note, however, that as hydroelectric technology becomes more advanced, smaller hydroelectric dams become more financially feasible as generators of electricity. Additionally, geothermal and solar energy continue to be investigated. However, geothermal energy is very localized and solar energy still must overcome consumer concerns such as initial costs and reliability.

Historic Energy Mix

After reviewing Figure 1-14 it should be evident that it may take many years before a society uses energy resources other than fossil fuels as their major energy source. A long time is required to develop new energy resources before they can provide a significant contribution to the total energy supply. This lag time is portrayed in Figure 1-14. The horizontal axis represents years, and the vertical axis the percentage that each energy resource is able to contribute. For example, in the year 1890, approximately 38 percent of the energy came from wood, 57 percent from coal (95 percent – 38 percent), and the remaining energy came from oil and natural gas.

Forty to 50 years were involved to change from wood to coal as a major resource (1850 to 1900). It also took nearly 50 years to change from coal to oil and natural gas (1900 to 1950). Although improved technology can shorten this time somewhat, 30 to 50 years will probably still be needed to change from one dominate resource to an alternative resource. Such a time increment would, therefore, be necessary to switch from fossil fuel dominance to alternative fuels.

Lead Times

Lead time is a term used to describe how long it takes to get new energy resources and technology to the customer. Lead time is defined as the length of time required from initial planning to final operation of any plant or mine. For any given energy resource to be developed, a lead time of 3 to 12 years usually occurs before energy production occurs. Figure 1-15 indicates the approximate lead times necessary to develop various plants and mines to produce additional energy.

These lead times are often much longer because of increasing environmental and social pressures, economic constraints, and political influences. Considering lead time, it becomes apparent that any significant change in energy resources will take a substantial length of time.

Future Energy Resources—Energy Issue 1-6

What will be the energy resources of the future? How long will it take for these future resources to gain significant use (10 to 20 percent) in the United States? How long will it take for these future resources to gain meaningful use in developing countries?

HISTORIC ENERGY SCENE

Energy Awareness

Historically, the United States really has not devoted much attention to energy consciousness within society. However, such consciousness was first sparked by the energy crisis of 1973 when increased funding, research, and

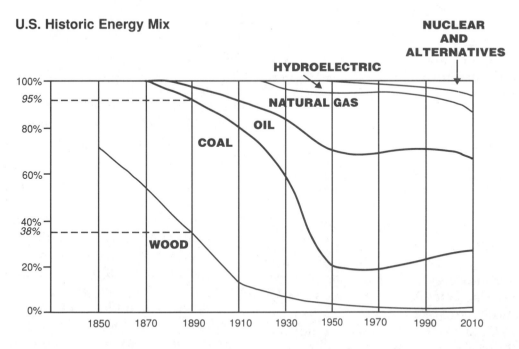

U.S. Historic Energy Mix

FIGURE 1-14 This historic energy mix suggests that to change from one major form of energy to another may take 50 years or so.

Anticipated Lead Times for Plants and Mines

	Number of Years
Surface Coal Mines	3–6
Underground Coal Mines	4–7
New Oil Refinery Construction	4–7
Alaskan Pipeline	4–7
Coal Gasification Plant	5–8
Shale Oil Mines	5–8
New Oil/Gas Production	3–10

FIGURE 1-15 Shown are various lead times necessary to develop different energy technologies. (Adapted from the U.S. Bureau of Mines)

education were initiated. The energy crisis, although inconvenient at the time, served as a starting point for United States' energy awareness. Since that time, however, awareness has subsided because of increased supplies of hydrocarbon fuels. Availability of supplies and their cost have a direct relationship to energy consciousness in a country. Nevertheless, understanding the energy crisis of 1973 helps one to see the total complexity of energy in our society. The Middle East Crises and the Iraq invasion of Kuwait in late 1990 again served as a spark to make the United States and other countries very aware of the need for energy.

The Energy Crisis

To begin, the countries of the Middle East provide most of the world's oil. Figure 1-16 shows the world crude oil flow. The thickness of the lines represents the amount of oil being shipped.

The following is a brief synopsis of events that occurred during the oil embargo of 1973. Beginning in October of 1973, the Organization of Petroleum Exporting Countries (OPEC) raised the price of crude oil from $3.00 to $5.12 per barrel. At the time, imported oil accounted for approximately one-third of the United States oil supply. Therefore, this price increase was immediately felt by consumers, most frequently at the gasoline pumps.

Besides this increase, during the same month the Organization of Arab Petroleum Exporting Countries (OAPEC) reduced crude-oil production by 5 percent and halted all crude-oil sales to the United States. Although oil was available at higher prices, the United States could not get its oil from OAPEC suppliers. This meant that approximately 33 percent of the United States' oil supply was not available at any price.

Then, in November of 1973, OPEC cut crude-oil production by an additional 25 percent. This cut in production was accompanied by OPEC's increasing the price of crude oil from $5.12 to $11.65 per barrel. At the time, imported crude oil accounted for approximately 36.6 percent of total United States' domestic oil consumption. Consumers felt the effect of OPEC's price hike when gasoline rose from approximately thirty cents per gallon to more than sixty cents per gallon.

To deal with the increased crude-oil prices and the reduced production rates, various occurrences and factors came together to offset the threatening energy crisis.

1. The Trans-Alaska Pipeline Act was passed to build the Trans-Alaskan pipeline, shown in Figure 1-17.
2. In 1973, the Emergency Petroleum Allocation Act (EPAA) and the Federal Energy Office (FEO) were created to assist in the allocation of petroleum products.
3. In 1974, the vehicle speed limit was reduced to 55 mph to improve fuel mileage and reduce United States' dependence on foreign oil. However, despite the savings, the Reagan Administration

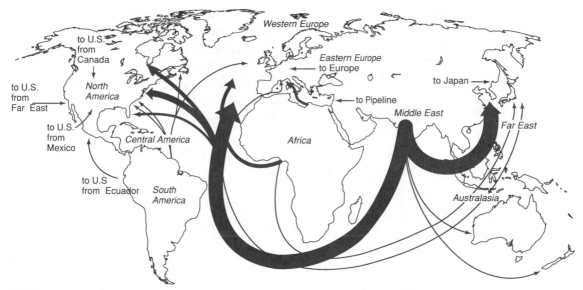

FIGURE 1-16   Most oil comes from the Middle East, as indicated by the thickness of the arrows. (Adapted from the U.S. Bureau of Mines)

increased the speed limit to 65 mph on interstate highways.

4. In 1974, the Energy Supply and Environmental Coordination Act (ESECA) was implemented.

5. In 1974, after the initial shock to U.S. consumers began to wear off, OPEC lifted the embargo.

6. Natural gas became very difficult to purchase in 1975. Many price controls had been placed on natural gas in preceding years to help control rising inflation. However, the controls served as a disincentive to United States producers to drill for new wells. All natural gas price controls have since been removed.

7. In 1978, there was a major coal strike in the eastern part of the United States. Certain areas of the United States were without coal for more than 90 days while oil continued in short supply.

Factors in the Energy Crisis

The energy crisis of the 1970s could be described as a political and economic incident, and a result of four major factors.

1. The OPEC oil embargo, decreased production, and increased prices.

2. Price controls on natural gas, discouraging producers from drilling new wells.

3. The federal government's lack of established energy policies at the time of the oil embargo.

4. Consumers' continued requirements and demands for additional energy to power their conveniences and products.

> *Oil Embargo—Energy Issue 1-7*
>
> *What were the economic and social effects that resulted from the oil embargo of 1973? What effects might occur today if another oil embargo as severe as that of 1973 were imposed on the United States?*

Trans-Alaskan Pipeline

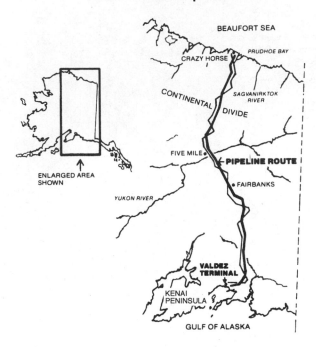

FIGURE 1-17 The Trans-Alaskan Pipeline transfers oil from Prudhoe Bay in the Beaufort Sea to the Valdez Terminal in the Gulf of Alaska. From there, the oil is transported to the United States by large oil tankers.

TODAY'S ENERGY PICTURE

Influencing Factors

The United States and other countries have learned much since the oil embargo of 1973. Gasoline and petroleum prices are sufficiently high to spur continued development of oil and natural gas reserves. This has tended to increase our oil and natural gas supplies. However, if prices of oil and natural gas are reduced because of a plentiful supply, these incentives will again be decreased. This was the case in Oklahoma and Texas during the late 1980s. Solar energy is also much closer to being competitive with other resources. Additionally, coal has taken on an increasing role as a major energy resource.

Nevertheless, dependence upon foreign oil persists. Depending upon the particular year, the United States has been between 17 and 50 percent reliant upon foreign supplies. The Middle East oil cartels influence this percentage. For example, when the price of a barrel of oil is reduced by OPEC, United States' producers buy more foreign oil to keep their costs down and accumulate more profit. Figure 1-18 shows how the wellhead price of world crude oil has changed and is likely to change in the future.

The level of inflation of the United States' economy is another factor causing producers to seek additional lower-priced foreign oil. As inflation increases in the United States, producers typically purchase more foreign oil because it becomes competitively priced.

Doubling Times

Today, energy is frequently discussed in reference to *doubling times*. Doubling time is a measure of the length of time required for anything growing at a specific rate to double in size. For example, if $1000.00 were placed in a bank at 7 percent interest, the $1000.00 would double in 10.2 years. This means that every 10.2 years, the total value of the investment would be doubled (ignoring inflation). After 10.2 years the $1000 would be equal to $2000.00. After a second 10.2 years, the $2000.00 would be equal to $4000.00, and so on. Appendix J, entitled "Exponential Growth," illustrates doubling times for certain rates of increase. Doubling times are found by dividing the rate of increase (in percent) into a constant of 72 (for our example of 7 percent, 72/7 = 10.2).

In terms of energy supplies, if energy use is increasing by 3 percent per year, total energy

World and United States Oil Prices since 1979 with Projections to 2010

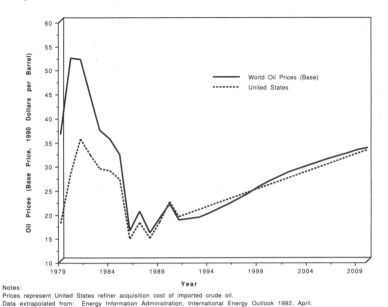

FIGURE 1-18 This graph illustrates how the price of a barrel of world crude oil changes. The price of oil in the United States is also shown for comparison.

Notes:
Prices represent United States refiner acquisition cost of imported crude oil.
Data extrapolated from: Energy Information Administration, International Energy Outlook 1992, April.

usage will double every 24 years (72/3 = 24). Are there adequate resources in the earth to handle such increased demand? Electrical energy demand is increasing at an annual rate of approximately 7 percent. This means that every 10.2 years electrical power plants must double their production capability. Clearly, continued increases in energy consumption will have serious consequences. In this connection, the following questions arise:

1. Can our society exist with fewer products and less convenience?
2. How long will oil and natural gas last if use of these resources continues to double?
3. What can we do to improve techniques of energy conservation through society?

Such questions should become part of any future energy scenario. In addition, education should play an important role in planning for future energy needs. One could conclude that technology alone is not usually the solution to society's ills. Often, the major questions regarding technology are not technical, but human questions involving choices and values. These human questions help to explore the ways in which technology can serve human purposes and how it is altering society. Regarding energy utilization, how the technology is applied depends upon what society thinks it is, what its limitations are perceived to be, what its role is, and how that role is assessed. These then should become issues addressed in educational programs that attempt to make citizens better consumers of energy and more aware of the sociocultural and environmental problems of energy utilization.

MAKING DECISIONS ABOUT ENERGY

Becoming Energy Literate

In today's society, consumers, business people, and governmental personnel are continually making decisions about energy and its associated technology. For example, logical decisions must be made about new and emerging energy forms. The use of existing resources and their effects on our society must also be evaluated. How does one make sound decisions about these issues? A person who can make sound and logical decisions based upon adequate information about energy issues can be called *energy literate*. One decision-making model used is shown in Figure 1-19.

Any decision about energy can be viewed by considering several aspects. These include sociocultural values and political, environmental, technical, and economic influences. The more these factors are considered, the more the "blinds open" and the more the problem can be observed. If we, as decision makers, view energy by considering these factors, the decision will be clearer, and more justifiable.

Sociocultural Values

All energy issues are closely tied to the sociocultural values of society. These are values in which people believe. For example, energy is in high demand today because of the value placed on material things and creature comforts. Our society highly regards materialism, which has contributed to an energy problem within our society. As energy decisions are made, the sociocultural factor must surely be discussed to understand the whole picture.

Political Influences

The political influences on energy include any decision that is controlled or decided by rules, regulations, policies, governments (city/state/federal), and so on. For example, the decision to build more nuclear power plants may be affected by political rules, safety standards, and agencies in effect in a particular geographical location. Further, zoning ordinances may restrict the location of a plant.

Environmental Influences

The environmental effects of energy decisions can be very great. For example, if solar photo-

Decisions About Energy Technology

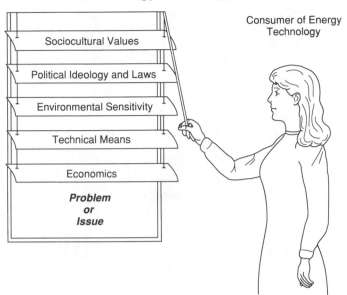

Consumer of Energy Technology

Sociocultural Values

Political Ideology and Laws

Environmental Sensitivity

Technical Means

Economics

Problem or Issue

FIGURE 1-19 A possible decision-making model for energy issues would consider: sociocultural values, political ideology and laws, environmental sensitivity, technical means, and economics.

voltaic cells will be used more extensively, what environmental effects will be felt at the production site? If coal energy resources are to be increased, what are the acid-rain effects? All energy decisions must be filtered through the environmental screen.

Technical Influences

When making decisions about energy, technical aspects must also be considered. Ask the question, *Can the technology accomplish what is needed for the energy issue?* One often hears that the technology is ready, so why isn't the product on the market? For example, solar energy can be collected and stored. However, it may not produce economic incentives to secure societal adoption.

Economic Influences

In the United States, society is based upon supply and demand (the components of capitalism), among other economic factors. Many decisions are made about energy based upon the dollar. Questions like *Is this a good investment?* or *Will I spend fewer dollars if I buy this type of energy?* play an important part in energy decisions.

Combining these influences—sociocultural, political, environmental, technical, and economic—helps to make complex decisions much easier within the energy field. All five parts must be researched and considered if the decision is to be sound and justifiable. To help see the interrelation of the parts, consider the following energy issue: *Should battery-powered cars continue to be promoted and developed within society?* Using the five parts, the following questions could be researched and answered in each of the areas to help make the best decisions. Obviously, both yes and no answers help to make the decision.

Sociocultural

Will society generally accept the battery-powered vehicle? Will people object to having a battery charge instead of a fill-up? Will people mind having more charges than a single fill-up? Will people desire more speed, performance, and power? Will an insurance company insure the vehicle? Are the passengers covered by insurance if they are in an accident and suffer acid burns? Are the speed limits too high for this vehicle? How might the adoption of this type of transportation affect speed limits?

Political

Will the regulatory agencies and government adopt legislation that encourages the promotion of battery-powered vehicles and conservation of fossil fuels? How will this legislation affect the economy and individual use of energy? How will political action groups (PACs) attempt to alter legislation to protect industrial, business, or auto enthusiast interests?

Environmental

Is the vehicle as safe as other cars in crash tests? Is there less pollution (carbon monoxide, nitrogen oxide)? How will the adoption of electric vehicles affect the demand for electricity, and would any increases cause further environmental demise? What environmental damage will acid spills cause on road surfaces?

Technical

Can an efficient battery-powered car be technically produced? Can it be designed to last as long as a gasoline-powered vehicle? Can it be designed to pull heavy loads? Can it achieve sufficient distance?

Economic

Will driving a battery-powered vehicle be less costly than driving a gasoline-powered vehicle? Will the insurance be more or less? Will the company that manufactures the vehicle make a suitable profit?

Obviously, more questions could be asked. However, as these questions are answered, the decisions then become more clearly defined regarding the success or failure of any energy issue being studied.

Making Energy Decisions—Energy Issue 1-8
Using the decision-making model in Figure 1-19, determine the appropriate energy sources and technologies for the developing nation of Botswana. In this analysis, consider the following question: Should the people of this country be restricted in the use of fossil fuels because of global ecological damage, more specifically greenhouse effects?

ALTERNATIVE FUTURES

Resource Recovery (Tire-to-energy)

A relatively new technical process is emerging within energy technology called *resource recovery*. Resource recovery is taking waste products from our society and using them to produce either a new product or different forms of energy. One process that is currently being examined is that of tire-to-energy resource recovery.

Each year our society produces over 200 million waste tires. In California alone there are more than 22 million tires disposed of each year (see Figure 1-20). Currently, these tires are either placed in landfills or dumped in illegal tire piles. Since tires are not biodegradable, they take up valuable space and tend to migrate to the top, ruining the landfill. Tire piles can catch fire, emitting noxious smoke and a hazardous oily runoff. Because they trap water, tires in piles

FIGURE 1-20 Each tire contains energy equivalent to 2.5 gallons of oil. The recovery of this resource may lessen some of the environmental effects of landfilling and provide society with usable energy. (Courtesy of Oxford Energy Company)

may also serve as a breeding ground for disease-carrying insects. Some piles have been estimated to contain over 15 million tires.

From research it has been found that waste tires can be an excellent high-Btu fuel for recycling into energy. Each tire, for example, contains energy equivalent to 2.5 gallons of oil, enough to heat a medium-sized house for a day.

One recycling process uses tires to produce electrical energy. In a plant recently constructed by the Oxford Energy Company, 700 tires per hour are burned, generating about 15 megawatts of electricity (see Figure 1-21).

In the operation, (review Figure 1-21) tires are fed into a large furnace where they are burned. Furnace temperatures of over 2000° F and state-of-the-art pollution controls ensure complete and clean combustion. The thermal energy in the furnace will boil water and produce steam. The steam is then used to operate a turbine generator to produce electricity.

FIGURE 1-21 At the Modesto Energy Project, tires are fed into a large furnace where they are burned. The thermal energy in the furnace creates steam, which is used to produce electricity. (Courtesy of Oxford Energy Company)

POINT/COUNTERPOINT 1-1

There are numerous issues in the study of energy supplies that are controversial. Often, data can be found that support almost any idea, concept, or belief. Point/Counterpoint is designed to help you explore both sides of various energy issues. Often, such controversial topics are excellent sources of discussion and debate.

TOPIC

Sustainable - No Growth - Drawdown

Theme: Should society encourage one of the following options for economic and social development: Sustainable Development, No Growth, or Drawdown? What are the consequences of the development options selected regarding energy utilization?

Sustainable development can be defined as the harmonious existence of economic growth and environmental sensitivity. Through the attainment of sustainable development goals, economic and social aims are balanced by the constraints of nature. Sustainable development requires compatible consumption, living patterns, and time management; low energy consumption and large-scale use of renewable energy systems; appropriate technology; reduction, reuse, and large-scale recycling efforts; land-use plans based on environmental concerns; a respect for species and environmental diversity; and, participatory management by citizens (i.e., majority over individual concerns).

No growth can be defined as the maintenance of the status quo regarding development. With this as a development strategy, citizens are limited to the resource extraction and material goods and services already in place. Emphasis is placed on maintaining the quality of life as it currently exists. The disparity between developed and developing countries is maintained, with no effort to reduce the pain and suffering of the less fortunate. Note that no growth is considered mostly a short-term fix. As resources continue to be exhausted, an alternative development strategy is to be selected. The no-growth development strategy requires immediate plans to control population growth; continued dependency on fossil fuels and increases in renewable energy as resources decline; reuse and large-scale recycling efforts; an increased emphasis on the anthropocentric paradigm (i.e., the well-being of nonhuman beings and other living things do not count; emphasis is placed on human survival; and an increased emphasis on centrally planned governments with less citizen involvement.

Drawdown can be defined as the process by which the dominant species depletes the resources faster than they can be replaced. As the drawdown continues, resources must be borrowed from other places until complete resource depletion occurs. The result of drawdown occurs when the population has exceeded the capacity of its environment in one life-giving respect or another. In such an instance, nothing can be done until the population is reduced to the level at which the resources can recover. Drawdown requires a blind faith in science and technology to solve human wants, needs, and problems; the current beliefs regarding resource depletion and development of new goods and services; continued population growth; emphasis on the anthropocentric paradigm; and less interaction in governmental affairs by citizens.

SUMMARY & REVIEW

Summary/Review Statements

1. Energy demand is directly related to the production and consumption of goods.
2. The United States consumes about 28 percent of the world's energy resources while constituting only about 5 percent of the world's population.
3. The oil embargo of 1973 had a significant effect on the United States in terms of energy awareness.
4. Energy and power are often confused. Energy is the ability to do work; power is the rate of work being done.
5. Energy can take on a variety of forms: chemical, thermal, mechanical, nuclear, radiant, and electrical.
6. Technology is currently used to convert energy from one form to another.
7. When converting energy from form to form, some energy is always transformed into a less useful state.
8. Coal, oil, and natural gas are considered hydrocarbons or fossil fuels.
9. Use of coal and solar energy is expected to significantly increase in the next 20 years.
10. Because of societal and technological complexities, it may take between 30 and 50 years or more to change from one major fuel resource to another.
11. To make energy decisions, the environmental, political, technical, economic, and sociocultural aspects of each energy issue must be studied.

Discussion Questions

1. In your opinion, what would be the best way to reduce the energy needs of society? Give examples and support them.
2. What is the importance of studying doubling times? How does the concept of doubling times relate to the study of energy?
3. Discuss the importance of entropy to the study of energy technology.
4. Consider the use of biomass energy sources to replace traditional fossil fuels. Give examples of how each of the five decision making factors (sociocultural, economic, technical, environmental, political) helps to make a sound and energy-literate decision.

chapter 2

Energy Demand and Consumption

PURPOSES

Goals of Chapter Two

The purpose of this chapter is to determine how societies demand and consume energy. A knowledge of energy consumption is a prerequisite to developing solutions to many energy problems studied today. If one knows the extent and patterns of energy demand and consumption, then informed decisions can be made to effect positive changes in energy use throughout one's life. Thus, the general goal of this chapter is to develop a sound knowledge of the demand and consumption of energy. At the completion of this chapter, you will be able to

1. Define the sectors within our society that consume the energy being supplied.
2. Compare the relative percentages of energy consumed by each sector within society.
3. Examine the growth patterns and trends of the energy consumed within society.
4. Examine how energy flows from resources to consuming sectors within society.
5. Identify the various applications in which consumers utilize the energy supplied within society.

6. Illustrate the reasons for the growth in electrical energy demand.

Terms to Know

By studying this chapter, the following terms can be identified, defined, and used:

Cogeneration
Commercial Sector
Embargo
Energy Consumption
Energy Demand
Energy Sector
Industrial Sector
MB/DOE
Nonenergy Sector
Residential Sector
Sectors
Transportation Sector

INTRODUCTION

Although the United States consumes one quarter or more of the world's energy resources, some of this energy is used to supply goods and products to other countries. Currently, United States workers produce about 25 percent of the world's goods. The United States also provides almost one-half of the world's food exports. All of these uses of energy are considered *energy consumption*. This means that energy consumption, although very high, is being used to fulfill not only the needs of people within the United States but also those of consumers in other countries. However, United States consumers still use more energy per capita than consumers in most other countries.

Supply and Demand Relationship

In the study of energy, supply and demand play important roles. In fact, all products sold within our country are based upon common laws of supply and demand. Supply can be considered the energy that is put into a system. Energy resources, such as coal (see Figure 2-1), oil, natural gas, nuclear power, solar energy, wind, hydroelectricity, and so on, are considered part of the energy supply. Note that energy supplied to society may be influenced by many factors. Political constraints, technical and economic influences, and so on, play important roles in determining energy supplied to society. Energy supply and demand must be in balance. If they are not in balance, fuel prices may fluctuate drastically.

Quite often in the past, energy supply and demand have been unbalanced. The oil embargo of 1973, for example, reduced the supply of oil. An embargo is a political measure of one country to stop shipping a product to another country. When the ratio between supply and demand is not balanced, energy prices change—as illustrated in Figure 2-2. If there is a surplus of a resource, such as oil, prices may be reduced to help sell the product. On the

FIGURE 2-1 Coal is considered one of the major energy resources in the United States. (Courtesy of Union Pacific Corporation)

other hand, if there is a shortage of oil, prices will probably increase because consumer demand for the product is great. To keep prices stable, supply and demand should be balanced as closely as possible.

It is important to note that high energy prices tend to induce more conservation. This is because the high prices provide greater incentives to improve efficiency. Low economic growth also reduces the demands for energy-related services—particularly regarding energy consumption in the industrial and transporta-

Relationship Between Energy Supply and Demand

FIGURE 2-2 There is a direct correlation between energy supply and demand. Supply and demand of energy must be in balance to keep prices stable.

tion areas. This is because low economic growth is associated with sluggish industrial activity, unemployment, and reduced levels of travel in all modes of transportation.

Supply and Demand Balance—Energy Issue 2-1

List several consequences of supply and demand of energy not being in balance. What are some possible ways to augment energy supply while reducing energy demand? (Answer in terms of energy availability, cost of energy, technological advancements, and societal/environmental consequences.)

Demand and Consumption

The demand for energy and the energy consumed by a society may not be equal. **Energy demand** represents the resources that a society requires for a certain standard of living. Consumption is defined as the amount of energy that a society consumes. A society may, in fact, demand more energy than it consumes. It would be unlikely for a society to consume more energy than it demanded. If energy systems in our society are working correctly, the energy consumed will be equivalent to the energy demanded. According to energy projections of the United States Energy Information

Administration, the United States is projected to consume about 106 quadrillion Btu's of energy resources (in all forms combined) by 2010. This would be an increase of about 25 percent above the 1990 level of energy use.

High-Grade and Low-Grade Energy

Energy can be said to be either high grade or low grade. High grade is considered very condensed energy (i.e., having a high Btu content) and is associated with high temperatures (600 to 3000° F) during conversion processes. For example, coal energy that is eventually used to turn a turbine to produce electricity is considered high-grade energy. Similarly, the petro-

Consumption is defined as the amount of energy that a society consumes. (Michael Weisbrat/Stock, Boston)

leum used for combustion in a gasoline or diesel engine is high-grade energy (1400° F). Low-grade energy is considered less dense and is usually associated with lower temperatures (80–600° F). For example, the type of energy required for central heating (70–80° F) in a house is low grade. However, if fuel oil is used to heat the house, the fuel oil is considered high grade.

Energy Consumption—Math Interface 2-1

Being aware of how much energy one uses per day allows one to realize the need for saving energy. Solve the following problems, which illustrate the energy being consumed by people in society.

1. If all the countries in the world consumed a total of 67 million barrels per day of oil and the United States consumed about one quarter of this energy, how many barrels of oil would the United States consume in one year?

   United States consumption equals

   _____.

2. In the same year described in question 1, the United States population was about 255 million. Using the answer from question 1, how much energy (in millions of barrels) was consumed per person, per day in the United States in this year?

   Consumption in barrels per day, per person equals

   _____.

3. If a barrel of oil is equal to 42 gallons, how many equivalent gallons of oil are used per day by each person in the United States (using the figure from question 2)?

   Equivalent gallons per person, per day equal _____.

4. Using the same procedure as illustrated in questions 1–3, determine the oil used in gallons per day by people in Japan if there were 124 million people and 5.5 million barrels per day of oil consumed in that country.

   Equivalent gallons per person, per day equal _____.

SECTORS WITHIN SOCIETY

There are several major energy consuming subdivisions or *sectors* within most industrialized societies. These sectors, designated as end-uses of energy, are the demanding and consuming parts of our society. Figure 2-3 illustrates the various energy sectors. They identify where the energy is going or being consumed. Awareness of these sectors helps consumers and government representatives to establish policies for augmenting energy efficiency and for encouraging energy to be used wisely. The challenge for society is to produce and use energy more efficiently while preserving resources for future generations. See Figure 2-3 for sector titles.

Energy Sectors

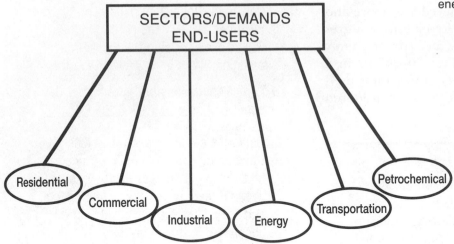

FIGURE 2-3 There are six major sectors that demand and consume energy in most societies.

Residential Sector

The *residential sector* of society uses energy to operate residential dwellings including houses or other buildings where people live. Residential usage involves energy-consuming devices used directly in the dwelling for day-to-day living. Some prominent energy-consuming applications of this sector include:

1. home heating;
2. domestic water heating;
3. cooking;
4. clothes drying;
5. refrigeration;
6. air-conditioning; and,
7. use of other appliances—including televisions, freezers, dishwashers, washing machines, lighting fixtures, computers, and other small appliances.

Residential applications commonly use electricity, natural gas, or oil; however increasingly, there is use of renewable energy sources for heating purposes. Renewable energy sources include solar panels, wind turbines, and *ground-source heat pumps.* Ground-source heat pumps (also called *geothermal heat pumps*) transfer heat between the air in buildings and the ground via piping that is installed outside the building and a short distance below the earth's surface.

Frequently, the energy that is used in the residential sector can be of a lower grade. For

Residential usage involves energy-consuming devices used directly in the dwelling for day-to-day living.

example, space or central heating requires temperatures between 60 and 80° F. Domestic water heating requires temperatures between 80 and 150° F. However, not all energy utilized in the residential sector is low grade. Electricity used for lighting and air-conditioning, for example, requires high-grade energy.

The selection of home heating fuel in new construction is an influencing factor in determining the future pattern of energy consumption by the residential sector. Traditionally, heating accounts for more than half of all residential energy use. Since 1985, the use of natural gas as the main heating fuel for single-family dwellings has risen nationwide. Electricity's share for heating has continued to decline. These two trends are expected to continue for the rest of this decade, especially if additional gas pipelines are introduced to new areas. However, the consumption of natural gas for heating will begin to decline as the efficiency of all heating equipment improves and as natural gas prices rise relative to electricity prices. This equilibrium should occur in the early 2000s. The projected rise of renewable energy sources will also displace some natural gas in the residential sector. For example, the use of wood is expected to show a slight increase in use through the year 2010. This increase will occur because of more efficient woodstove designs, a general decrease in the use of other traditional means of heating, and a need by consumers to lower home-heating costs.

Further, technical improvements are also expected in both heating and cooling. Heat-pump water heaters and condensing furnaces are projected to be fully developed and widely available before the turn of the century. With augmented federal efficiency standards for heating, cooling, water heating, and refrigeration technologies coming into effect, all-new

and replacement purchases of appliances should help to raise the overall efficiency level of energy use in the residential sector.

Commercial Sector

The *commercial sector* of society is that part that uses energy for business operations. The commercial sector includes restaurants, office buildings, hotels, motels, and educational buildings. Some major energy-consuming applications of this sector include

1. heating for the buildings;
2. water heating;
3. cooking for school cafeterias;
4. refrigeration;
5. air-conditioning; and,
6. other end uses, including those for lighting, appliance use, and computer operations.

The residential and commercial sectors are often combined in studies and reports concerning different sectors of energy consumption. This is done because the type of energy used for each is similar. The aggregate energy consumption of the residential and commercial sectors during 1990 was estimated at 16.9 quadrillion Btu. This represented approximately 26 percent of the U.S. end-use energy total.

Computers and Energy Use—Math Interface 2-2
Research has illustrated that the vast majority of the time the United States' 32 to 38 million personal computers are turned on, they are not actively in use. Furthermore, approximately 30 to 40 percent of these computers are left running at night and on weekends. It has been projected that computer equipment is the fastest growing electricity load in the

commercial sector. Computer systems are believed to account for 5 percent of commercial electricity consumption at this writing and potentially 10 percent by the early 2000s. In 1992 the United States Environmental Protection Agency (EPA) formed an alliance with computer manufacturers to promote the introduction of energy-efficient personal computers to reduce electrical consumption and air pollution caused by their power generation. Working with the Apple, Compaq, Digital, Hewlett-Packard, IBM, NCR, Smith Corona, and Zenith Data Systems, the EPA has proposed a system wherein computers "powerdown" when not in use. This powerdown feature could effect a 50 percent reduction in the energy used by computers. This voluntary program has been designated the Energy Star Computer Program.

The EPA estimates that the energy saved as a result of this voluntary effort would prevent carbon dioxide emissions of 20 million tons. This amount of pollution is equivalent to the emissions from five million automobiles. Reductions in energy use would also lead to the elimination of 140,000 tons of sulfur dioxide and 75,000 tons of nitrogen oxides. These two pollutants are responsible for acid rain deposits.

Conduct a computer audit of a school or building to determine the number of computers operated and how many hours per day they are left running. On the back of each computer and monitor find the wattage consumed. Calculate the potential energy that would be saved if these computers had a power-down feature as described in this exercise. After completing these calculations, develop and implement a strategy to educate computer users about the benefits of turning off their machines when not in use.

The total energy consumption in the residential and commercial sectors changes as the number of occupied buildings increases (based on the additions of new buildings, minus demolitions and abandonments). As newly constructed houses and commercial buildings are added, the effects on energy consumption are mixed. New buildings often require significantly less energy than older buildings. However, new buildings tend to incorporate a greater range of energy-consuming devices—such as air-conditioning, office equipment, and entertainment equipment (e.g., radios, televisions, and so on).

Because construction standards are more stringent than in years past, new buildings are more efficient than those already in place—even with their greater use of energy-consuming devices. It is estimated that 30 percent of all residential units and 44 percent of the United States' commercial buildings will be of post-1987 construction by 2010. This may represent a substantial savings in energy consumption as new buildings replace the older, less energy-efficient residential and commercial structures.

Industrial Sector

The United States industrial sector consumed approximately 27 percent of all the energy consumed in 1990. This sector includes both manufacturing and nonmanufacturing industries. The *industrial sector* consumes energy for heat and power. This includes a wide range of end uses—such as powering equipment, process heating, using feedstocks, and using boilers. The nonmanufacturing industry consists of agriculture, mining, and construction. The manufacturing industry consists of the following:

1. primary metals and fabrication industries;

2. chemical and allied industries;
3. stone, clay, glass, and concrete industries;
4. paper and allied industries;
5. food and related industries;
6. transportation equipment-manufacturing industries;
7. electronic product industries; and,
8. other manufacturing industries—such as heat processing, lumber making, and so on.

The industrial sector normally requires a higher quality of energy than do the residential and commercial sectors. A significant amount of industrially consumed energy is considered high-grade energy. Required temperatures for industrial processes often exceed 1500° F. Predominant uses of energy within this sector include the manufacture and use of

1. processing steam equipment;
2. electric drives, motors, and clutches;
3. electrolytic manufacturing processes;
4. direct heating and drying equipment; and,
5. refrigeration equipment.

Energy and Agriculture—Energy Issue 2-2

The industrial sector uses energy for agricultural purposes. List and discuss five ways in which energy is used for agricultural purposes in society. What would be some of the consequences to agricultural activities of a 5 percent energy tax imposed on all energy sources?

Petroleum Refining Industry

The petroleum refining industry uses large amounts of still gas and petroleum coke. These are byproducts and residues of the refining processes. These account for about two-thirds of all the energy consumed in refining. Natural gas constitutes about three-fourths of the remainder.

Paper and Pulp Manufacturing

As noted in chapter 1, the pulp and paper industry utilizes large amounts of wood and other wastes recovered in its various processes to burn as fuel. These supply a little more than half of the energy for this industry. Close to three-quarters of the remaining fuel comes from coal and natural gas.

Rubber, Plastic, and Chemical Industries

The rubber, plastic, and chemical industries require large amounts of conventional fuels to produce heat and power. These industries additionally use substantial amounts of natural gas and petroleum as the chemical raw material from which they manufacture final-use chemicals, fertilizers, and plastic resins. The natural gas and petroleum used by these industries to produce consumer products are known as *feedstocks*. Feedstocks represent the raw materials for the production of these products. These feedstocks (equivalent to about 3.5 quadrillion Btu if used as a fuel) account for almost 60 percent of all energy consumed by the chemicals industry. Electricity and natural gas account for more than 80 percent of the remaining consumption.

Primary Metals Industry

The primary metals industry (e.g., aluminum, iron, steel, and other metals) consumes a large amount of one predominant fuel that is unique to this industry. This fuel, used as a feedstock to produce coke (a fuel made from coal to make steel), is metallurgical coal. Once used as a feedstock, it is consumed in the iron and steel industry for heat and power. More than 40 percent of all energy consumption in primary metals comes from this special form of coal. Approximately 90 percent of the remaining energy consumed is in the form of natural gas or electricity. Figure 2-4 provides a graphic illustration of the energy consumed in the most energy-intensive industries of petroleum refining, pulp and paper manufacturing, chemical and plastics production, and primary metals manufacturing. In this graphic representation, base consumption refers to the amount of fuel (in quadrillion Btu's) used in the manufacture of various products. *Base plus renewable* refers to the amount of fuel and renewable energy sources used. The renewable energy sources are in the form of waste products that result from manufacturing. An example of this form of renewable energy would include the burning of bark, spent cooking liquor, and lignin in the manufacture of paper and pulp. The addition of feedstocks would include fuel and chemical raw materials (e.g.

petroleum, natural gas, and so on) needed to manufacture a product. The addition of primary inputs to electricity means that this is the total energy content of all the fuels that went into the manufacture of this specific product, including the generation of electricity at a utility company.

Making Energy Conclusions—Energy Issue 2-3

Interpreting statistics about energy consumption often depends upon the type of data and how it was collected. Energy data can be presented in different ways that may be misleading or provide only a partial inventory of the energy being used. For example, in reviewing Figure 2-4 several different conclusions can be reached about the overall energy consumption of various industries. If one were interested in ranking these industries from the most energy intensive to the least energy

Energy Consumed in Energy-Intensive Industries (Quadrillion Btu)

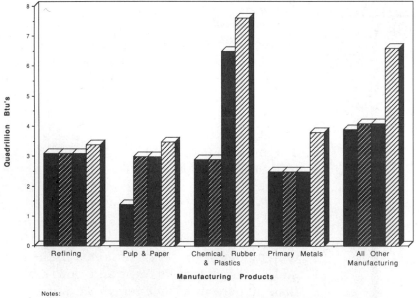

Base Consumption

Base Plus Renewable Energy Sources

Also Adding Feedstocks

Counting Primary Inputs to Electricity

FIGURE 2-4 Energy-intensive industries require and use a vast amount of energy resources.

Notes:

Data extrapolated from: Energy Information Administration, 1992 Annual Energy Outlook with Projections to 2010.

With the average annual projected GDP growth rates to the year 2010 ranging from a low of 1.8 percent to 2.7 percent, the industrial sector's expected energy consumption in 2010 will increase to a range of 28.0 to 30.8 quadrillion Btu. This means that the United States industrial sector is projected to consume between 0.5 and 1.0 percent more energy each year while increasing the value of its output. Energy-intensive industries are anticipated to continue growing more slowly than those less energy-intensive. The statistics attributed here should help to illustrate that improved energy-use technologies will need to be adopted, especially as energy prices rise.

Electrical Energy Sector

The electrical energy sector, also called the *energy sector*, includes the industries involved in the production of electricity from various resources. The use of electrical energy is a fact of life in the United States. Our reliance upon electricity has been dramatically demonstrated on two occasions: the East Coast power outage of 1968 and the New York City outage of July 13, 1977.

The New York blackout occurred when lightning struck a power station. People were stranded in elevators, water did not get pumped into high-rise apartments, street lights would not operate, subways did not run, and air-conditioning systems could not function—all because of the outage. Although considerable looting was evident, no major personal injuries took place. However, this event and similar occurrences underscore the current dependence upon electricity in society.

As noted in chapter 1, the electric utilities are both major producers and consumers of energy. They are consumers because it requires energy to operate the electrical power plants. The utility companies will play a pivotal role in the United States' energy future. Almost 20 percent of the energy consumed by users is anticipated to be in the form of electricity by 2010. The electric utilities are expected to use approximately 40 percent of all the energy consumed in 2010 to meet growing electrical needs.

During the next 10 to 15 years, electrical demand is projected to account for approximately 35 to 50 percent of the growth in aggregate end-use energy consumption. Some 90 percent of this growth will be in the commercial sector—primarily for room heating, water heating, lighting, and cooling. Significant increases will be seen within the residential sector in electricity use for lighting, room heating, and water heating.

Figure 2-5 illustrates the electrical growth (in kilowatt-hours) in all sectors since 1981 (with projections to 2010). One kilowatt-hour is defined as the electricity equal to 1,000 watts used for one hour or the equivalent. For example, a simple hair dryer operated on the 1,000-watt level for one hour would consume 1 kilowatt-hour. Estimates show that the consumption of electricity is projected to grow

Electrical Generation and Consumption with Projections to 2010

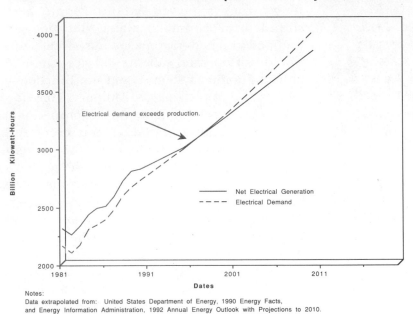

FIGURE 2-5 Electrical demand is projected to outpace electrical production by the turn of the century.

Notes:
Data extrapolated from: United States Department of Energy, 1990 Energy Facts, and Energy Information Administration, 1992 Annual Energy Outlook with Projections to 2010.

by 2 percent each year until 2010. However, note that electrical generation (production) is only forecasted to increase by 1.6 percent each year according to this graph. This means that unless new construction or energy conservation measures occur, electrical demand will outpace production.

> **Electrical Energy Demand—Energy Issue 2-4**
>
> *Electrical energy demand is growing at a faster rate than can be met by electric utilities. What will be the consequences of this demand if no new power plants are built in the future, and what are some alternatives to building new power plants?*

Electrical Prices to Consumers

Consumers perceive electricity as clean and convenient. Electricity's general share in end uses is anticipated to increase since the price of electricity is projected to rise more slowly than the costs of other energy sources. In fact, numerous state regulatory bodies often structure electricity prices in order to mitigate the consequences that cause large price increases in other areas. For instance, regulators often require that capital costs of expensive generating facilities be phased in over several years to avoid the sudden price impacts on consumers.

In the United States, electric prices on average are predicted to remain stable until the turn of the century. (Figure 2-6 illustrates the expected increase in costs per kilowatt-hour for electricity to the year 2010.) This is despite the fact that the price of all fuels used to generate electricity is expected to increase. The reason for the stable price is that the capital costs per unit of output are projected to decline. The decline is expected because utilities are extending the operation life of many existing plants instead of building new plants. Thus, the return on invested capital and the depreciation expenses are lower than those for a newly built plant. However, after the year 2000, as new plants are built by utilities to enter service, the revenue related to the total return on invested capital is projected to grow more slowly than the growth in electrical energy costs.

Strategies for Meeting Electrical Demand

The growth in the demand for electricity is forecasted to create a need for new generating resources as the turn of the century approaches. Utilities are taking different strategies to meet this demand, from building new power plants to reducing electrical requirements through investments in conservation and *demand-side management* (DSM) programs. DSM programs are designed to modify the hourly, daily, or seasonal variations in electrical demand for a utility and increase the efficiency of production and consumption. These pro-

grams may include informational efforts, time-of-use and *interruptible* rate programs, and rebates to customers who use more efficient appliances and equipment. For example, many homes today utilize an interruptible rate program with nighttime domestic water heating reductions. Under such a program, the electric water heater might have a day rate of 6.7 cents per kilowatt-hour and a night rate of 3.2 cents per kilowatt-hour. Thus, the consumer saves money by using the lower rates for washing clothes at night.

Utilities anticipate to augment expenditures on DSM programs over the next decade from around $1 billion annually to about $3 billion annually. This effort is being made because of the proven benefits of DSM programs. Successful programs have reduced the need for expensive new generating plants and have led to decreases in energy consumption. The economic benefit to consumers lies in the ability of utility companies to provide the same

Expected Increase in Cents per Kilowatt-Hour of Electricity with Projections to 2010

FIGURE 2-6 Electrical prices are expected to remain stable until the turn of the century.

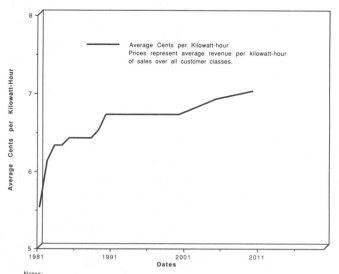

Notes:
Data extrapolated from: United States Department of Energy, 1990 Energy Facts, and Energy Information Administration, 1992 Annual Energy Outlook with Projections to 2010.

level of energy services at a lower cost. In addition, there are environmental benefits from DSM initiatives. Effective DSM programs often lower fossil fuel consumption by the utility. This reduction of fossil fuel consumption saves resources and reduces the hazardous emissions of carbon dioxide, sulfur dioxide, and nitrogen oxides. However, some new plant construction will be required before the turn of the century, even with the use of DSM programs.

When a utility company finds that it must build a new power plant, the technology it chooses is determined in large part by how the new unit will be used. Additional factors may include the relative price of fuels, the type of generating capacity already in the region, and the electrical demand expected in the future.

Since demand fluctuates during the day and because electricity cannot be stored easily, some plants operate almost continuously. These units handle what is known as the *base load*. Other electrical generation plants will be called on only when demand is at its highest peak, called *peak load* capacity. Large coal and nuclear power plants are generally best suited for baseload operation. This is true because, while expensive to build, these power plants use low-priced fuels. This allows these plants to be relatively inexpensive to operate. Electric power plants based on combustion turbines, however, are less expensive to build but use more expensive fuels (i.e., oil and natural gas). This in turn makes them more expensive to operate and maintain. Since the cost per kilowatt-hour is much higher for peak load plants, these power units are operated only for short periods when peak demand is reached.

Many of the new power plants added before 2000 will be used for intermediate and peak load operation. Some 45 to 50 percent of the 80 to 100 gigawatts of capacity added via new plants by the end of the century are anticipated to be fueled by natural gas or oil. An additional 130 to 200 gigawatts of capacity will be added after 2000 to the production facilities that already exist. Some 40 or 50 percent of these power facilities are projected to burn coal.

Fuels for Electrical Energy

The addition of nuclear energy has gradually been added to the sources of electrical energy production. In fact, the U.S. Department of Energy suggests that nuclear energy will continue to provide electrical energy production in future years. However, no new nuclear energy power plants are currently under construction or planned for the near future. High costs of construction, safety standards, consumer concerns, and disposal of spent fuel associated with nuclear energy will surely continue to be debated.

Aside from nuclear energy, other resources also provide fuel for the generation of electricity. Figures 2-7 and 2-8 list major energy resources used to produce electrical energy. The bottom or horizontal axes of each figure show the years from 1973 to 2013. The vertical axis represents electrical energy being produced in million kilowatt-hours. Nuclear energy made very little impact on the total electrical energy supplied in the 1960s. However today nuclear energy's contribution to the generation of electricity is greater and is projected to continue at current levels for the foreseeable future.

Natural gas and petroleum use has declined in the past few years, while use of coal has grown significantly. Increases in production of coal and nuclear power will directly change the amount of electrical energy available. This is because both coal and nuclear energy are the main resources used by public utilities to produce electricity, and changes in the production of coal and nuclear power directly affect base load operations.

Energy Sources of Electric Utility Generation (Typically Base Load) with Projections to 2010

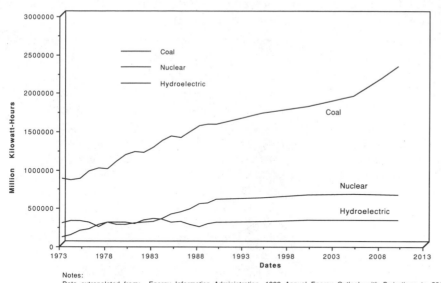

FIGURE 2-7 The major energy sources for electric utility generation are coal, nuclear, and hydroelectric. These sources are primarily used for base-load power plants.

Notes:
Data extrapolated from: Energy Information Administration, 1992 Annual Energy Outlook with Projections to 2010, and Energy Information Administration, September 1992, Monthly Energy Review.

Energy Sources of Electric Utility Generation (Typically Peak Load) with Projections to 2010

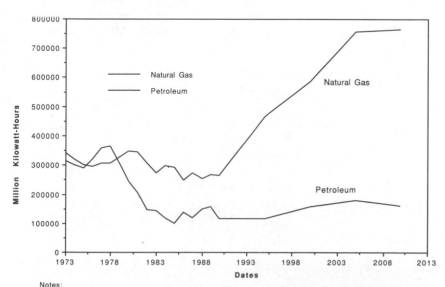

FIGURE 2-8 Additional energy sources of electric utility generation are natural gas and petroleum. These sources are primarily used for peak-load power plants.

Notes:
Data extrapolated from: Energy Information Administration, 1992 Annual Energy Outlook with Projections to 2010, and Energy Information Administration, September 1992, Monthly Energy Review.

While coal and nuclear power represent the two major sources of fuel for electrical generation, renewable and nondepletable energy technologies are important to the energy mix. Renewable and nondepletable energy technologies currently used to produce electrical energy include hydroelectric, geothermal, wood, biomass, municipal solid waste, solar, and wind power. Of these sources, hydroelectric accounts for approximately 84 percent of this generation; municipal solid waste and biomass provide 11 percent; geothermal, solar, wind, and wood supply the remainder. These technologies will be presented in more detail in chapters 8 and 9. Figure 2-9 illustrates the growth of renewable and nondepletable energy technologies to produce electricity. Note in this graph that renewable energy technologies are expected to grow by an average annual increase of about 1.3 percent to the year 2010. This growth is expected to come from municipal solid waste (roughly 42 percent), geothermal (24 percent), wind (13 percent), hydroelectric (11 percent), and other energy sources (10 percent).

Electricity and Solid Waste—Energy Issue 2-5

The rapid growth anticipated in power generation from municipal solid waste is a result of improvements in combustion technologies, pollution controls, and the problems associated with the handling of municipal solid waste. These problems include costly and stringent landfill regulations and the difficulty of finding new sites that will meet EPA and local environmental guidelines. In a classroom setting, debate the pros and cons of generating electricity from municipal solid waste. In this debate contemplate the various implications of this alternative—including resource depletion, changing consumer purchasing habits, air and water pollution, and ownership of this renewable electricity-generating facility.

Renewable and Nondepletable Energy Technologies in the Production of Electricity with Projections to 2010

FIGURE 2-9 Renewable energy sources are expected to play an even more significant role in the production of electricity in the future.

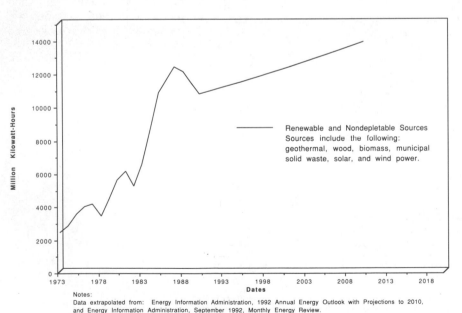

Renewable and Nondepletable Sources
Sources include the following: geothermal, wood, biomass, municipal solid waste, solar, and wind power.

Notes:
Data extrapolated from: Energy Information Administration, 1992 Annual Energy Outlook with Projections to 2010, and Energy Information Administration, September 1992, Monthly Energy Review.

FIGURES 2-10 and 2-11   These vehicles are examples of how energy is consumed within the transportation sector.

Transportation Sector

The *transportation sector* of society is that which moves people and goods. Figures 2-10 and 2-11 illustrate two examples of transportation vehicles that consume energy in this sector. Since our society relies heavily upon many forms of transportation, this sector is important. Sixty-four percent of all the petroleum products used in this country in 1990 went into providing transportation services. This share is expected to increase in the future. However, efficient vehicles will help to moderate this increase. More importantly, note that the trans-portation sector consumes about 12 percent more petroleum than is produced domestically. This deficit in petroleum production must be obtained from foreign suppliers. This foreign dependency increases the United States' trade deficit and makes consumers more vulnerable to price increases in various goods and services. In total, United States oil consumption is anticipated to increase by about 3.2 million barrels per day by 2010. Four-fifths of this increase will be used to transport people and goods. Depending upon the future predictions reviewed, the United States could consume between 43 and 107 percent more petroleum for transportation than is expected to be produced domestically in 2010. Most of this petroleum would be used in automobiles and light trucks. These alarming statistics help to illustrate the United States' dependence on foreign oil.

Some major end uses that demand and consume energy for the transportation sector include

1. automobiles;
2. aircraft;
3. trains;
4. trucks;
5. ships and barges;
6. buses;
7. military and government vehicles;
8. space transportation;
9. conveyors, elevators, escalators; and,
10. recreational vehicles.

The transportation sector requires high-grade energy in the form of gasoline, diesel fuel, and jet fuel to operate propulsion systems. Generally, combustion temperatures exceed 2000° F for such applications.

The transportation sector is almost entirely oil-dependent (i.e., 97 percent of the energy used is from petroleum) in the United States. While other end-use sectors have substituted other energy sources (namely coal, nuclear, and natural gas) for petroleum, the transportation sector is not projected to make this shift on a large scale until sometime after 2010. The principal sources of transportation energy are motor gasoline for automobiles and light trucks, diesel fuel and motor gasoline for freight trucks, jet fuel and aviation gasoline for aircraft, diesel oil and electricity for trains, and diesel and residual oil for domestic and international ships. Figure 2-12 provides a graphic representation of energy use in the United States' transportation sector as of 1990 and provides projections for 2010. By observing these charts one can conclude that the United States is not planning to make any significant policy decisions to reduce significantly the amount of energy directed to the transportation sector (particularly in the mode of personal consumer travel—autos/light trucks). Note also that the most energy-efficient modes of travel [i.e., marine (barges, ships, and so on) and rail] consume the least amount of energy in the transportation sector.

Fuel Pricing and Efficiency

Historically, gasoline prices affect the total fuel consumption of automobiles and light trucks. Consumers tend to respond to rising gasoline prices by driving less and by changing their attitudes over time to demand greater fuel efficiency. The response to high fuel costs, however, is limited in the short term. This is because automotive manufacturers need time to design, make production plans, and retool before manufacturing more energy-efficient automobiles. In the long term, higher gasoline prices, use of alternative fuels, adoption of more efficient vehicles, and the changing of consumer attitudes should increase potential fuel savings. An examination of several trends in the transportation sector helps to illustrate the need for a long term approach.

Since the 1950s, the United States transportation system has been dominated by petroleum-powered automobiles and light trucks. During the 1980s vehicles on the road grew almost twice as fast as the percentage increase of workers in the country (i.e., 17.4 percent increase in vehicles to 9.7 percent increase in workers) from 1980 to 1990. Figure 2-13 illustrates the actual and projected travel mode trends of commuters. By reviewing this chart,

Energy Use in the United States Transportation Sector 1990 (Actual) and 2010 (Projected)

FIGURE 2-12 Energy use in the transportation sector is expected to increase by 2010.

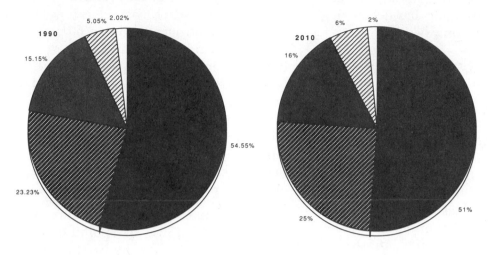

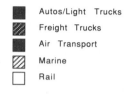

Autos/Light Trucks
Freight Trucks
Air Transport
Marine
Rail

Notes:
Data extrapolated from: Energy Information Administration, 1992 Annual Energy Outlook with Projections to 2010.

Travel Modal Trends of Commuters with Projections to 2000

FIGURE 2-13 The predominant mode of travel in the United States is the automobile, with over 110 million people expected to be driving alone by 2000.

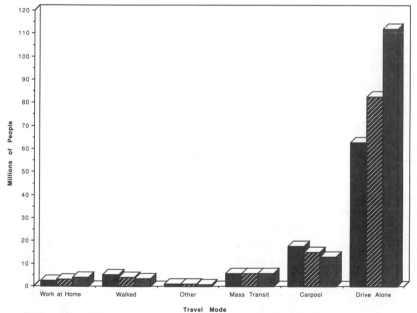

Millions 1980
Millions 1990
Millions 2000

Notes:
Data extrapolated from: The Council on Environmental Quality, Environmental Quality 23rd Annual Report

one can conclude that commuters who drive alone represent the fastest growing travel group. Additionally, carpooling has declined, and most carpools involve only two people.

Governmental Policies in the Transportation Sector

The transportation of people and goods is essential to the well-being and competitiveness of a country. The United States' transportation system plays a vital role in this society's well-being. At the same time, the transportation sector consumes approximately 23 quadrillion btu's of energy per year, uses large land areas, and produces a major portion of the U.S.' pollution problems. The transportation infrastructure has further contributed to habitat stress, ecosystem disruption, and transportation accidents involving hazardous cargoes. Despite these problems, in recent years greater emphasis has been placed on bringing the United States' transportation system into harmony with the environment. The 1990 Clean Air Act Amendments and the Oil Pollution Act of 1990 contain numerous provisions that will change the way the nation moves people and goods. Regarding transportation issues, the 1990 Clean Air Act Amendments provide provisions that include

1. Setting national limits on gasoline volatility. This will reduce the release of hydrocarbons, which contribute to smog.
2. Introducing cleaner reformulated gasolines that will reduce hydrocarbon and toxic emissions by 15 percent in nine cities with the worst pollution.
3. Increasing the oxygen content of gasoline in 39 cities with high levels of carbon monoxide pollution. (Increasing the oxygen content of gasoline reduces carbon monoxide pollution.)

4. Reducing the emission limits of automobiles and light trucks. This reduction would reduce hydrocarbon emissions by 30 percent and nitrogen oxides by 60 percent (relative to current standards).
5. Requiring, in 1998, that new fleet vehicles (e.g., taxis, delivery service vehicles, and so on) in 22 cities meet tailpipe standards that are more stringent than those required for conventional vehicles. These guidelines further provide incentives for fleet owners to purchase vehicles fueled by electricity, propane, alcohol, or natural gas.

Each of these provisions is an attempt to encourage reductions in pollution and resource depletion in the transportation sector. Additionally, the 1991 Intermodal Surface Transportation Efficiency Act (ISTEA) established new sources of funding for environmental initiatives. ISTEA initiatives include funds to mitigate transportation congestion, improve air and water quality, protect scenery and historic sites, and provide bicycle paths and green areas. Each of these initiatives is seen as an effort to minimize some of the environmental and social problems that result from the transportation sector.

Automobile fuel efficiency also became a major political issue in 1992. Congress, during deliberations on national energy legislation and in the presidential campaign, investigated the costs and benefits of proposed major increases in *corporate average fuel economy* (CAFE) standards. (These standards set limits that auto manufacturers must meet [miles per gallon] to earn credits under the Alternative Fuel Act of 1988.) Some lawmakers proposed amending the CAFE standard for automobiles from the present 27.5 miles per gallon to 40 miles per gallon by the year 2000. During President Clinton's administration, it was proposed that the standard reach 45 miles per gallon by 2010.

Figures 2-14 and 2-15 provide an overview of passenger vehicle travel, fuel consumption, and automobile fuel efficiency since 1969 with projections for the future. Based upon data on the charts, passenger vehicle travel will be increasing annually by 1.01 percent. Automobile fuel consumption, while lower than its high mark in 1978, has been steadily rising. Nonetheless, automobile fuel efficiency has continued to increase since 1976. Several social and technical changes that may lead to improved fuel economy in the future include

1. Meeting the proposed tailpipe emission standards mandated by the 1990 Clean Air Act Amendments;
2. Changing consumer preferences from larger to smaller vehicles;

3. Increasing taxes on fuels or roadway use; and,
4. Increasing emphasis on alternative fuels, fuel efficient cars getting improved mileage ratings, and mass transit.

Petrochemical Sector

One additional sector that is sometimes referenced to help identify where energy is used is the petrochemical or nonenergy sector. If not separately categorized, this sector is considered part of the industrial sector. However, in this text it has been separated and identified as a distinct sector. The petrochemical sector utilizes fossil fuels, mostly oil and natural gas, to produce various plastic and synthetic products. Note that the petrochemical sector does not use energy in a way that is distinct from other industrial manufactured products. It utilizes petrochemicals as the primary component for end-use products.

Automobile Fuel Consumption and Passenger Vehicle Travel with Projections to 2000

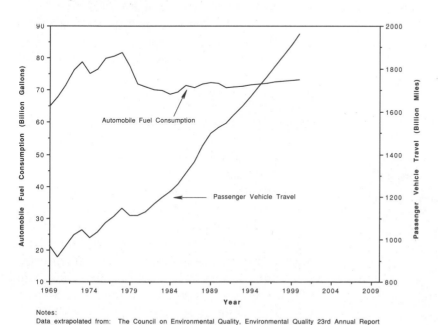

Notes:
Data extrapolated from: The Council on Environmental Quality, Environmental Quality 23rd Annual Report

FIGURE 2-14 Passenger vehicle travel will continue to increase into the next decade. Note that fuel efficiency has continued to increase since 1979.

Automobile Fuel Efficiency with Projections to 2000

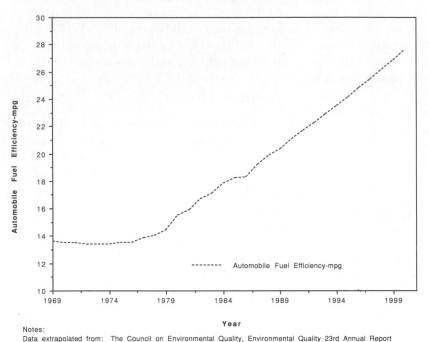

Notes:
Data extrapolated from: The Council on Environmental Quality, Environmental Quality 23rd Annual Report

FIGURE 2-15 Automobile fuel efficiency (measured in miles per gallon) has continued to increase since 1979. This increase has been a result of federal regulations, improved technology, and lighter and smaller vehicles.

Products such as alcohol, fertilizer, plastic, polyester, and so on, are considered petrochemicals. Generally, this type of energy application requires the use of high-grade energy. Figure 2-16 shows an example of a petrochemical plant that makes alcohol and plastic. More will be said about this type of industry in chapter 6, which covers natural gas in all its uses.

The Petrochemical Industry—Energy Issue 2-6

Realizing that the petrochemical sector requires the use of high-grade energy, develop a list of the benefits and a list of the negative consequences of diverting more energy to this sector. To make these lists more manageable, concentrate on the areas of transportation and agriculture. Finally, attempt to answer and justify the question:

"Should more energy sources (i.e., coal, natural gas, and oil) be redirected to this sector for the production of consumer goods?"

Percentages Within Each Sector

As U.S. government agencies have attempted to determine how energy is being used, they have generally investigated what end-use sectors utilize energy and the percentages of energy growth from these sectors. Figure 2-17 shows the relative consumption of each sector within our society. Note that these figures change annually. In the United States the energy sector consumes the most energy. Although figures vary, many sources indicate that the energy sector consumes about 32 percent of the energy available in our society.

FIGURE 2-16 This petrochemical plant is used to process natural gas into alcohol and plastic. These products are considered part of the petrochemical sector. (Courtesy of Conoco Incorporated)

A variety of influences may produce changes in how energy is consumed in society. Any of the sectors may increase or decrease its energy consumption depending upon societal changes. For example, as more products require plastic and synthetic materials, the petrochemical sector may increase its energy demand. On the other hand, if a recession should occur and residential building declines, the residential and commercial sector energy demand might be reduced. As more energy-efficient homes are built and less efficient homes are made more energy efficient, energy consumption may also decrease within the residential sector. Figure 2-18 illustrates the growth of energy consumption per capita in the United States since 1950. Observe from the graph that United States consumers have become more energy dependent to maintain their standard of living.

Growth in Sectors

The total United States' energy demand is projected to grow at a rate of approximately 2 to 4 percent annually. Each sector will increase or decrease at a different rate depending upon many factors. These factors include governmental regulations, availability of resources, supply and demand relationships, and others. Energy growth rates can be controlled by several factors, including

1. Better conservation measures in all sectors of society;
2. A growing awareness of the energy problems, leading to a more efficient use of energy;
3. A reduction in the need for too many products and conveniences;
4. The balance between supply and demand, as a regulator of energy prices (energy price increases tend to lead to decreased demand);
5. Government energy policies and regulations (tax incentives, tax increases, and the like);
6. A reduction in population growth; and,
7. A redefinition of the concepts of quality of life, progress, and an appropriate standard of living.

GDP and Energy Growth—Energy Issue 2-7
What effect does an increasing GNP or GDP have on the growth rates of energy within each consuming sector? Considering the content of chapters 1 and 2, what are the possible methods of augmenting the GNP or GDP without increasing energy use?

Energy Consumption by End-Use Sectors

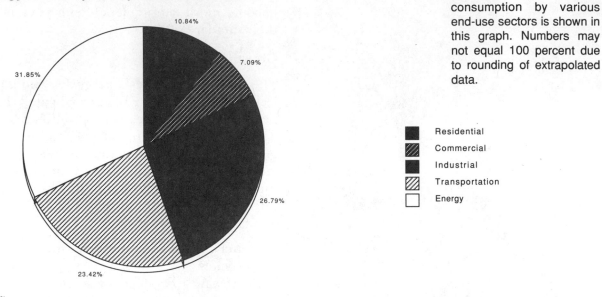

Notes:
Data extrapolated from: Energy Information Administration, 1992 Annual Energy Outlook with Projections to 2010.

FIGURE 2-17 Energy consumption by various end-use sectors is shown in this graph. Numbers may not equal 100 percent due to rounding of extrapolated data.

- ■ Residential
- ▨ Commercial
- ■ Industrial
- ▨ Transportation
- □ Energy

United States Energy Consumption Per Capita since 1950 with Projections to 2000

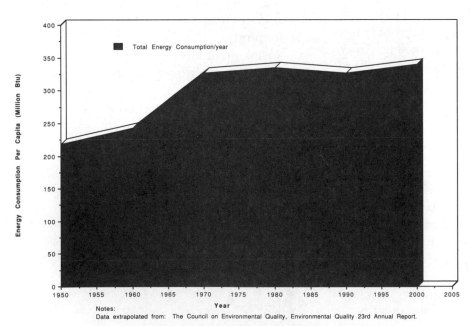

Notes:
Data extrapolated from: The Council on Environmental Quality, Environmental Quality 23rd Annual Report.

FIGURE 2-18 Total energy consumption is expected to increase during the next decade.

Figure 2-19 illustrates the projected growth rates of each sector. The horizontal axis shows the various energy sectors. The vertical axis represents energy consumption in quadrillion Btu's per year. This data is presented with projections to 2010. The residential and commercial sectors are projected to increase, but at a slightly slower rate than has been observed in past years. This estimate does not assume fewer houses and other buildings will be constructed. It relies upon expected use of better energy-conservation measures. New types of insulation, more insulation in older homes, and conservation awareness are all factors that will contribute to the slower growth rate.

Many energy analysts believe that the transportation sector offers the greatest opportunity for energy-saving measures. Data clearly indicates that people in the U.S. have become dependent upon the automobile. Automobile use in vacations, social, and recreational pursuits has continued to play an important role in family life. Energy consumption within the transportation sector is anticipated to increase by 1.1 percent annually until 2010. Figure 2-20 reviews these trends with projections for the future.

While gas mileage of automobiles is continuing to improve each year (figures of 30 to 50 miles per gallon are now possible), this augmented efficiency has been offset by growth in automobiles used and increases in distances traveled annually by all consumers. As noted in Figure 2-20 (see line in figure), since 1983 daily vehicle miles traveled per household has increased from approximately 32 miles to 46 miles. This change in driving patterns reflects the rise of the suburbs, increased automobile

Energy Consumption by End-use Sectors with Projections to 2010

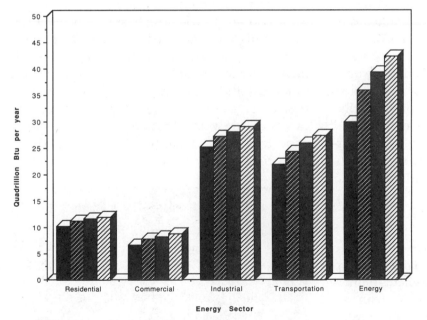

FIGURE 2-19 The energy sector is projected to increase significantly by 2010, using approximately 42 quadrillion Btu's per year.

Notes:
Data extrapolated from: Energy Information Administration, 1992 Annual Energy Outlook with Projections to 2010.

Personal Travel and Automobile Use with Projections to 2000

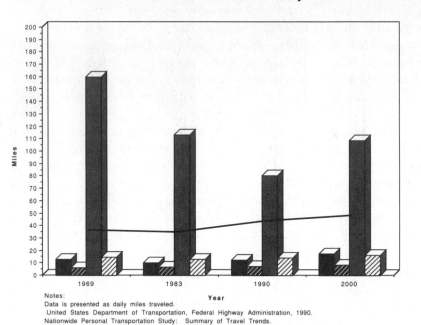

Notes:
Data is presented as daily miles traveled.
United States Department of Transportation, Federal Highway Administration, 1990.
Nationwide Personal Transportation Study: Summary of Travel Trends.

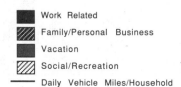

FIGURE 2-20 The automobile continues to play an important part in consumers' lives despite the numerous environmental and social problems of resource depletion, road congestion, and pollution.

use for work, and declines in the use of mass transit. Obviously, changes are needed in the transportation sector if the United States is to overcome its dependency on foreign oil. The opinion of many transportation experts is that higher fuel prices and CAFE standards would increase the fuel savings within this sector. This opinion is based upon the premise that new technologies would be developed and consumer attitudes would change to the purchasing and use of more fuel-efficient automobiles. The current CAFE standard of 27.5 miles per gallon is lower than the fuel efficiency already achieved by most domestic automobile manufacturers.

The government representatives in the state of California have taken a national leadership role in encouraging the development of new automobile technology. California's clean air plan designates that 10 percent of all new vehi-

cles sold after 2003 must have "zero emissions." Under this criteria, the most likely type of vehicle would be electric, which can easily meet this zero emission standard. Another technical advance promoted in California to better air quality and lessen fuel demand is the use of alcohol and natural gas vehicles for business, utilities, and government agencies. However, the prospects for wide adoption of these vehicles is expected only after substantial increases in gasoline prices and increases in technical efficiency.

Energy Demand Increases—Math Interface 2-4
Energy usage has continued to increase every year. Simple calculations can help to illustrate how rapidly energy use is increasing and how long it takes to double energy consumption and demand. Solve the following growth-rate problems:

Other factors that affect transportation growth include new programs designed to encourage airlines to cancel flights low in passenger numbers, car pooling in large cities, increased use of mass transportation systems, air-quality management programs in major urban areas, and conscientious conservation by people who are driving.

The industrial sector's growth is also forecasted to continue, but at a slower rate than in previous years. Energy consumption within this sector is projected to increase by 0.7 percent annually until 2010. Industrial growth depends upon how many and what type of products consumers buy. Every product requires energy for manufacturing. Since energy demand is linked to industrial growth, the increased use of energy-consuming manufacturing technologies tends to increase energy consumption. Inflation and recession will directly affect energy consumption in this sector.

Energy conservation in the industrial sector has other influences that moderate the advances. This is because the cost of improving the efficiency of motors, processing equipment, and heating/drying equipment is quite high. However, some conservation improvements are currently in use. For example, an effort has been initiated to capture waste heat from various conversion processes. (This is called *waste heat management* and, in some cases, cogeneration.) However, some savings of energy will always be offset by increased demand for new products.

The petrochemical *(nonenergy) sector* reflects the use of gas, oil, and sometimes coal as raw material for manufacturing such products as asphalt, waxes, and fertilizers. Projections indicate that this sector will increase in energy consumption. Many scientific groups believe that natural gas, coal, and oil should be redirected to petrochemical energy consumption. Energy loss is considerably less when such fossil fuels are utilized for these purposes instead of for other uses. This is because once the energy is used in manufacturing, the result is a product. This product retains an energy value. That is, the product may later be converted by a chemical, thermal, mechanical, or other process to extract the energy contained within it. However, plastic products produced from the petrochemical industries pose a significant environmental problem for disposal. For example, most plastics retain their form in a landfill for many years (in some cases thousands of years) after disposal. Thus, a strong move has been made to recycle these products.

The electrical energy sector's growth rate will continue to increase. As earlier noted, energy consumption within this sector is predicted to increase 2 percent annually until the year 2010. Most of this sector's growth is expected to be from increases in electrical demand. This demand will be met by fossil fuels, nuclear power, and by renewable energy sources.

ENERGY FLOW DIAGRAM

Diagram Description

Chapter 1 dealt briefly with the United States' energy supply. The link between that supply, the demand, and consumption of the sectors of society is illustrated in Figure 2-21.

The left side of the diagram indicates the supply of energy made available to the United States. For example, oil, coal, natural gas, nuclear, hydroelectric, and so on, are all shown. Both domestic and imported supplies are included. The thickness of the lines indi-

cates the quantity of fuel in **MB/DOE** (Million Barrels per Day Oil Equivalent). The middle boxes indicate the demanding and consuming sectors of society. On the right side of the flow diagram, the useful and rejected energy is shown. The *useful energy* is the amount that is used for its intended purposes; the *rejected energy* is that amount lost (entropy), as discussed in chapter 1.

The box labeled "electrical energy generation" represents all the power plants used to convert coal, oil, natural gas, nuclear energy, and hydroelectric power into electricity. The electricity produced from such power plants is further supplied to the individual sectors for their intended use.

The height of each box indicates relative percentages of energy supply and demand. As

U.S. Energy Flow

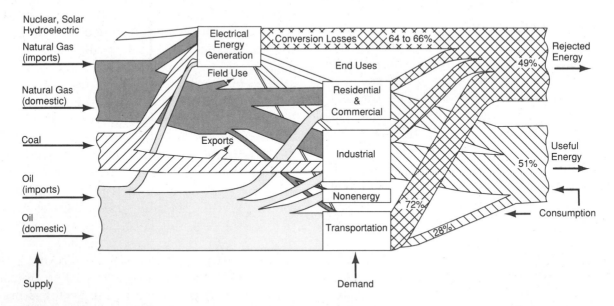

FIGURE 2-21   The total United States' energy flow is depicted in this diagram. Energy supplies such as coal, oil, natural gas, and so on, are placed into different demand sectors. The thickness of the lines represents the amount of energy in MB/DOE.

societal energy demands increase, the height of each of these boxes will increase at an annual rate of approximately 2 to 4 percent.

Total Energy Flow

Most energy is supplied by coal, oil, and natural gas. Nuclear power, hydroelectricity, and alternative energy sources (such as solar and wind) account for a much smaller proportion.

A portion of each resource is converted into electrical energy. As Figure 2-21 indicates, fossil fuels (such as coal) contribute the largest amount for generation of electricity.

The main disadvantage of converting coal, oil, and natural gas into electrical energy is that, in existing power plants, approximately 64 to 66 percent becomes unusable energy. This is entropy. Thus, only about 34 to 36 percent of the energy is useful to the various sectors. This point is important; for example, when a home uses electrical energy for heating. Electrical energy for home heating is advertised as being 100 percent efficient. However, such efficiency is only at the stage of converting from electrical energy to thermal energy. When the chemical energy of coal is converted into electrical energy and subsequently into thermal energy, the total conversion is only 34 to 36 percent efficient. This does not account for transformer, line losses, and so on.

New technology is now available for fossil-fuel electrical-generating facilities defined as *cogeneration systems*. **Cogeneration** is a process in which wasted or rejected heat is used to produce either electrical or mechanical energy. Cogeneration systems can achieve close to 55 percent efficiency. If all electrical-generating power plants used cogeneration, the electrical production system would be more efficient.

Tracing Energy Flow

To get an idea of how to trace energy flow, use the following example of oil:

1. Crude oil comes from both foreign and domestic resources;
2. Crude oil is used mostly in the transportation sector;
3. Small amounts of oil go into the nonenergy sector, industrial sector, and residential/commercial sectors; and,
4. A small amount of oil is used in electrical energy generation (power plants).

Energy Diagram Conclusions

The following conclusions can be made by observing the energy flow diagram (Figure 2-21):

1. Most energy used in the transportation sector is derived from oil. Automobiles consume about 52 percent of this fuel. The average number of cars per family in the United States is 1.77. Trucks, boats, planes, trains, and so on, consume most of the remaining oil used in this sector.
2. Because of the characteristics of the automobile and other propulsion technology, a great deal of energy is rejected in the transportation sector. For example, most propulsion devices are approximately 25 to 30 percent efficient.

This means that only 25 to 30 percent of the energy in the fuel is used to move the vehicles forward. The remaining energy from combustion is removed by the exhaust and cooling system or used in overcoming friction in the engine and drivetrain. This loss is considered part of the rejected energy shown in Figure 2-21.

3. Coal is utilized mainly in the energy sector, where it is used to generate electricity. Additional amounts of coal are used in the industrial sector to produce steam for processing operations. If a coal shortage were to exist, the electrical energy conversion sector would be most affected, thus reducing electricity provided to the industrial and residential sectors.

4. Natural gas feeds into the energy, industrial, residential, and commercial sectors. A small amount of natural gas is used for liquid-propane powered vehicles (Liquified Natural Gas, also called LNG). If a natural gas shortage were to occur, these four sectors would be affected most. For example, during the winter of 1977/1978, one of the coldest in recorded history, U.S. industrial, residential, and commercial consumers were significantly affected by a natural gas shortage. In fact, many industries, businesses, and homes were without energy at intervals throughout that winter.

5. The fact that there is little rejected energy in the petrochemical, or nonenergy, sector suggests that more energy might be channeled into this sector. However, environmental concerns such as the effects from the disposal of plastic and synthetic products may not allow this to occur unless a stronger recycling program is initiated.

6. Perhaps the most distressing fact revealed in Figure 2-21 is that a sizable quantity of energy is not directed into useful work. Approximately 65 percent of the energy put into electrical generation is rejected (entropy). About 72 percent is rejected in the transportation sector. Consequently, the transportation sector and electrical generating sector are obvious places for technical improvement. Combined entropy losses in all sectors result in 49 percent energy rejection, leaving 51 percent for useful work. Society, therefore, loses almost as much as it uses.

ENERGY SYSTEMS

Just as production, communication, and transportation technologies can be studied by using the systems approach, so can energy technologies. The material covered in this text clearly shows that energy technologies are an extensive part of any society. A systems approach helps to identify, organize, use, and evaluate specific energy technologies much more easily.

All technologies have system components. These components include inputs, processes, outputs, feedback loops, and effects. (See Figure 2-22.)

Input

The input can also be called the command or objective. The input is that which needs to be accomplished by the energy technology. The input must help accomplish and complete the desired result or output. The input can also be considered a statement of the problem. To understand the input, use the following energy technology as an example: Statement of the problem: To generate an electric energy capacity of 500 megawatts from coal. This is called the input, the command, or the objective.

Process

The process is the technical concept or principle used to help accomplish the desired output or result. The process requires numerous resources for correct operation, including any of the following:

People	Tools	Time
Information	Energy	
Materials	Capital	

The processes in any energy technology are primarily designed to convert and change one form of energy to another. Some common examples include:

engines	boilers
transmissions	solar collectors
nuclear power plants	wind generators

Using the electrical energy example previously stated, a coal fired power plant is that technology used to process coal into electricity. The coal power plant requires the use of each of the above resources of people, information, materials, tools, energy for operation, capital (or money), and time.

Output

The output is considered the result of the process. The output is often called the industrial application. In the previous example, the output is to have available the 500 megawatts of electric energy generated from coal.

Effect

As technology is being studied, greater importance is being placed on the technological effect of using energy. Every energy technology always has some form of negative or positive effect on people and society. These effects must be studied to determine the overall impact of the energy technology system. Effects are often the social, environmental, economic, and individual consequences of the technology.

Systems Model

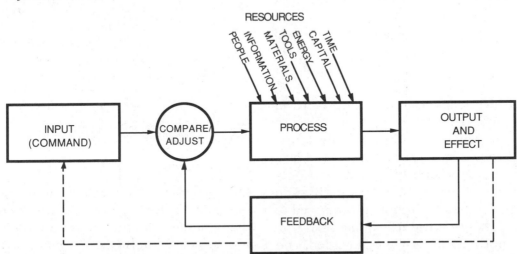

FIGURE 2-22 The systems approach provides an opportunity to examine various energy technologies. Note the dotted line from output and effect to input. As the consequences of certain technologies become evident (both positive and negative) new technologies or social changes may be developed to take advantage of or minimize the effect of the implemented technology.

For example, some negative environmental effects of converting coal to electricity include producing acid rain, scarring the land through strip mining, and depleting natural resources. Positive effects may include more electricity, more conveniences, and new and innovative products.

Feedback

Feedback is the technology used for monitoring the complete system. It is also considered a control system. Feedback takes on the form of a loop. The feedback loop exists so that the question can be asked: *Was the result or output correct and accurately accomplished?* Note in Figure 2-22 that two feedback loops were used. The first loop is from the output and effect of the proposed technology to compare/adjust. In the given example of the coal-fired power plant, the computer used for controlling and monitoring systems in the plant constitutes the first feedback loop. For example, sensors are used to check temperatures, steam-flow rates are monitored to control steam valves and traps, pressures are monitored to control coal input amounts, and so on. Thus, the first feedback loop is merely a method to monitor a proven technological system.

A second feedback loop in Figure 2-22 can be seen; it goes from the output and effect to the input or command (as shown by the dotted line). This feedback loop can be viewed as a method to monitor or redesign new technological systems. During this feedback portion of the loop, the result (output) is compared to the desired result (input). Sometimes the results do not match what was intended, and the input must be adjusted and changed to achieve the desired effect. For example, when coal-fired electrical power plants had to meet stricter legislation involving emissions, new technologies had to be developed and implemented for the existing power plants. In some cases, such technologies had to be approached from entirely different perspectives (a different input), each with new challenges and opportunities. For example, in the power plant mentioned earlier, the input is to develop technologies to burn fuel more cleanly. This input could have also taken a second form—to clean the spent gases as they move out of the smokestack. An example of an opportunity that could result from this challenge is to sell (for profit) any residue from the cleaning of the spent gases. Therefore, the second feedback loop can be viewed as a means to examine an implemented new or proposed technology (a different input) in view of social or environmental challenges or opportunities.

The systems approach of input, process, output, effects, and feedback works throughout all the energy technologies. Although the coal-fired electrical plant example was very broad in nature, the system's model can also be used to help study specific energy technologies—such as engines, solar collectors, wind energy systems, bioconversion systems, battery storage systems, drilling rigs, natural gas storage systems, and others.

ALTERNATIVE FUTURES

Cogeneration

One promising new electrical generation technology is called *cogeneration*. Cogeneration is the simultaneous production of power using the rejected thermal energy of a primary process. The result is that the same fuel is used twice. One such system is called the *combined cycle unit* and nears 55 percent efficiency. Combined cycle generating units are those that

Cogeneration (Combined Cycle)

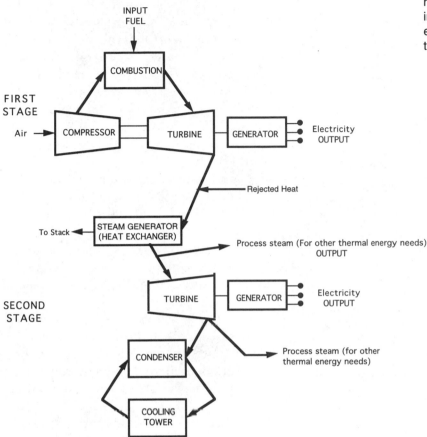

FIGURE 2-23 Cogeneration technologies, as illustrated here, offer improved efficiency to numerous electrical generation and manufacturing operations.

use exhaust heat to operate a second turbine/generator set, thus making the entire system more efficient. Figure 2-23 shows a combined cycle cogeneration unit. The first stage has a compressor, combustion chamber, and turbine that are used to drive the electrical generator. The rejected heat from the turbine (1st stage) is extracted through a steam generator (heat exchanger).

The extracted thermal energy is then used to operate a second turbine (2d stage) and generator. In this system, energy from the input fuel is utilized twice, thus improving efficiency. Note that steam can also be extracted for other thermal energy needs. Any remaining heat is sent to a condenser and cooling tower for removal.

POINT/COUNTERPOINT 2-1

TOPIC

Electrical Energy Growth

Theme: Electrical energy demand is increasing each year. It is projected to outpace supply before the end of the century. Where should the resources for this electrical energy come from? Possible resources could be from coal, oil, natural gas, nuclear, hydroelectric, solar, wind, and so on. Research this issue and attempt to answer the following questions in a classroom discussion or debate.

Discussion/Debate Questions:

1. Can the demand for electrical energy be reduced? If so, how could this be done without declines in the GNP or GDP? Should electrical energy usage be taxed? What consequences would this have on lifestyles and on the economy?

2. If nuclear energy is used more to provide electricity, how could consumer confidence be increased? What mandates would be needed to ensure safer power plant designs and disposal of spent fuel? What are the advantages and disadvantages of this power source (i.e., environmental and social)?

3. If coal is to be used to generate more electrical energy, how would the acid rain problems be resolved? What could be done about the problems associated with strip mining? What additional costs from making environmentally friendly changes would be passed to consumers?

4. If oil and natural gas will be used for electrical energy production, what might this do to the future availability of these resources for other energy sectors?

5. If hydroelectric power is to be used for electrical energy production, what are some of the social and environmental consequences?

6. Economically, will the electric power companies be able to build the additional plants that will be needed? What are the consequences if consumers do not want more power plants, but demand more electricity?

7. What can citizens of the United States learn from other countries regarding their electrical usage and demand?

SUMMARY & REVIEW

Summary/Review Statements

1. Six sectors—industrial, commercial, residential, electrical energy, transportation, and petrochemical—utilize energy today.
2. The United States constitutes approximately 5 percent of the world's population, yet consumes about 25 percent of the world's energy.
3. Energy supply and demand must be carefully controlled to keep prices constant.
4. If there is too great a supply of energy, prices will generally fall.
5. If there is too great a demand for energy, prices will generally rise.
6. Approximately 50 percent of all energy put into our society is lost as entropy.
7. Oil is the primary energy source for the transportation sector. The transportation sector consumes about 12 percent more petroleum than is produced domestically.
8. Electrical demand is expected to increase at a faster rate than can be supplied by utility companies.
9. Energy demand is increasing at a rate of approximately 2 to 4 percent annually.
10. Production of coal energy resources is projected to increase in the next 20 years. This projected growth will occur primarily in response to anticipated increased demand for electrical energy.
11. Electric utilities are both major producers and consumers of energy.
12. Cogeneration is using wasted or rejected heat to operate a second generation system.
13. The petrochemical industry, also called the nonenergy sector, uses natural gas and oil to manufacture plastics and other synthetic products. However, a strong recycling program is needed to reduce damage to the environment.

Discussion Questions

1. List ways in which energy demand could be reduced in the industrial sector.
2. The transportation sector utilizes and wastes large amounts of energy. What are several methods that could be used to reduce energy consumption in the transportation sector?
3. Which of the demand sectors are anticipated to grow most in the future? State why the growth is expected.
4. Why is the demand for electric energy growing so fast? What are some methods that our society could incorporate that would reduce electrical energy demand?

chapter 3

Energy Terminology

PURPOSES

Goals of Chapter Three

This chapter is designed for students who have a familiarity with, but not an exhaustive knowledge of, the science of physics. The terminology and mathematics employed provide insights and a better understanding of the applications of energy technology. Many of the terms defined in this chapter will also be useful when reading the remainder of this textbook and other literature on energy technology. The overall purpose of this chapter is to define and utilize various terms and principles of physics associated with the study of energy technology. At the completion of this chapter, you will be able to

1. Define and apply mechanical energy terminology.
2. Define and apply thermal energy terminology.
3. Understand concepts of matter, energy, and space relationships.

Terms to Know

By studying this chapter, the following terms can be identified, defined and used:

Accelerated Motion
Acceleration
British Thermal Unit (Btu)
Conduction
Convection
Deceleration
Degree-Day (DD)
Efficiency
Horsepower
Manometer
Radiation
Speed
Torque
Velocity

Introduction

The study of the physical nature of the universe has always been a fundamental pursuit of scientists and researchers. Physics is a basic science that encompasses a range of subject matter from subatomic particles to beyond the galaxies and the large arena of the nature of the universe. From this perspective, the physical sciences deal specifically with the nonliving aspects of nature. Thus, the study of physics is concerned with the concepts of matter, energy, space, and the relationships between them. This chapter illustrates several of these concepts and their associated relationships.

MECHANICAL ENERGY TERMINOLOGY

There are numerous concepts used to represent mechanical energy systems. The following selected list of concepts and definitions will help you better understand mechanical energy systems. In addition, a variety of mathematical computations has been provided to illustrate various applications of physics to mechanical energy.

Speed

It can be said that everything in the physical world is in motion, from the largest galaxies to the smallest subatomic particles. The study of the motion of objects allows better understanding of their behavior and helps us learn how to control them. Concepts that relate to *displacement* and distance to time should be introduced along with motion or *kinematics*.

The simplest type of motion that an object can experience is uniform motion in a straight line. When an object covers the same distances in successive units of equal time, it is said to move with constant speed. Thus, **speed** is the ratio of the distance traveled by an object to the time it takes to travel that distance. This is irrespective of the direction the object may take. *Average speed* is the ratio of distance traveled to the time elapsed, which represents an average value of the speed over time. Algebraically, average speed is represented as

$$Average\ Speed\ =\ \frac{distance\ traveled}{time\ elapsed}\ \ or\ \ \bar{v} = \frac{s}{t}$$

In the final equation, s represents the distance traveled and t represents the time elapsed. The bar over the symbol v denotes that the speed represents an average value for the interval t. The above mathematical equation can be illustrated with the following example. A salesperson drives 335 miles from St. Cloud, Minnesota, to Thunder Bay, Ontario, in 7 hours. What is the salesperson's average speed? Solve for the average speed as follows:

$$\bar{v} = \frac{s}{t}\ or$$

$$\bar{v} = \frac{335\ miles}{7\ hours}\ ;\ therefore,$$

$$\bar{v} = 47.86\ miles\ per\ hour$$

Speed is a scalar quantity that is independent of direction. A scalar quantity is specified by its magnitude. It consists of a number and a unit. Examples of scalar quantities are speed (20 miles per hour), distance (10 miles), and volume (18 liters). Note that Appendix A includes a complete list of English and metric conversions used in energy technology. In the example of the salesperson driving from St. Cloud to Thunder Bay, it was unnecessary to know either the speed of the car at any instant or the direction of its travel. Similarly, the average speed of a car traveling from Illinois to Massachusetts is a function only of the distance measured by the odometer and the time taken to make the trip. It makes no difference if the driver takes the scenic or direct route in determining the average speed of the automobile.

The scalar quantity by which that speed is most conveniently measured depends upon what objects are in movement. When the movement taking place is that of a car, train, or airplane, miles per hour or kilometers per hour are used. If the measurement is that of the speed of the subatomic particles of nuclear energy, then it is proper to express their speeds in terms of feet per second, kilometers per second, or even fractions of the speed of light. Many energy-related problems often require that a distinction be made between the scalar quantity speed and its directional counterpart, velocity.

Velocity

The terms *speed* and *velocity* are often used synonymously, but there is a clear difference between the two. Speed requires only one facet of information to define it completely—that is "how much." When a person says that a vehicle was moving at 30 miles per hour, this reveals its speed (i.e., how much). Velocity, however, requires not only how much but also "in what direction." Thus, velocity is a vector quantity that describes a distance traveled in a specific direction. *Velocity* may also be defined as the ratio of displacement (which considers both distance and direction) to a time interval. Displacement, in the study of motion, is the distance between the starting and ending positions of a moving object during an interval

of interest. Vectors such as velocity consist of a number, a unit, and an angular orientation—such as 10 meters per second, 30 degrees.

To illustrate this difference between the scalar quantity of speed and the displacement (or angular orientation) of velocity, consider the following example. Assume that a salesperson is traveling by automobile to three locations and ending where beginning—as illustrated in Figure 3-1. The salesperson's displacement has been represented by a solid line with arrows to illustrate the direction of travel. Note that the salesperson's displacement is represented by the vector D (D_1, D_2, and D_3). This displacement is completely independent of the salesperson's path (represented by the dotted lines S_1, S_2, and S_3). The car's odometer may show that a scalar distance of 900 miles was traveled, although the vehicle displaced only 700 miles. Also, note that this displacement has a constant direction of 130.2° (from the starting point to Location 1);

Displacement is a Vector Quantity

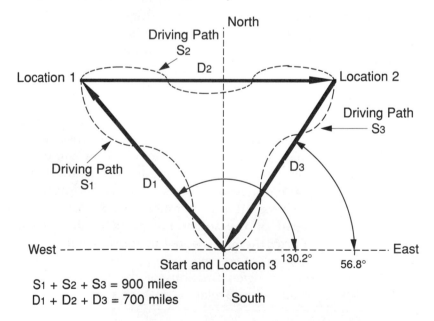

$S_1 + S_2 + S_3 = 900$ miles
$D_1 + D_2 + D_3 = 700$ miles

FIGURE 3-1 Displacement is completely independent of the path taken. While a car traveling to the three locations may record 900 miles on its odometer ($S_1 + S_2 + S_3$), the vehicle only displaced 700 miles, represented by $D_1 + D_2 + D_3$.

whereas, the direction of the automobile at any instant during the travel could take much wider or narrower angles, depending upon the route taken by the road.

To better understand the relationship between speed and velocity, it may help to apply the concepts. The following examples have been used to illustrate the difference between speed and velocity. Suppose the distance (s in Figure 3-2) is 600 miles, and the displacement (D_1) is 450 miles at 45°.

If the driving time is 12 hours, the average speed can be determined by the following example, where:

$\bar{v} = velocity \quad s = distance \quad t = time$

$\bar{v} = \dfrac{s}{t}$ or

$\bar{v} = \dfrac{600 \text{ miles}}{12 \text{ hours}}$; therefore,

$\bar{v} = 50$ miles per hour

However, the average velocity must consider the displacement magnitude and direction. The average velocity can be obtained by computing the ratio of displacement (D) to time (t). This is illustrated, using the same examples as

---

$\bar{v} = \dfrac{D}{t}$ or

$\bar{v} = \dfrac{450 \text{ miles, } 45°}{12 \text{ hours}}$; therefore,

$\bar{v} = 37.5$ miles per hour, 45°

Therefore, if the path of a moving object is anything other than a straight line, the difference between speed and velocity is one of direction and magnitude.

Velocity is a Vector Quantity

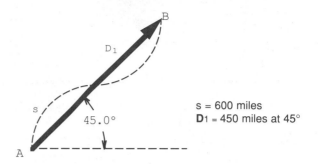

s = 600 miles
D_1 = 450 miles at 45°

FIGURE 3-2 Velocity is a vector quantity involving displacement, which takes into account direction. Speed, however, is not a function of direction.

Acceleration

Another important quantity that must be considered is that of *acceleration*. When a driver in a parked automobile begins to move the vehicle in a direction, the car is said to be accelerating. The driver has changed the speed of the automobile in a positive manner, increasing it with time. *Acceleration* can be defined as the time rate at which velocity changes.

In most cases, the velocity of a moving object changes as the motion continues. This type of motion is referred to as *accelerated motion*. Acceleration is called positive when the velocity increases, and negative when the velocity decreases. Mathematically, positive acceleration or an increase in speed is represented by positive numbers, while negative numbers are used to denote a reduction in speed. When speed is reduced, this type of acceleration is sometimes called *deceleration*. An example of accelerated motion is a car accelerating from 0 miles per hour to 55 miles per hour in 20 seconds. Mathematically, acceleration can be expressed as

$$Acceleration = \frac{change\ in\ velocity}{time\ interval}\ \text{or}$$

$$a = \frac{v_f - v_o}{t}$$

In this equation, v_o represents the body and its velocity at the beginning of the time interval. The body and its velocity at the end of the time interval are represented by v_f. The time is represented by the symbol t. To illustrate this equation, review the following example. Assume a boat moves with constant acceleration from point A to point B, as illustrated in Figure 3-3.

Acceleration of a Boat

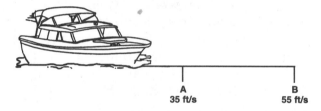

A
35 ft/s

B
55 ft/s

FIGURE 3-3 The boat moves with constant acceleration from point A to point B. At point A, the boat's speed is 35 ft/s. At point B, the boat's speed is 55 ft/s.

The boat's speed at point A is 35 feet per second. At point B, the boat's speed is 55 feet per second. If the increase in speed requires four seconds, the acceleration can be calculated as follows:

$$a = \frac{v_f - v_o}{t} = \frac{55\text{ft/s} - 35\text{ft/s}}{4\text{s}} = \frac{20\text{ft/s}}{4\text{s}} = 5\text{ft/s}^2$$

The answer is read as five feet per second per second or five feet per second squared. This means that every second the boat increases its speed by five feet per second. Having noted

that the boat had already acquired a speed of 35 ft/s when it reached point A, in one more second the boat has reached a speed of 40 ft/s. At two seconds the boat reaches a speed of 45 ft/s, and at three seconds the boat would be moving at 50 ft/s—assuming constant acceleration.

It should be further noted that the mathematical formula for determining acceleration can be used to solve for different quantities. Final speed, for example, could be determined from the original formula using the following equation:

$$v_f = v_o + at$$

In this equation, final speed (v_f) is equal to the initial speed (v_o) plus the change in speed (at).

One classic example of acceleration is shown in Figure 3-4. In this figure the gravity of the earth is the cause or force of acceleration. If a mass is released from a point above the earth's surface, gravity will accelerate the body as it falls at approximately 32.8 ft per s^2. The acceleration by gravity means that during each second, the velocity of a falling mass will increase by approximately 32.8 feet per second. This assumes zero air resistance to the falling body or mass.

Speed, Velocity, and Acceleration—Math Interface 3-1

Mathematics problems involving speed, velocity, and acceleration help to provide a greater understanding of motion. Solve the following applications problems:

1. A train reduces its speed from 70 to 15 miles per hour as it enters a rail-

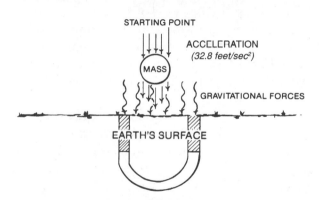

FIGURE 3-4 Gravity causes an object to be accelerated toward the earth's surface.

Force

Related to acceleration is the concept of force. *Force* is the cause of acceleration of a moving body or mass. In the acceleration example of Figure 3-4, gravity was considered the force. Force is defined as the pushing or pulling action of one object upon another. Force causes acceleration and deceleration. Examples of various common forces include muscle, wind, gravity, thermal expansion, and mechanical energy (such as that produced by a gasoline or diesel engine). Therefore, force can be thought of as that component which changes the acceleration or direction of a moving object or mass.

To conceptualize force and its relationship to acceleration, imagine that a concrete block is placed on a flat surface. Assume that a person pushes the concrete block from the side, and assume there is no friction. This push or force results in the block accelerating across the surface. If a second identical block were placed on top of the first and the same force applied, the acceleration would be half as great. If a third identical concrete block were added and the same force applied, the acceleration would be reduced to one third of its original value. Mathematically, this relationship is expressed as

$$Force = mass \times acceleration \quad \text{or} \quad F = ma$$

The primary unit of measure for force is called the newton (N). Using this measurement, mass is measured in kilograms (kg), and the acceleration unit in meters per second per second (m/s^2). One newton represents that force required to move a 1 kg mass at an acceleration of $1\ m/s^2$. Force may also be represented in pounds (i.e., pounds of force).

Problems of force, mass, and acceleration have many applications in energy-related fields—for example, in the production of automobiles and trucks. Automobile and truck engine designs are matched to the performance requirements and weights of the vehicles. If a vehicle were designed to pull light loads and maintain high fuel efficiency, then the engines manufactured would not need to generate much force. This is true because the mass of the vehicle would not be great, in order to maintain its high fuel efficiency. However, one could not expect to use this engine for a large truck that is needed to move heavy loads. A large truck and the weight of its load would have a far greater mass than that of the fuel efficient small truck. Thus, the engine designed for this larger vehicle would have to generate a greater force to meet its performance requirements.

Understanding the difference between the weight of a body and its mass is important since the terms are often applied differently depending on what is being investigated. When studying chemistry for example, weight is measured in kilograms. In the study of thermodynamics, mass is often represented in pounds. These inconsistencies result from the use of four different systems to describe mass and weight: the metric absolute (SI), the British absolute, the metric gravitational, and the British gravitational (Bgs) (also known as the English system of measurement). To help remove some of this confusion, this text only refers to SI and Bgs units. Thus the pound (lb) will always refer to weight, and the kilogram (kg) will always refer to the mass of an object.

Distinguishing Between Mass and Weight

The *weight* of a body refers to the force by which that object is pulled vertically downward by gravity. Weight, then, is simply the gravitational force acting on an object. To illustrate this, drop a pencil from eye height. If the pencil accelerates from gravitational force alone, it will increase its speed as it reaches the earth's surface. This relationship between weight and acceleration can be illustrated by

$$Weight = mass \times acceleration\ due\ to\ gravity \quad \text{or}$$
$$W = mg; \quad \text{also then}$$
$$m = \frac{W}{g}$$

Note from these mathematical expressions that the mass of an object (m) is equal to its weight (W) divided by the acceleration of gravity (g). Weight, additionally, has the same units as those of force, and the acceleration of gravity has the same units as those of acceleration. Therefore, when using the SI unit of measurement, weight in newtons is equal to mass in kilograms times gravity. The numerical value of gravity in the SI unit of measure is 9.8 m/s^2. When using the Bgs unit of measure, weight in pounds is equal to mass in slugs times gravity. Note that a slug is a unit of mass that is accelerated at the rate of one foot per second per second when acted upon by a force of one pound of weight.

The mathematical constant for gravity in the Bgs unit of measure is equal to 32 ft/s^2. (To determine other conversion factors between SI and Bgs systems of measurement, refer to Appendix A of this text.)

The following problems will help to demonstrate the relationships between force, mass, and acceleration.

Problem 1: Find the mass of an object from its known weight of 420 lbs.

Solution: Mass is equal to weight divided by gravity.

$$m = \frac{W}{g} = \frac{420\ lbs}{32\ ft/s^2} = 13.125\ slugs$$

Problem 2: What force would be needed to move a 100 lb object at an acceleration of 60 ft/s^2?

Solution: To find the force, first determine the mass of the object from its weight of 100 lbs. Then, determine force after the mass is known.

$$m = \frac{W}{g} = \frac{100 \text{ lbs}}{32 \text{ ft/s}^2} = 3.125 \text{ lbs/ft/s}^2$$

$$F = ma = 3.125 \text{ lbs/ft}^2\text{/s } (60 \text{ ft/s}^2) = 187.5 \text{ lbs}$$

Force, Mass, and Acceleration— Math Interface 3-2

Everyday life is full of examples where the equation $F = ma$ is put into practice. Solve the following problems to gain a greater understanding of the relationship between force, mass, and acceleration:

1. A 1000 lb wrecking ball is supported on a crane by a cable. Find the tension in the cable if a) the wrecking ball is accelerating upward at 9 ft/s² and b) downward at 12 ft/s². (Hint: First draw a diagram of the forces and the direction of the acceleration. For the first part of the question, if the wrecking ball is moving upward there is a negative acceleration. Similarly, a movement downward would create a positive acceleration.)

   Tension equals (a) _____

   Tension equals (b) _____

2. A farmer working in a field must move a large boulder weighing approximately 500 lbs. The tractor the farmer is using weighs 2000 lbs and accelerates at 15 ft/s². Although an example of this has not been given in the book, use your problem-solving skills to determine the following: If the drag of the ground on the boulder produces a force (i.e., friction) of 750 lbs, what total force must the tractor produce to move itself and the boulder?

   Total force equals _____

Pressure

In everyday language, people often use the terms of force and pressure interchangeably. However, there is a difference between the two. Before defining pressure, an example will be used to show a need for a distinction between force and pressure. Imagine a woman who weighs about 125 lbs standing on one foot on a wooden floor. If the woman is wearing a pair of tennis shoes and is standing with all her weight on her one heel, one would not expect to see any damage to the floor. Now imagine the same woman standing on the same floor and in the same position on one heel of a pair of spike heels. While the downward force is the same in each case, one might now find a dent in the wooden floor where the woman stood. This would be true because in the tennis shoes the woman's downward force was distributed over a much wider area than the one-quarter square inch of the spike heel. The idea of force and area is important when describing the type of situation presented above. Therefore, *pressure* can be defined as a force acting upon a unit of area. Algebraically, pressure is written as

---

$$Pressure = \frac{Force}{Area} \quad \text{or} \quad P = \frac{F}{A}$$

When force is measured in pounds and area is measured in square inches, a new term, pounds per square inch (psi), represents pressure. The equivalent metric unit would be newtons per square meter. Newtons per square meter are referred to as *Pascals*. Pressure in the metric system can then be stated as

$$Pascals \ (Pa) = \frac{newtons}{meters \ squared}$$

(Since Pascals are relatively small units, kilopascals (Kilo = 1,000) are often used.) See Appendix C.

To illustrate this mathematical relationship, the example of the woman in tennis shoes or spike heels will be used to determine the pressure being exerted on the floor. For this example it will be assumed the woman in the tennis shoes has her weight all on one foot, displaced over a surface area of 4 square inches; and the spike heels weight, all on one foot, is displaced over one-quarter of an inch surface area.

Solution: For the tennis shoes, pressure is equal to the force exerted divided by the surface area, as follows:

$$P = \frac{F}{A} = \frac{125 \ lbs}{4 \ in^2} = 31.25 \ lbs/in^2$$

Solution: For the spike heels, pressure is equal to the force exerted divided by the surface area, as follows.

$$P = \frac{F}{A} = \frac{125 \ lbs}{0.25 \ in^2} = 500 \ lbs/in^2$$

Obviously, the wooden floor must be able to resist a great deal of pressure if some people wear spike-heeled shoes. The application of

pressure is, therefore, an important consideration in deciding the strength of materials used in buildings and other technology.

Liquids and Gases

The concept of pressure is also useful when working with liquids and gases. Yet, there is a marked difference in the way a force acts on a liquid or gas (fluid) as compared to a solid. A solid is a rigid body that can absorb a force without a change in its shape. Like a solid, a fluid can sustain a force, but it can do so only in an enclosed surface or boundary. Additionally, the force from a fluid on the walls of the container is perpendicular to its walls, as demonstrated in Figure 3-5.

The concept of pressure is also useful when working with liquids and gases.

If holes are bored into the sides and bottom of a barrel of water, as illustrated in Figure 3-5, it can be observed that the force exerted by the water at the holes is perpendicular to the barrel's surface.

Fluids also exert pressure in all directions. This point can be easily illustrated with the following example. Suppose that a small pool float is pushed below the surface of the water. As the float is pushed down, the water exerts an upward pressure on the bottom of the float. The greater the force applied to the top of the float to submerge it, the greater the pressure exerted on the bottom of the float by the water. Also, if the float is pushed underwater and to the left, a force will be exerted from under the float and to the right to counteract the original force applied. Other pressures exerted by fluids are similar to those exerted by solids.

Force Exerted by a Liquid

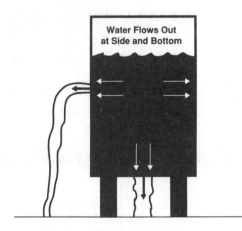

FIGURE 3-5 The force exerted by a liquid (or gas) is perpendicular to its retaining surfaces.

Just as solid objects exert greater forces at their base, fluids exert greater pressures as their depth increases. This is simply the result of the weight of the overlying liquid. The weight of the overlying fluid is proportional to its density, and the pressure at any depth is also proportional to the density of the fluid. Symbolically, this relationship is expressed as

---

Pressure (weight per unit area) = Density × depth

or $P = Dh$

To illustrate this equation, the following problem is offered:

Problem: A small freshwater dam is proposed to generate hydroelectric power. If the water has a depth of 120 feet at its retaining structure, what pressure will be exerted to the base of the dam?

Solution: The weight of the overlying fluid is proportional to its density. In this example, the weight density of water is 62.4 lb/ft^3. Therefore,

$P = Dh$ or

$P = 62.4 \text{ lb/ft}^3 (120 \text{ ft})$

$P = 7488 \text{ lb/ft}^2$

Psig and Psia Scales

The pressure presented in the previous section examined only the fluid itself. This pressure did not consider the downward force exerted from the column of air that extends from the top of the atmosphere to sea level. This force, known as *atmospheric pressure*, is defined as the average pressure exerted by the atmosphere at sea level. This average pressure measures 14.7 lb/in^2 and may also be called *1 atmosphere.*

There are two pressure scales commonly used in industry. They are called *gauge* and *absolute pressure*. Gauge pressures measure the pressure of a gas or liquid above or below the pressure of the surrounding atmospheric air. Absolute pressure is defined as the pressure of the atmosphere, also called barometric pressure. To show the relationship between the two: at a barometric pressure of 14.7 psi, a gauge pressure will read zero. The most common gauge pressure unit is referred to as psig, or *p*ressure per *s*quare *i*nch on a *g*auge. Most pressure gauges, such as the one illustrated in Figure 3-6, read psig pressures.

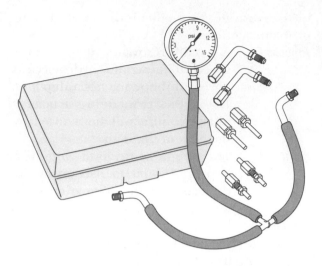

FIGURE 3-6 This gauge measures psig pressures. Although the gauge reads in psi, it means psig readings.

In many mechanical energy calculations, the absolute pressure of the confined gas or liquid is required. Absolute pressure is read as psia, or *pressure per square inch absolute*. The two scales illustrated in Figure 3-7 demonstrate the difference between gauge and absolute pressures. Note that 0 on the gauge scale (left) is equal to 14.7 psi on the absolute scale.

Thus, to convert from gauge pressure to absolute pressure, add 14.7 units to the gauge pressure. For example, if the pressure in a system were 12 psig, then the equivalent absolute pressure would be 26.7 psia (12 + 14.7). Also, note that the psig scale can be read below 0, as in the case of an absence of pressure. Below 0, the pressure is read as negative pressure (called a vacuum). Negative pressures can be read as positive numbers on the psia scale. Often, working with all positive pressure units (using the psia scale) is easier than working with negative and positive pressure units on the psig scale.

Measuring Pressures

Absolute pressure, also called barometric pressure, is measured with the use of a barometer and is usually expressed in terms of centimeters or inches of mercury. Other instruments can also be used to measure such pressures. One such instrument is called a *manometer*, as shown in Figure 3-8.

A *manometer* consists of a U-shaped tube containing a liquid (usually mercury or water). Because of the difference in densities, water is used for measuring lower pressures while mercury is used for higher pressures. When both ends of the U-shaped tube are open, the liquid inside seeks its own level. This is caused by 14.7 lb/in² of pressure being exerted at each of the open ends. If one end is connected to a pressurized chamber, the liquid rises in the open tube until the pressures are equalized.

Difference Between Absolute and Gauge Pressures

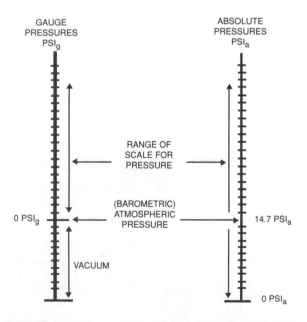

FIGURE 3-7 These two scales illustrate the difference between psig and psia readings.

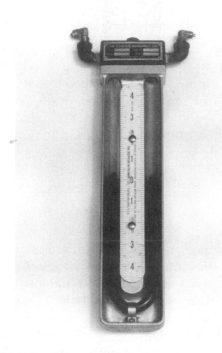

FIGURE 3-8 This is an example of a manometer used to measure pressures and vacuums.

The amount of change can be measured on an inch or a centimeter scale found within the center of the U-shaped tube. Gauge pressure can then be determined by taking the difference between the two levels of liquid (i.e., the difference between the absolute pressure in the chamber and the atmospheric pressure at the open end).

Figure 3-9 illustrates a manometer with a pressure being applied to the left side. Note the fluid is pushed down on the left side a distance of 1 inch. The fluid also rises 1 inch on the right side. The total reading for this pressure would be read as 2 inches of water.

Measuring barometric or absolute pressure is not the only use for a manometer. Common examples of pressures measured by a manometer in a gasoline, diesel, or turbine engine or aircraft include

exhaust manifold pressure
intake manifold pressure
turbocharger boost
crankcase pressure
altimeter (aircraft)

There are also charts available to convert inches of mercury or water to either psig or psia scales. The following conversion factors show the relationships:

1 inch of water = 0.0735 inches of mercury
1 inch of water = 0.0361 psig
1 inch of mercury = 0.491 psig
1 inch of mercury = 13.6 inches of water
1 psig = 27.7 inches of water
1 psig = 2.036 inches of mercury

The manometer is most often used in low-pressure applications. Where high pressures are encountered, Bourdon gauges are typically used. The Bourdon gauge is often found on high-pressure equipment—such as compressors, water and steam lines, boilers, and similar equipment. Bourdon gauges, as illustrated in Figure 3-10, are normally fitted with a dial and

Manometer

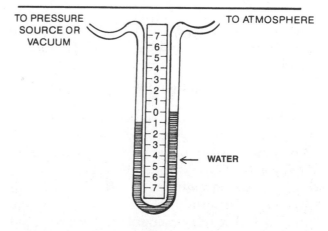

FIGURE 3-9 This manometer is read by applying a pressure to the left side and measuring in inches how far the liquid has moved. In this case the liquid moved one inch on each side or a total of 2 inches of water.

Bourdon Gauge

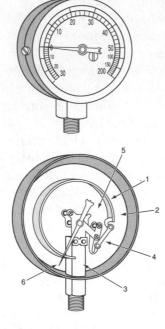

FIGURE 3-10 Shown is a Bourdon pressure-vacuum gauge. The left and bottom sides of the dial shows readings from 0 to 30 inches of mercury. The right and top sides shows a pressure scale from 0 to 200 lb/in². A cutaway view illustrates the interior working parts of the gauge.

pointer. These gauges can be calibrated to read in pounds per square inch, kilopascals, inches, or millimeters of mercury.

As illustrated in the cutaway of Figure 3-10, the circular tube (1) is made of a thin piece of flexible brass. One end of this brass tube (2) is free to move while the other end (3) is fixed. When the Bourdon gauge is connected to a line, gas (or liquid) enters the tube; the resulting pressure straightens the brass tube. This pressure causes the free end (2) to move outward. A linkage (4) is connected to a gear (5) that rotates the pointer (6). This pointer then moves across a scale that has been calibrated to show the measured pressure units.

Work

Work has a precise definition in the study of mechanical energy and energy technology. This definition has been refined by scientists in the study of physics as the result of applying a force to a body or mass through a certain distance when both the force and the movement are in the same direction. Thus, work is actually energy in transition. For work to be done, three specific aspects are necessary:

1. There must be a force applied to an object.
2. The force must act through a distance. (This distance is called *displacement.*)
3. The force must have a component (i.e., vector) along its displacement.

The calculation of the component of the force in its direction is an exercise in trigonometry. As such, one multiplies the force by the cosine of the angle (*cos θ*) between the direction of the acting force and the direction of the motion that it causes. Mathematically, work (*W*) is equal to the product of the displacement (*s*) of an object from the influence of a force (*F cos θ*) and can be represented by the following equation:

$$W = (F \cos \theta)s$$

The illustration in Figure 3-11 represents an automobile being pulled along a straight, level street by a steel cable that is slightly above the level of the street. It could be assumed this cable is connected to a tow truck at an angle of 10° to the street. Further assume that 2750 pounds of force are applied to the cable (i.e., pulling the automobile).

To determine the amount of work that must be done by the tow truck to move the automobile a distance of 20 feet, the equation for work is utilized:

$W = (F \cos \theta)s$ or
$W = (2750 \text{ lb})(\cos 10°)(20 \text{ ft})$; therefore,
$W = 54,164.43 \text{ ft lb}$

As shown earlier, force is measured in pounds, and distance is measured in feet. Thus, work is measured in foot-pounds (ft lb) in the Bgs unit of measure. Work would be measured as newton-meters in the SI system. When newtons are multiplied by meters, the resulting measure is called joules (J). Two conversion factors helpful in solving work problems are:

1. $1 \text{ J} = 0.7376 \text{ ft lb}$ (the work done by 1 newton acting through a distance of 1 meter)
2. $1 \text{ ft lb} = 1.356 \text{ J}$ (the work done by 1 pound acting through a distance of 1 foot)

Note that if a force is applied and the object does not move, no work has been done. In this case, the applied force would not have been sufficient to move the object because the force would have been perpendicular to the displacement ($\cos 90° = 0$). To illustrate what would happen if the force occurred perpendicular to the displacement, solve for work using the example previously used in Figure 3-11:

$W = (F \cos \theta)s$
$W = (2750 \text{ lb})(\cos 90°)(20 \text{ ft})$
$W = (2750 \text{ lb})(0)(20 \text{ ft})$
$W = 0$

While a great deal of energy may be expended, no work has been performed in this example. Plainly, the amount of work will be greatest when the angle of the force is 0° to the movement. Using the towed car as an example, if the angle were zero rather than 10°, more work would be performed. In this instance, the component would have its maximum value of 1 ($\cos 0° = .9998477$). Worded in another way, the force has been applied directly in line with its displacement. This reduces the equation for determining work to

$Work = Force \times Distance$ or
$W = Fs$

Power

Power is the measure of work being done over time. More specifically, power is considered the rate at which work is being done. For example, in an automobile, work is done when the vehicle is displaced by pushing (rear-wheel drive) or pulling (front-wheel drive). The pushing or pulling motion is accomplished by some mechanical means. A vehicle, for example, may move 60 miles over a period of 1 hour. The amount of power determines the speed at which this movement is accomplished. Note that power is required to overcome resistance,

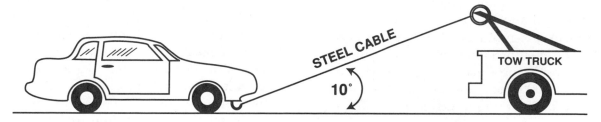

FIGURE 3-11   An automobile is being pulled by a tow truck at a specific angle. Therefore, the work required can be calculated.

friction, and gravity. Without these to overcome, no power would be needed to keep machines working once they were set in motion.

The unit of power in the SI system is defined as the number of joules being performed per second, or J/s. Joules per second can also be called watts (W) and are a measure of both electrical power and mechanical power. For a complete listing of power and energy units, refer to Appendices N and O.

Horsepower

The unit of power in the English system of measurement is called *horsepower*. The term *horsepower* has commonly been used in reference to engines or motors. One horsepower (in the English gravitational system) is defined as the energy required to lift 550 pounds of weight 1 foot in 1 second. One horsepower is also defined as lifting 33,000 pounds of weight 1 foot in 1 minute (33,000 foot-pounds of work per minute). Generally, the movement of the weight is in a straight-line direction.

To get an idea of how much work is called for to produce a certain horsepower, use the following example: A person weighs 200 pounds and must climb a ladder that is 55 feet tall. *Problem:* How much horsepower is needed to climb the 55-foot ladder in one minute? *Solution:* The person will lift 200 pounds a distance of 55 feet or 11,000 foot-pounds in 1 minute (200 lbs × 55 ft). This is about a third of a horsepower (1 horsepower is 33,000 foot-pounds in one minute). If the person were to climb the 55-foot ladder in half the time, the person would produce twice as much horsepower (11,000 foot-pounds in 30 seconds is equal to 22,000 foot-pounds in 1 minute).

Torque

The *moment of force* or *torque* is defined as the tendency to produce a change in rotational motion. Torque can, therefore, be thought of as a twisting force. A torque wrench is one device used to measure torque when tightening bolts and nuts. The twisting force from the torque wrench causes a rotational motion on either the bolt or nut.

Most mechanical energy conversion and power transmission systems have outputs measured in torque. Some twisting force (torque) applications include motors, gears, transmissions, engines, and wheels.

Torque is expressed in foot-pounds in the Bgs system of measurement. Newton-meters are the unit used to measure torque in the SI system. When measuring torque, time is not considered.

Torque can be measured directly from rotating energy converters, such as gasoline and diesel engines. This is done with the use of a *dynamometer*. A dynamometer is a loading device that applies a mechanical, electrical, or hydraulic (fluid) load (resistance) to a rotating shaft so that work can be accomplished. Figure 3-12 provides an illustration of a dynamometer.

When operating a fluid type dynamometer, the engine drive shaft is attached to a turbine wheel inside the dynamometer. As the engine is operated (i.e., increases its revolutions per min [rpm]), a fluid is allowed to enter the dynamometer turbine. As more fluid is added, the engine shaft speed is reduced by the additional load unless the throttle is open to compensate. A small torque sensor placed on the outside of the dynamometer housing measures the engine torque being produced. This sensor is normally positioned a certain distance (1 foot) from the center line of the main dynamometer shaft. As the torque sensor

FIGURE 3-12 This dynamometer is used to extract and measure the torque on engines. (Courtesy of Clayton Dynamometer)

records the twisting pressure, torque in foot-pounds is registered. Horsepower from an engine can be calculated from the torque measurement by using the following formula:

$$Horsepower = \frac{Torque \times revolutions\ per\ minute\ (rpm)}{5252}$$

Torque and rpm are read directly from the dynamometer. The number on the bottom part of the formula, 5252, is a constant and is determined based on the distance the torque sensor is placed from the center of the dynamometer shaft. For example, if an engine on a dynamometer produces 230 foot-pounds of torque at 3000 rpm, 131.38 equivalent horsepower was created (230 foot-pounds × 3000 / 5252).

Horsepower and Watts

Horsepower can be measured in watts or kilowatts in the SI system. In the United States, the use of the terms *watt* and *kilowatt* are used almost exclusively in connection with electrical power, while the term *horsepower* is primarily used for mechanical power. This practice is only a convention. It is perfectly possible to speak of a 0.1-horsepower light bulb (i.e., 75-watt light bulb) or a 141,740-watt automobile engine (i.e., 190-horsepower engine). The conversion factors for horsepower and watts are:

1. 1 horsepower = 746 watt = 0.746 kilowatts
2. 1 kilowatt = 1.34 horsepower

Mechanical Advantage

A machine is used to transmit a force into useful work in mechanical systems. These machines may be very simple in design, such as a screwdriver or chain hoist, or they may be very complex, like an automobile. Regardless of their complexity, machines are made to produce a *mechanical advantage*. Mechanical advantage (M_A) can be defined as the ratio of the output force (F_o) to the input force (F_i). Symbolically this is represented as

$$M_A = \frac{F_o}{F_i}$$

A simple machine, such as a chain hoist, can be used to illustrate mechanical advantage. Assume 20 pounds of force have been applied to a chain hoist at point A, for a distance of 3 feet—as shown in Figure 3-13. The weight lifted at point B is 120 pounds and is moved a distance of 0.5 feet. The mechanical advantage of this machine is then 20 to 120, or 1 to 6. This means that for every pound applied to the input at point A, 6 pounds can be lifted at point B if there is no friction.

When the final analysis of the total energy system is calculated, the total work put into the

A machine is used to transmit a force into useful work in mechanical systems.

Chain Hoist

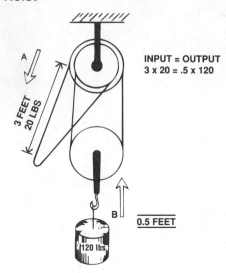

INPUT = OUTPUT
3 x 20 = .5 x 120

3 FEET
20 LBS

A

B

0.5 FEET

120 lbs.

FIGURE 3-13 This chain hoist enables 20 pounds of force to lift a weight of 120 pounds.

machine must equal the total work output by the machine (plus the energy needed to overcome the friction of the chains). This means that the force output cannot be increased without surrendering some distance. Note that the distance of the output has been reduced to 0.5 feet. Since the output force increased by a factor of six, the output distance must be decreased by a factor of six. This is because of the following:

Work Input = Work Output   or
Force × Distance Input = Force × Distance Output

The component reduced in the previous example was the distance that the 120 pounds were lifted. Using the above formula, note that 60 foot-pounds of work input (3 feet × 20 pounds) = 60 foot-pounds of work output (0.5 feet × 120 pounds). Using the formula for mechanical advantage, a range of speeds and output distances can be achieved within any mechanical energy system.

Levers

Perhaps the oldest and most useful machine used is the *lever*. A lever consists of a rigid bar that is pivoted at a certain point. The point at which the lever is pivoted is called the *fulcrum*.

Levers are used to change forces and distances moved in mechanical energy systems. Levers provide a form of mechanical advantage since the input and output work remain the same, yet different forces or distances can be achieved. Levers are used on such machines as robots, large cranes, tape players, steering systems, and so on. The lever may be used to increase a force, lift, or move an object in each of these machines.

The three types of levers used in mechanical energy systems are called *first-class*, *second-class*, and *third-class levers*. Figure 3-14 shows each type of lever for comparison.

Types of Levers

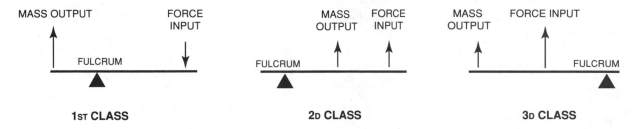

FIGURE 3-14 Three types of levers are commonly used on mechanical systems: first-, second-, and third-class levers.

First-Class Levers

The most common reason for using a first-class lever is to increase output forces while decreasing the distance moved. The amount of force gained is dependent upon the position of the fulcrum. If the pivot point is moved to the left in the drawing of the first-class lever, then more mass can be lifted. Common examples of first-class levers are the use of a teeter-totter or a crowbar (as illustrated in Figure 3-15).

First-class levers can also be used to gain distance instead of force. This can be accomplished by moving the fulcrum to the right. Once the pivot point moves past the center, then distance will be gained while losing force at the output.

1st Class Lever

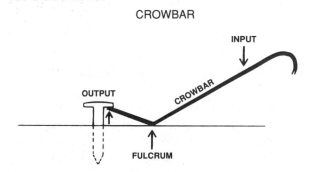

FIGURE 3-15 A crowbar is an example of a first-class lever.

Second-Class Levers

A second-class lever is used to increase output forces while allowing the output distance moved to decrease. An example of this class of lever, shown in Figure 3-16, would be a hand water pump or a hydraulic jack. Note that the arrangement of the input, output, and pivot points differ from that of a first-class lever.

Third-Class Levers

The characteristics of a third-class lever are such that input forces are greater than the output forces, but there is an increase in output distance. An example of this type of lever might be an automobile windshield-wiper mechanism, shown in Figure 3-17.

Application

The term *application* is often used in the study of energy technology to denote the use or function of an energy source. For example, solar collectors can have several applications—such as heating water, heating forced air, operating electrical appliances, heating swimming pools, or drying grain. Gasoline engines can be used in several applications, including such machines as automobiles, snow blowers, lawn mowers, and garden tractors. All energy converters must be put to the correct application to avoid losses of economy and efficiency. For example, engines must be correctly sized for the specific application.

2nd Class Lever

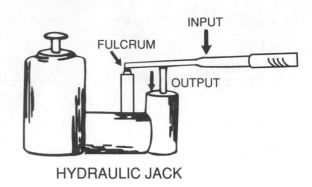

HYDRAULIC JACK

FIGURE 3-16   This hydraulic jack uses the principles of a second-class lever.

3rd Class Lever

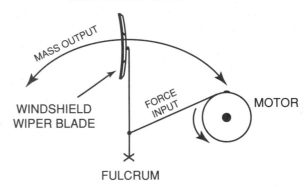

FIGURE 3-17   This windshield-wiper mechanism is an example of a third-class lever.

THERMAL ENERGY TERMINOLOGY

There are numerous thermal energy terms used throughout the literature of energy technology. These terms relate to heat, heat transfer, and heating degree days. The following section provides an overview of the terms and their relationship to thermal energy.

Heat and Temperature

The use of thermal energy is important in the conversion of various fuels to allow work to be performed. As noted in chapter 1, when the atoms and molecules of an object gain kinetic energy, there is an increase in the temperature of the object. This energy is known as *thermal energy*. Thus, thermal energy, or *heat*, is energy in transition. When measuring how hot or how cool a body is, its temperature is usually used.

Temperature is a measure of the average kinetic energy of the random moving molecules within an object. As such, temperature is a scalar quantity that determines the direction of thermal energy between any two objects (or systems) in contact. Temperature is usually measured on one of three scales; Celsius, Fahrenheit, or Kelvin. For a simple comparison, note the following:

	°C	°F	°K
Boiling point			
of Water	100	212	373
Water Freezes	0	32	273
Absolute Zero	−273.15	−459.67	0

Heat is always transmitted from a higher temperature area or mass to a lower temperature area or mass, as shown in Figure 3-18. In this example, one side is of a greater temperature; therefore, the transition takes place. Both sides will eventually equalize, and their temperatures will remain the same if no additional thermal energy is added. The greater the difference in temperature between the two masses, the faster the heat will transfer from the hotter to the colder area. Often, differences or changes in temperature are represented by the Greek letter Delta (Δ) and the symbol T, shown as ΔT.

Direction of Heat Transfer

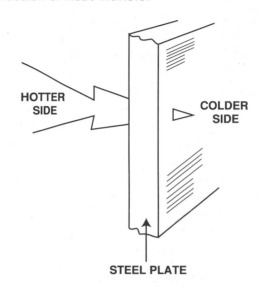

FIGURE 3-18 Heat transfer is always from a warmer area to a colder area.

The energy that makes up heat results from increased molecular movement. All substances are made of molecules, which are continually vibrating and moving. If thermal energy is applied to a mass, energy is added to these moving molecules. This addition of energy causes the molecules to increase their acceleration and vibrating motion. The greater the acceleration and vibration of the molecules, the greater the heat energy and temperature of the substance.

As long as there is molecular movement in a body or mass, heat is present. Molecular movement is said to stop at absolute zero. As will be demonstrated later in this text, this concept becomes important when studying the principles of operation involved in energy devices such as heat pumps and air-conditioners and in studying the principles of superconductivity.

Heat Transfer

Heat can be transferred from one substance to another in three ways: conduction, convection, and radiation. All three methods are integral to the study of heat-energy transfer. *Conduction* is defined as the transfer of heat energy by molecular collision. In conduction, thermal energy flows by the contact from a body or mass at a higher temperature to a body or mass at a lower temperature. Figure 3-19, for example, illustrates that if the coils on a range are at a higher temperature and the pan on the coils at a lower temperature, thermal energy is transferred from the coils to the pan through conduction.

The key consideration is that both bodies are at rest and are touching each other during the heat-transfer process. Thus, the thermal energy is transferred from molecule to molecule within and between materials.

Thermal energy is transferred through substances at different rates. Copper, for example,

Transfer of Thermal Energy through Conduction

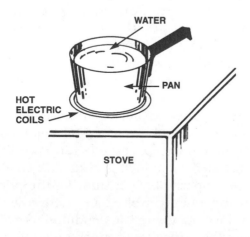

FIGURE 3-19 Heat is transferred by conduction from the hot electric coils on the stove to the cooler pan.

transfers thermal energy rapidly; on the other hand, fiberglass transfers thermal energy more slowly. Materials that rapidly transfer thermal energy are called *conductors*. Materials that slowly transfer thermal energy are poor conductors, the poorest being called *insulators*. Since fiberglass is such a poor thermal conductor, it is a popular product for thermal insulation applications.

The ability of a material to transfer heat depends a great deal upon its molecular structure. Generally, gases are poor conductors of heat since they have large spaces between their molecules. On the other hand, solids are much better conductors. However, some solids—such as cork, asbestos, brick, concrete, and glass—are poor conductors of heat. The ability of a material to transfer heat is expressed by an experimentally determined quantity called *thermal conductivity*. This concept is presented in more detail in chapter 8.

Convection is defined as that method which transfers thermal energy from a body or mass through a moving fluid (liquid or gas). Convection will convey the thermal energy away from a hot body to a colder substance. Note first that warm air on earth rises. It rises because it is less dense than colder air. Convection can be illustrated by reviewing the operation of a hot air balloon: As air is heated in a hot air balloon by the burning of a fuel, the air in the balloon will become less dense causing the balloon to lift into the higher, cooler atmosphere. This same principle is utilized in convection heating.

Convection heating can be demonstrated by an electric heater. As the warm air rises from the heater, the cool air moves lower since it is more dense (see Figure 3-20). Thus, thermal energy is being transferred by convection as the warmer air rises and the cooler air drops.

Convective Loop

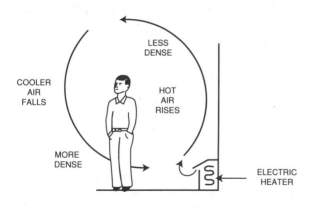

FIGURE 3-20 A convective loop of heat transfer is produced because hot, less dense air rises, while colder, more dense air falls.

The effect is that the air cycles around and around causing a *convection current* or a *convective loop* to be developed.

As noted in chapter 1, the movement of energy by **radiation** allows heat to be transferred from one substance to another by electromagnetic waves. Electromagnetic waves are generally measured by their frequency. Any object that has heat in it gives off electromagnetic waves at a specific frequency. When an electromagnetic wave strikes another object, some of its energy is absorbed and converted to thermal energy. Typically, the hotter the object, the higher the frequency (short wavelengths). The colder the object, the lower the frequency (long wavelengths). For a more detailed description of electromagnetic waves, refer to chapter 8 on solar energy resources.

Radiant or electromagnetic waves are transmitted at the speed of light (186,000 miles per second). The energy is emitted from atoms and molecules because of internal changes in the molecular structure. The total amount of thermal energy that is transmitted through radiation increases with increases in temperature.

Radiant energy, unlike conduction, does not transfer a great deal of heat to the medium (e.g., air) through which it passes. However, radiant energy is best transferred through a vacuum. For example, the radiant energy that is transferred from the sun to the earth moves through the vacuum of space. In this process, the electromagnetic waves are either absorbed by objects on earth or are reflected back into space. Note that lighter and brighter colors reflect more energy, but darker colors can absorb more. This fact is important when designing solar collectors.

British Thermal Unit (Btu) and Calories

British Thermal Units (Btu's) provide a method of measuring amounts of thermal energy. A Btu is a thermal unit of measurement in the Bgs system. A calorie is a thermal unit in the SI system of measurement. Btu's will be used throughout this textbook since the calorie is less frequently used by the energy industry within the United States.

One Btu is the amount of thermal energy necessary to raise the temperature of 1 pound of water 1 degree Fahrenheit under standard atmospheric conditions. This is close to the amount of thermal energy released by a wooden match burned completely to ash. A calorie is the thermal energy required at a pressure of 1 atmosphere to raise 1 gram of water 1 degree Celsius. Since the calorie is such a small unit, 1000 calories or one kilocalorie is often used. One Btu is equal to 252 calories. One kilocalorie is equal to 3.968 Btu's.

To differentiate between the 'calorie' used as a measure of thermal energy and 'calorie' used as a measure of the energy in food, a convention will be used. This convention will refer to the energy in food as Calories, with an uppercase 'C.' This is also known as the large calorie. The food Calorie is equal to 1,000 standard calories.

Heat energy and Btu's can also be converted into mechanical energy. Theoretically, 1 Btu of heat energy will produce 778 foot-pounds of mechanical energy. Btu's are used in numerous energy applications. The cooling capacity for air-conditioning, for example, is measured in Btu's. The cooling capacity is the amount of thermal energy that an air-conditioning unit can remove from the air in an hour. The removal of 12,000 Btu's in 1 hour is defined as 1 ton of refrigeration. Poor home insulation may lead to a loss of 20,000 Btu's per hour during the cold winter months.

Degree-Days

A *degree-day (DD)* is a measuring standard used for determining the severity of winter or summer. There can be either heating or cooling degree-days. Heating degree-days are used in the winter; cooling degree-days are used in the summer. Degree-days represent important data for determining the total energy needed by the population of a city during each heating or cooling season. The degree-day is the difference between a fixed setpoint and the averages of the high and low temperature of the day. The fixed setpoint is 65° F. This is the temperature at which a residential home normally needs neither energy for heating or air-conditioning. This temperature is the base. To find the average of the high and low temperatures, the sum of the highest and lowest temperatures is divided by two. The mathematical formula for calculating a degree-day (DD) is:

$$Degree\text{-}Day = 65 - \frac{High + Low\ Temperature}{2}$$

To illustrate the use of a degree-day, assume a high temperature of the day at 60° F and a low temperature of 25° F, the heating degree-day can be calculated by:

$$Degree\text{-}Day = 65 - \frac{High + Low\ Temperature}{2} \quad \text{or}$$

$$Degree\text{-}Day = 65 - \frac{60 + 25}{2} = 22.5$$

Therefore, the heating degree-day for this day would be 22.5. The sum of degree-days for a heating or cooling season serves as an indication of the total energy needed. By using the total heating or cooling degree-days, one can learn which seasons were colder or warmer than normal. This data is used by the United States Department of Energy to determine increases or decreases in energy use for a given season.

An example of normal heating and cooling degree-days for a typical midwestern city is shown in Figure 3-21. Degree-Day data for other geographical areas can be obtained from the local office of the National Weather Bureau. Figure 3-22 shows a map of the United States with nine zones. The total heating degree-days as derived from the National Weather Bureau are shown for each zone.

Applying Thermal Energy Terms— Math Interface 3-4

There are many thermal energy terms and applications in the study of energy technology. These terms and applications can be better understood by solving mathematical problems. Solve the problems below:

1. A thermal energy system produces 1,260,000 calories. What is the equivalent amount of Btu's being produced?

Btu's equal _____.

2. What are the degree-days produced if the high temperature of the day is 50° F and the low temperature of the day is 12° F?

   Degree-days equals _____.

3. A hot-water tank for a commercial building has 8000 pounds of water in it. The water must be heated from 70° F to 120° F in a period of one day. How many Btu's must be put into the water during this time? How many calories?

   Btu's equal _____.

   Calories equal _____.

Average Degree-Days

Month	Heating Degree-Days	Cooling Degree-Days
September	68.1	
October	326.1	
November	716	
December	1078	
January	1159	
February	930	
March	723	
April	298	
May	68	
June		300
July		600
August		400

FIGURE 3-21   Both heating and cooling degree-days are used to determine the severity of summer and winter energy use.

EFFICIENCY TERMINOLOGY

Efficiency

As society becomes more energy conscious, the need for increased efficiency in conver-

sion technology becomes a key issue in designing energy systems. The more efficient an energy system becomes, the less energy it will consume. Energy losses are best reduced by improving the efficiency of the energy conversion system and changing consumer habits.

First Law Efficiency

Efficiency (also called *First Law Efficiency*) is the ratio of output to input work in an energy system. Efficiencies are expressed as percentages and are always less than 100 percent. This less-than-perfect efficiency is a result of numerous practical difficulties. The losses resulting from friction and from heat, through conduction and radiation, prevent machines from obtaining their maximum efficiency. Thus, some energy will always be downgraded because of friction and heat losses. As discussed in chapter 1, this is called entropy losses. The most efficient machines are those that reject the least amount of thermal energy for a given work output. The mathematical formula for calculating First Law Efficiency is

$$Efficiency\ Percentage = \frac{output\ energy}{input\ energy} \times 100$$

To illustrate the use of First Law Efficiency, assume that 500 Btu's were absorbed by a solar collector and that 300 Btu's were able to be removed for use. The efficiency of the solar collector could be calculated by

$$Efficiency\ Percentage = \frac{output\ energy}{input\ energy} \times 100 \quad \text{or}$$

$$Efficiency\ Percentage = \frac{300\ Btu\ output}{500\ Btu\ input} \times 100 = 60\%$$

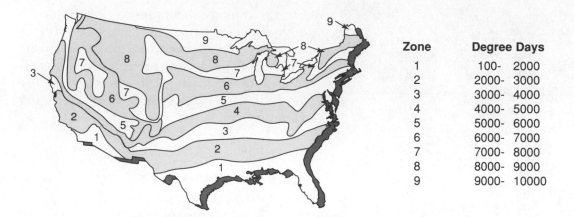

Zone	Degree Days
1	100- 2000
2	2000- 3000
3	3000- 4000
4	4000- 5000
5	5000- 6000
6	6000- 7000
7	7000- 8000
8	8000- 9000
9	9000- 10000

FIGURE 3-22 This chart shows the average heating degrees accumulated over a period of one year at different zones throughout the United States.

This means that 40 percent of the input energy has been downgraded due to poor insulation, reflection from the solar collector, and other problems. Similarly, the efficiency of a home heating system is reduced when heat escapes through windows, doors, and poorly insulated walls and ceilings. A typical gasoline engine ranges in overall efficiency from 22 to 30 percent. This indicates that for every 100 units of energy input, only 22 to 30 units yield power. The remaining units are downgraded in their usefulness through exhaust, friction, cooling, and so on.

Mechanical Efficiency

Another way to measure efficiency is to measure how efficient the mechanical systems are within a machine. *Mechanical efficiency* is the relationship between the theoretical amount of work required to do a certain job and the actual amount of work used. To illustrate mechanical efficiency, assume that a machine (a pump), required to do a certain job (say irrigation), uses 205 horsepower and the theoretical horsepower required to do the same amount of work is 185. The mechanical efficiency of this engine can be calculated. The formula and how to apply it are shown using the irrigation pump example:

$$\text{Mechanical Efficiency Percentage} = \frac{\textit{Theoretical horsepower}}{\textit{Actual horsepower}} \times 100 \quad \text{or}$$

$$\text{Mechanical Efficiency Percentage} = \frac{185 \text{ horsepower}}{205 \text{ horsepower}} \times 100 = 90\%$$

The losses on any mechanical system are primarily due to friction and to heat, through conduction and radiation. If these can be reduced on a mechanical system, the mechanical efficiency increases. Alternatively, the efficiency will decrease if friction and heat loss increase in a mechanical system.

Volumetric Efficiency

Efficiency can also relate to how easily air flows in and out of an internal combustion engine or, for that matter, any energy converter that draws air in or out. As an engine increases in revolutions per minute, its intake valves are open for an increasingly shorter amount of time. This means that the amount of air per time period going into an engine has been reduced. Volumetric efficiency measures this condition. The formula for measuring volumetric efficiency is

$$\text{Volumetric Efficiency (as a percentage)} = \frac{\text{actual air used}}{\text{maximum air possible}} \times 100$$

To illustrate volumetric efficiency, assume that at a certain engine speed, 40 cubic inches of air/fuel mixture enters each cylinder. However, to fill each of the cylinders, 55 cubic inches should enter. Using these two numbers, one could determine the volumetric efficiency of the engine:

$$\text{Volumetric Efficiency (as a percentage)} = \frac{\text{actual air used}}{\text{maximum air possible}} \times 100$$

$$\text{Volumetric Efficiency (as a percentage)} = \frac{40}{55} \times 100 = 73\%$$

One way to increase the volumetric efficiency is to improve the ease with which air and fuel can enter the internal combustion engine/converter. This is known as *scavenging*. Factors that will affect the volumetric efficiency on internal combustion engine/converters are:

1. Exhaust restrictions
2. Air-cleaner restrictions
3. Carbon deposits on cylinder walls and valves
4. Shape, design, and number of valves
5. Amounts of restriction in the intake and exhaust ports by curves. (Ports can be polished to reduce friction.)

Thermal Efficiency

Another form of efficiency is called *thermal efficiency*. Thermal efficiency measures how effectively thermal devices convert heat energy in fuel into power. This type of efficiency considers all the losses—including thermal losses, mechanical losses, and volumetric losses. Therefore, thermal efficiency is sometimes called *overall efficiency*. It is the most common form of efficiency used to compare thermal converters. Thermal efficiency is measured by using the following mathematical formula:

$$\text{Thermal Efficiency Percentage} = \frac{\text{actual output}}{\text{heat input}} \times 100$$

When using this formula, the units of measure for input and output must be the same. The following example illustrates the use of the formula for calculating the thermal efficiency of an engine. In a gasoline engine, the heat input is expressed in Btu's. A gallon of gasoline, for example, has approximately 110,000 Btu's. The output of an internal combustion engine is expressed in horsepower. To find the actual output in the same unit, note that 1 horsepower is equal to 42.5 Btu/min. Therefore the formula can now be shown as:

Thermal Efficiency Percentage =

$$\frac{actual\ output}{heat\ input} \times 100;\ therefore,$$

Thermal Efficiency Percentage =

$$\frac{horsepower \times 42.5\ Btu/min}{110,000\ Btu's/gal \times gal\ per\ min} \times 100$$

Combining Efficiencies

Large energy systems are likely to contain several smaller energy systems, each of which has its own efficiency. Consider, for example, a typical automobile. The energy from gasoline eventually is used to turn the back or front wheels of the vehicle. Look at the overall efficiency from the gasoline engine to the power at the wheels. The overall efficiency (including the engine, transmission, and differential) could be calculated mathematically, as could the individual efficiencies of each. This is further illustrated in Figure 3-23.

The individual components of the engine and drivetrain form the efficiency of the total system. Total vehicle efficiency is calculated by multiplying the individual efficiencies by one another and multiplying by 100. Using the data in Figure 3-23, for example, would yield a total efficiency of 25.65 percent ($0.3 \times 0.9 \times 0.95 \times 100$). Thus, although each component has a certain efficiency, the overall efficiency of the total vehicle is less than any one component.

This calculation can also be made when using electrical power plants that produce electricity for electric heating in a home. Often, electric heating is misleadingly advertised by some manufacturers as being 100 percent efficient. Yet, when converting electrical energy into thermal energy, it is only 33 percent efficient when the coal is converted to electricity and then to thermal energy. For example, when an electrical power plant is 33 percent efficient, and an electrical heater in a home is 100 percent efficient, the total efficiency of converting coal into electricity and finally into thermal energy in the home yields no more than 33 percent efficiency ($0.33 \times 1.00 \times 100$).

Energy Efficiency Ratio—EER

Many energy systems and machines use different units of energy to measure input and output values. Therefore, a true indication of a system's efficiency or performance cannot be

Energy System

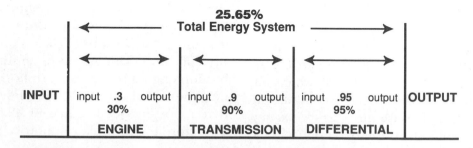

FIGURE 3-23 When different components are used in a vehicle, each with different efficiencies, the total efficiency of the vehicle is less than the lowest efficiency.

determined unless conversions are made to place the values in the same units of measure. This is the case for an air-conditioner in which the input is in watts and the output is in Btu's. To handle this problem, the *energy efficiency ratio*, or EER, has been established in the construction industry to measure an air-conditioning system's efficiency.

A ratio is defined as a relationship between two similar values, such as Btu's per hour and watts. The energy efficiency ratio is calculated by the formula

$$EER = \frac{Btu / hour\ output}{Watts\ input}$$

For example, a 5,000-Btu per hour air-conditioner needs 500 watts of electricity to operate at full capacity. The EER of this energy system is then

$$EER = \frac{Btu / hour\ output}{Watts\ input} \quad or$$

$$EER = \frac{5,000\ Btu/hour}{500\ watts} = 10.0$$

The average EER throughout the air-conditioning industry is between 8 and 11. This means that the air-conditioner can remove 8 to 11 Btu's of heat with each watt used. Obviously, the higher the EER, the more efficiently the unit performs.

Coefficient of Performance—COP

Another term used in the construction industry is the *coefficient of performance*, or COP. The term *coefficient* is a numerical measure of a physical or chemical property that is a constant for a system under a specified condition. The COP is, therefore, used as a measure of the performance effi-

ciency of various energy systems. An example of an energy system that utilizes the COP is the heat pump. A heat pump can extract heat from outside a home to use inside during winter. The heat pump is able to extract heat even in outside temperatures as low as 10° F. It is also able to remove heat inside the home to the outside during the cooling season (i.e., summer). Although the input and the output terms are different, the COP provides a clearer indication of the performance when input watts are converted to output Btu's. For conversion purposes, we shall use 1 watt of electrical energy to equal 3.414 Btu. The formula for COP is:

$$COP = \frac{Btu\ output / hour}{Watts\ input \times 3.414}$$

A heat pump that has a COP of 3 means that for each Btu put into the heat pump (from the operating electricity) 3 Btu's can be extracted by the heat pump and used in a residential dwelling. The extra Btu's come from the energy in the air surrounding the heat pump. Average COP ratios are about 1 to 3. The actual COP will depend upon the temperatures, type of fluid used to transfer the energy, and exact type of system and technology used.

Calculating Efficiencies—Math Interface 3-5

To better understand the terms and equations defined in this section, solve the following problems (refer to Appendix Q for certain formulas):

1. A solar collector had 1800 Btu's striking its surface during a one-hour period (input). The solar system was able to put 1200 Btu's into a house during this

same time (output). What is the First Law Efficiency of this solar system?

Efficiency equals _____.

2. An air-conditioner requires 1500 watts to operate properly. During operation, 7,000 Btu's are removed from the house per hour. How many Btu's are removed from the house for each watt of electricity used? State the answer as EER.

EER is equal to _____.

3. A vehicle with a gasoline engine has the following efficiencies:

Basic Engine Efficiency . . . 30 percent

Tire Efficiency (Rolling). . . . 90 percent

Drive System (Transmission and Differential) . . . . 91 percent

Rear Axle Efficiency . . . . . . 94 percent

What is the overall efficiency of this vehicle?

Efficiency overall equals

_____ percent.

4. A residential heating system uses a heat pump. At an outside temperature of 20° F, the heat pump is able to place 8300 Btu's into the house per hour. If the heat pump is consuming 1200 watts of power during this time, what is the COP of the heat pump?

COP of the heat pump is

_____ .

ALTERNATIVE FUTURES

Greenhouse Gases

One of the major environmental problems causing international debate is that of *global warming*. Global warming, or the *greenhouse effect*, results when certain atmospheric gases (e.g., carbon dioxide, methane, chlorofluoro-carbons, nitrous oxide, and ozone) trap solar rays that would otherwise be radiated back into space. Greenhouse gases are emitted by the burning of fossil fuels, industrial processes, changes in forest cover and land use, agricultural practices, and breakdown of organic wastes. These emissions inevitably produce carbon dioxide (CO_2) gas, leading to an increase of CO_2 in the atmosphere. CO_2 is a gas that traps heat, thus causing global warming.

Since the Industrial Revolution, the burning of fossil fuels and deforestation have led to a 25 percent increase in carbon dioxide in the atmosphere. In addition, there has been a 100 percent increase in methane and numerous additions of engineered chemicals, such as chlorofluorocarbons (CFCs).

The combination of these gases has trapped about two watts per square meter of additional radiant energy. While this energy increase may appear to be trivial (two watts would be the equivalent power needed to light a small holiday tree ornament—one for every square meter), it has added thermal energy to the atmosphere. What is unclear to scientists is how these extra two watts of heating return to the atmosphere in a temperature rise. This is because scientists are dealing with unverifiable assumptions of how this heat is distributed in the atmosphere and to the earth's surface.

Despite the unverifiable assumptions made by scientists, the Intergovernmental Panel on Climate Change (IPCC) has concluded that a global warming of several degrees is likely by the middle of the next century. The IPCC's first report on global warming showed that the projected average global temperature rise would be between 0.2 and 0.5 degrees Celsius per decade through the next century. At this rate of temperature rise, global warming would be greater than any experienced in human history. A four-degree warming is nearly equivalent to the temperature difference between the conclusion of the last ice age and the present interglacial epoch—a time that completely altered the ecological makeup of the earth.

Assessments of the ecological damage from this expected temperature increase have ranged from slightly beneficial to cataclysmic. More recently, however, most scientists studying this problem note that a temperature increase of the magnitude projected could seriously disrupt the ecological systems. Reductions in agricultural output, increases in hurricane activity, flooding of low-lying coastal areas, extinction of plant and animal species, worldwide climatic changes, and unknown effects to human and animal health are all possible.

If the scientists studying this dilemma are correct, agricultural areas could become deserts, and the melting polar ice caps could cause the oceans to rise by three feet or more. This rising of the oceans would cause flooding of the coastal areas where half of the world's population lives.

These theories of the greenhouse effect have caused chemical industries and the public utilities to debate their roles in the contribution to global warming. Chemical industries, for example, vigorously denied that a chemical as useful as CFC could have such a negative impact on the atmosphere. This chemical was viewed as being practically synonymous with progress. CFCs were being used in a large range of products including air-conditioners,

aerosol sprays, fire-fighting chemicals, insulation, and refrigeration. Industrial uses of the chemical included industries such as agriculture, electronics, health care, plastics, telecommunications, and transportation. From an industrial perspective, literally billions of dollars in investment capital were at stake.

Because of statistics and research such as these presented here, changes are beginning to show an effect on the greenhouse gases. For example, automobile manufacturers are now using less dangerous chemicals in their air-conditioning systems. Instead of using Freon 12 (a CFC) as a refrigerant, manufacturers are now using an HFC (hydrofluorocarbon) called HFC-134. This type of refrigerant does not harm the ozone layer as does R-12.

As noted earlier, the major gases known to contribute to global warming are carbon dioxide, chlorofluorocarbons, methane, and nitrous oxide. Figure 3-24 lists these gases and some of their characteristics. The second column in Figure 3-24 shows the atmospheric concentration of various greenhouse gases. For example, there are 351.3 ppm (parts per million) of carbon dioxide gas in the atmosphere. The third and fourth columns show the annual increase and the life span of each type of greenhouse gas, respectively. The relative greenhouse efficiency (column 5) is a number that represents the degree of danger from each of these gases as compared to carbon dioxide. Chlorofluorocarbons, for example, are present in minute concentrations. However, they are 15,000 times more harmful to the atmosphere as compared to carbon dioxide. Also, CFCs remain in the atmosphere between 75-111 years. The current greenhouse contribution (column 6) is also shown for each gas. On the right side of the table, the principle sources that produce the gases are shown, including their uses in products or services.

Major Greenhouse Gases and Their Characteristics

1	2	3	4	5	6	7
Gas	Atmospheric Concentration (ppm)	Annual Increase (percent)	Life Span (years)	Relative Greenhouse Efficiency CO_2-1	Current Greenhouse Contribution (percent)	Principal Sources of Gas
Carbon Dioxide (Fossil Fuels)	351.3	.4	x[1]	1	57 (44)	
(Biological)					(13)	Coal, Oil, Natural Gas, Deforestation
Chlorofluoro-carbons	.000225	5	75–111	15,000	25	Foams, Aerosols, Refrigerants, Solvents
Methane	1.675	1	11	25	12	Wetlands, Rice, Fossil Fuels, Livestock
Nitrous Oxide	.31	.2	150	230	6	Fossil Fuels, Fertilizers, Deforestation

[1]Carbon Dioxide is a stable molecule with a 2–4 year average residence time in the atmosphere

FIGURE 3-24   This chart shows the major greenhouse gases and their characteristics. (Courtesy of Worldwatch Institute)

POINT/COUNTERPOINT 3-1

TOPIC

The Greenhouse Effect

Theme: Global warming has become a major environmental concern. While there is debate over the severity and rate of climatic changes that may happen, preventive action needs to occur if there are to be reductions of greenhouse gases emitted into the atmosphere.

Currently, the burning of fossil fuels for energy is one of the major sources of greenhouse gas emissions. The United States alone emits over 5 billion tons of carbon dioxide annually. Since global warming is caused by human activity, it is a problem that can be solved. The solution to global warming will depend upon the establishment of governmental policies and the collective action of individual citizens. After further researching the issue of global warming, attempt to answer the following questions in a classroom discussion or debate:

1. What shift in energy priorities is needed to develop national interests to long-term global survival?
2. It has been estimated that the investments needed to counteract the effects of global warming will be in the billions of dollars. These funds would account for research, development, and implementation of new chemical substitutes for industrial products and services, cleaner burning fuels, and improved fuel efficiencies for the transportation sector. Who should pay to discover the effects of and possible responses to global warming? Should the industrialized countries who have been using greater quantities of fossil fuels than developing nations shoulder the financial responsibility or should there be an equal sharing of the costs by the world community?
3. What effect, if any, would the rapid development and use of solar energy and other carbon-free energy sources have on global warming? What transitional steps would be needed to increase the rate of adoption of these technologies?
4. What immediate actions can consumers take (e.g., the purchasing of products and services, reductions in energy consumption, and so on) to help reduce further increases in global warming? Regarding this question, specifics are needed—for example, switching to energy efficient fluorescent lighted bulbs. For each specific suggestion made, estimate the potential reductions in emissions of a greenhouse gas for a given population if this suggestion were adopted.
5. Develop an action plan that would allow citizens to work at the local, state, and federal level to promote energy efficiency and develop policies that address the economic and environmental consequences of global warming.

SUMMARY & REVIEW

Summary/Review Statements

1. Speed is the ratio of the distance traveled by an object to the time it takes to travel that distance.
2. Velocity is the ratio of displacement (which takes into account both distance and direction) to time.
3. Displacement is the distance between the starting and ending positions of a moving object during an interval of interest.
4. Acceleration is the ratio of a change in velocity of an object to a given unit of time.
5. When an object slows, its slowing is referred to as deceleration.
6. Force is the pushing or pulling action of one object upon another. As such, force is the cause of acceleration to a moving body or mass.
7. The mathematical relationship of force is expressed as mass times acceleration.
8. The weight of a body refers to the force by which that object is pulled vertically downward by gravity. Weight is equal to mass times acceleration due to gravity.
9. The mass of an object is equal to the ratio of the object's weight to its gravitational acceleration.
10. Pressure is a measure of a force applied to a given area.
11. A force applied to a fluid can sustain the force only in an enclosed surface or boundary.
12. The force from a fluid is applied perpendicular to the rigid body that surrounds it.
13. Fluids also exert pressures in all directions.
14. Fluids exert greater pressures as their depth increases.
15. Pressure is measured in psig or psia scales.
16. A manometer is an instrument containing water or liquid mercury in a U-shaped tube and is used to measure both vacuum and pressure. The readings are taken in inches or millimeters of water or inches of mercury.
17. Work is the result of applying a force to a body or mass through a certain distance when both the force and the movement are in the same direction.
18. Power is the measure of work being done over time.
19. One horsepower is the amount of energy required to move 550 pounds 1 foot in 1 second.
20. Torque or the moment of force is the tendency to produce a change in rotational motion.
21. Regardless of their complexity, all machines are made to produce a mechanical advantage. Mechanical advantage can be determined by calculating the ratio of the output force to the input force.
22. Levers are used to change forces and/or distances moved in any mechanical energy system.
23. Temperature is a measure of the average kinetic energy of the random moving molecules within an object.
24. Heat is always conducted from a higher temperature area or mass to a lower temperature area or mass.
25. The three ways in which heat can be transferred are conduction, convection, and radiation.
26. The units used to measure thermal energy are the Btu (Bgs unit of measure) and the calorie (SI unit of measure).
27. A degree-day is a measuring standard used for determining the severity of winter or summer.

28. First Law Efficiency is defined as the ratio of output to input in an energy system.
29. Thermal efficiency measures how effectively thermal devices convert the heat energy in fuel into power.
30. The total efficiency in any system is the product of all the component efficiencies multiplied by 100.
31. Global warming, or the greenhouse effect, is an environmental problem that results when certain atmospheric gases (i.e., carbon dioxide, methane, chlorofluorocarbons, nitrous oxide, and ozone) trap solar rays that would otherwise be radiated back into space. The burning of fossil fuels is a major contributor to global warming.

Discussion Questions

1. Using the definitions of *conduction*, *convection*, and *radiation*, identify examples of residential applications involving each method of heat transfer.
2. Which is numerically larger: the weight of an object or its mass?
3. Give an example in which records of heating degree-days are used in relationship to energy conservation.
4. Give two examples of applications calling for first-, second- and third-class levers.
5. A large piece of ice floats in a glass of water so that the level of the water is at the top of the glass. Will the water overflow the glass when the ice melts? Explain the answer.
6. A homeowner who has used a 4-horsepower log splitter purchases a new 6-horsepower model. Upon using the new log splitter the homeowner proclaims, "I now have twice the power that I had with the 4-horsepower splitter." Is the homeowner's statement correct? Why does the homeowner believe there is an increase in power?
7. Heat can flow by radiation and conduction. In what ways are they similar? In what ways are they different?
8. A bicycle has two gear ranges. In the low gear the pedals make two complete revolutions while the power wheel turns one revolution. In the high gear the pedals make one revolution while the power wheel turns three revolutions. What are the advantages and disadvantages of each gear, and under what road conditions would each be useful?

Introduction to Nonrenewable Energy Resources

Section 2 investigates the use of nonrenewable energy sources. As noted in chapter 1, nonrenewable energy sources have a rate of formation that is so slow that it is meaningless in terms of their replacement. As such, nonrenewable resources are exhausted or depleted once they are used.

Section 2 has four associated chapters to provide a more complete review of nonrenewable energy sources. Chapter 4 examines the use of coal. Chapters 5 and 6 investigate petroleum and natural gas, respectively. Chapter 7 is devoted to nuclear power and its associated technologies.

Each of the four chapters constituting Section 2 provide an historic analysis of the resources used and their consequences on society and the environment. Throughout these chapters three questions are addressed in relation to the study of energy technology. These questions are:

1. What are the limitations of the energy sources and technologies available for use?
2. How can nonrenewable energy sources be best utilized and managed for future generations?
3. How large are the resources and when will they be depleted?

chapter 4

Coal Resources

PURPOSES

Goals of Chapter Four

The goal of this chapter is to examine the use of coal as an energy source and its associated environmental and social costs. This chapter, therefore, first reviews coal usage from an historic perspective. The reserves, production rates, and technologies associated with coal's extraction, transportation, and conversion to useful work are then examined.

It is important to make reference to the fact that the United States does not have the necessary fossil fuel reserves (petroleum and natural gas) to meet its expected energy needs in the twenty-first century. However, the country does have approximately 23 percent of the world's coal reserves that could be used to avert some of the consequences of a future energy crisis. A knowledge of coal and its uses will help in determining how to use this resource in an environmentally and socially responsible manner. At the completion of this chapter, you will be able to

1. Define the characteristics of coal resources in current use.
2. Compare and contrast the types of coal relative to their respective geographic locations within the United States.
3. Analyze various coal mining and transportation technologies.
4. Identify the technology associated with coal-powered generating plants.
5. Identify the processes of coal gasification and liquefaction.
6. Analyze the environmental and social costs of coal use.

Terms To Know

By studying this chapter, the following terms can be identified, defined, and used:

Acid rain
Anthracite
Bituminous coal
Carbon content
Coalification
Coal slurry
Fluidized bed combustion
Gasification
Heating value
Lignite
Liquefaction
Overburden
Peat
Scrubber
Strip or surface mining
Subbituminous coal
Syngas
Synoil
Town gas

INTRODUCTION TO COAL

Coal Use in Britain

Coal has held a prominent place in the history of the industrialized world. Humans have mined and used coal for about 3000 years. Among the first peoples to use this resource on a large scale were the Chinese. Throughout Northern Europe and particularly in Britain, coal has been mined for a home-heating fuel since the thirteenth century. However, it was not until the seventeenth and eighteenth centuries that the use of coal became important for industrial-heating processes. Note that the use of coal for industry resulted from a shortage of wood for making charcoal. Charcoal was the principal heating fuel for home and industry, but large-scale production of charcoal had severely reduced Britain's wood reserves.

The deforestation that Britain experienced in the seventeenth and eighteenth centuries was a result of the expansion of agriculture, the increasing use of wood as a building material, and, especially, the use of wood as a fuel source. In particular, charcoal was used as a fuel for the smelting of ores to yield molten metal for industrial uses. This single industry (smelting) had a decisively negative effect on Britain's forests since the production of a single ton of metal required several tons of wood for making the charcoal used as furnace fuel. Dwindling sources of wood led Britain to seek substitute energy sources for its growing industrial base.

The fuel to which Britain turned was coal. This change proved to be an important stimulus for the Industrial Revolution. The access to the relatively inexpensive and large resource of coal led Britain to further its industrial base and helped in the development of the steam engine. With the coal-fired steam engine came the factory system with centralized steam power, new transportation systems, advances of precision in machinery, and an array of new machine tools (e.g., lathes, planes, and boring machines). Additionally, the steam engine made the mining of coal easier. The coal-fired steam engine, by removing the water that collected in the mines, allowed miners to go deeper into the earth. New coal uses and the expanding population resulted in more coal extraction and consumption.

Coal became the primary fuel in Britain during the nineteenth century; and as other countries were industrialized, the use of coal increased further. Numerous countries experienced the same type of deforestation as Britain before adopting coal as a basic energy source. Like Britain, the United States was not exempt from this process.

Coal Use in the United States

Coal was slower in supplanting wood as a fuel in the United States. This was because fuel wood was more abundant on the North American continent than in Europe during the nineteenth century. In fact in 1850, 90 percent of the energy consumed in the United States was wood generated. However, in the fifty years that followed, coal usage increased from approximately 10 percent to 71 percent of the energy supply.

This increase in the use of coal was a consequence of deforestation in the eastern United States. As more wood was harvested, it had to be transported from greater distances. The resulting higher fuel costs created the first energy crisis in the United States.

As the price for fuel wood increased, coal became the fuel of choice to meet the growing energy demands in the country. Coal mined in the East, in Pennsylvania and the mountains of the Appalachian region, became the principal production areas for this energy source. Coal mining and use was essential to the growth in industrialization and urban development of the United States from 1860 to 1900. It was during this period of expansion that the increases in population density and industrialization created an energy demand that fuel wood could not satisfy due to shrinking forests and the lower Btu content of fuel wood. The abundance of coal in the East solved these problems.

The dramatic growth of coal use (and fuel wood decline) is illustrated in Figure 4-1. Note how the use of coal continued to intensify until 1910. After that time, oil and natural gas began slowly to supplant coal as an energy source.

COAL IDENTIFICATION AND CHARACTERISTICS

Origin of Coal

Coal began to form nearly 350 million years ago. During this period of the earth's develop-

ment, plant growth exceeded losses to predators and decay. As plant material died, some of it was deposited into swamps. This material then underwent a decaying process of humification.

In the humification process the gases of carbon dioxide and methane leave the decaying plant material. Water is also driven from the decaying plant material leaving carbon. As humification continues, more gases and moisture are removed leaving a dark brown, jelly-like humus. As this carbon-enriched humus comes in contact with other plant remains, it forms peat.

Peat is a young form of coal that consists of partially decomposed plant and vegetable matter and inorganic minerals that have accumulated in a water-saturated environment. Peat contains substantial amounts of carbon and hydrogen. Under favorable conditions, peat can become covered by accumulating sediments from other decaying matter. This decaying matter protects the peat from contact with air. After each successive layer is added, heat and pressure may increase to a level that pro-

Consumption of Energy Sources in the United States Thru 1955

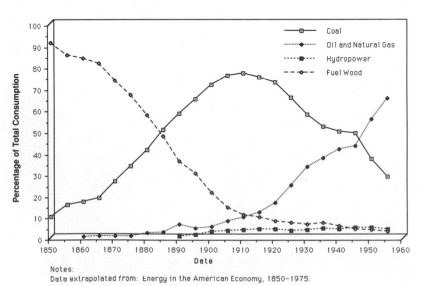

Notes:
Data extrapolated from: Energy in the American Economy, 1850-1975.

FIGURE 4-1 In 1850, approximately 90 percent of the energy consumed in the United States was from wood. Within the next fifty years coal exceeded wood as the primary energy source.

It has been estimated that the origin of coal began nearly 350 million years ago.

motes the process of coalification. *Coalification* is the process by which coal is formed.

In the coalification process, oxygen is removed in the form of carbon dioxide and water. This process is a result of increasing heat and pressure being applied to the peat from the addition of successive layers of organic and mineral material. Through millions of years, heat, and pressure, the peat is turned into various formations of coal.

Formations of Coal

The chemical and physical properties of coal are extremely variable. They depend upon numerous factors: the nature of the original vegetation, the age of the formation, and the geological conditions under which coalification took place. However, one aspect of coal formation is clear: the older the deposit, the higher the levels of carbon. It is the carbon and other volatile matter that allows coal to be used as an energy source. Although various types of coal can often be found together in a mining

operation, several different classes of coal can be recognized. These classes are peat (pre-coal), lignite, subbituminous coal, bituminous coal, and anthracite. The following offers a brief description of each class.

Peat

Although peat is seldom categorized as a coal, it is considered the first geological step in the formation of coal. Peat contains the preserved remains of vegetation (from which it was formed) and plant and animal materials that have decomposed beyond recognition. Peat is generally soft and spongy with a high water content (92 to 94 percent). Even after drying, the water content in peat may be as high as 50 percent. Since peat represents an early stage of decomposition, the carbon content is nearly 50 percent of its dry weight. The principal use of peat is for domestic heating fuel.

Although peat can be found in all 50 states, fewer than one-fourth of the states have deposits significant enough to be utilized on a commercial basis. These states include Florida, Maine, Michigan, Minnesota, North Carolina, and Wisconsin. Since peat can be considered a renewable resource if managed correctly, it will be presented in more detail later in this text.

Lignite

Lignite is believed to be the youngest of the different types of coal—formed around 150 million years ago. Lignite contains an amount of recognizable woody material from the original vegetation, and chemically it is not that different from peat. Like peat, lignite contains a large percentage of water and volatile material. Physically, lignite is brownish black and considered a soft coal. Lignite has been used as an industrial fuel and for the generation of electricity. Large reserves of lignite have been found throughout North and South Dakota, Montana, and Texas.

Subbituminous Coal

Coal classified as **subbituminous** was formed approximately 200 million years ago. This coal has less moisture and more volatile material than lignite. However, subbituminous coal is also considered a soft coal. It is dull black and is often called black lignite. This coal can be found in Alaska, Wyoming, and Montana. Subbituminous coal is used as an industrial fuel, a space-heating fuel, and an energy source for the generation of electricity.

Bituminous Coal

Bituminous coal can be found in Kentucky, Pennsylvania, Illinois, and West Virginia. This type of coal is about 300 million years old, has a high carbon content, and is more volatile than subbituminous coal, lignite, or peat. Bituminous coal is the most common type of solid fossil fuel. It is a soft dense coal that is black in color. This coal also has well-defined bands of bright and dull material intermixed within it. Bituminous coal is used primarily as an energy source for the production of electricity. Other uses include the making of coke for the steel industry and as a fuel source for space heating.

Anthracite

Anthracite was formed over 300 million years ago. Anthracite is a hard, brittle coal with a shiny, black appearance. This type of coal has a high carbon content and can be found mainly in northeast Pennsylvania. Anthracite is primarily used as a home-heating fuel since it is free from dust, burns cleanly, and burns longer than the other types of coal. Note, however,

that anthracite may not burn as hot as bituminous coals. Although the carbon content is higher, the heating value is slightly lower. Impurities in coal, therefore, affect its heating value.

Carbon Content-Ranking Coal

The nonrenewable resource of coal has a **carbon content** that varies according to the coalification process the organic material underwent during its formation. Time, heat, pressure, and organic matter were all variables in the formation of coal. Thus, these variables have had an effect on the heating value, impurities, and carbon content found in coal. Geologists use the carbon content as a means of ranking coal. The higher the carbon content, the higher the ranking given to the coal, and generally the more energy (heating value) is available per pound of fuel. The average carbon content of coal is approximately 50 percent. Figure 4-2 illustrates the rankings of coal based upon their carbon content and heating value. From this figure it can be seen that anthracite is the highest in carbon content (although slightly lower in heating value than one would expect), followed by bituminous coal, subbituminous coal, and lignite.

Heating Value

The **heating value** is a measure of how many Btu's can be produced by the burning of one pound or kilogram of coal. Thus, the heating value allows specific types of coals to be graded by their Btu output. Most often, the higher grades of coal will have a high carbon content and high Btu's available per pound. This can be seen in Figure 4-2. However, a specific increase in carbon does not necessarily magnify the heating value by the same amount. This means that the relationship is nonlinear. The

Ranking of Coal Based Upon Carbon Content and Heating Value

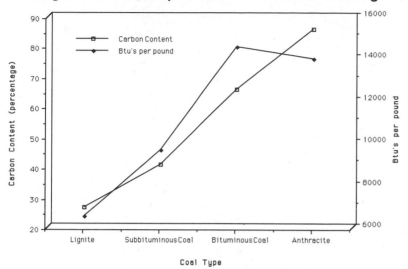

FIGURE 4-2 Carbon content (measured in percentage) and heating value (measured in Btu's per pound) are shown for different types of coal.

reasons for this nonlinear relationship are a result of various impurities in coal.

Impurities in Coal

In addition to classifying coal by carbon content and heating value, it is classified in terms of its impurities. The impurities in coal include moisture, ash, and sulfur. Figure 4-3 provides a graphic representation of two impurities in each of the four types of coal. Only the moisture impurities and ash content are reported in Figure 4-3. The ash content has been illustrated with two types of limits (i.e., a lower limit and an upper limit) since coals often have a varying range of ash content depending on the location and type of coal found. As can be seen in Figure 4-3, the moisture content decreases with the increasing rank of the coal. The ash content of coal is notably variable, ranging from a low of 4 percent to a high of 14 percent. Scientists hypothesize that this ash content is a result of the formation characteristics of the coalification process.

Another variable impurity in coal is sulfur.

The sulfur content of coals in the United States ranges from 0.5 percent to 4 percent depending upon the geological location of the resource. Generally, coal from the western regions of the United States has a lower sulfur content than coal in the East. The piece of coal pictured on the left side of Figure 4-4 contains fewer impurities than the one on the right and has a higher carbon content and heating value as well.

Impurities, such as sulfur, have a negative effect on the coal mining and processing equipment. Undesirable environmental effects are also produced by burning coal that has significant amounts of impurities. An example of some problems produced by these impurities include the following:

1. Moisture and sulfur cause corrosion to the equipment used to mine and process coal.
2. A high content of ash reduces the Btu's available per pound of coal.
3. Sulfur in coal enters the atmosphere and produces acid rain. Acid rain is environmentally hazardous to plant and animal

Impurities in Coal (moisture and ash)

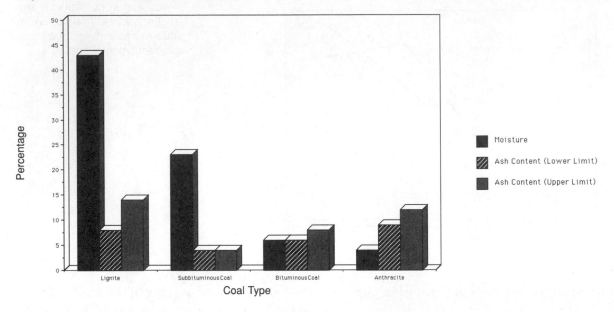

FIGURE 4-3 The impurities in coal include moisture and ash. These impurities are presented here for the various types of coal.

life and is also a corrosive agent to buildings and monuments. The effects of acid rain from the burning of coal will be presented later in this chapter.

In the final analysis of ranking and grading coal, anthracite has generally the highest value, followed by bituminous coal, subbituminous coal, and lignite. Since anthracite is high in carbon content, volatile matter, and heating value,

less coal is needed to produce the same amount of thermal energy than that needed if bituminous coal, subbituminous coal, or lignite were used. Additionally, whenever the poorer grades of coal are used (e.g., lignite and subbituminous coal) processing problems often occur. These processing problems include the addition of steps to remove impurities from the coal and the use of more coal by weight to achieve the needed amount of thermal energy.

FIGURE 4-4 Coal is considered a fossil fuel, black to brown in color. The coal on the right has more impurities, such as ash, than the one on the left.

Selecting Types of Coal—Energy Issue 4-1

A particular power company has both subbituminous and bituminous coal within 500 miles of its plant. The plant managers must decide which type of coal to use. Identify and justify the aspects that need to be considered when making the determination of which coal or combination of coals to use.

COAL AVAILABILITY

World Coal Consumption

The consumption of coal is forecasted to expand modestly in most parts of the world through 2010 (increasing approximately 1.7 percent per year). This modest increase is not surprising since coal currently has one of the slowest rates of growth among all the major fuels and is forecasted to lose a portion of its relative importance to natural gas. This is because known world natural gas reserves have continued to rise and because increased efficiencies in gas-fired technology have allowed natural gas to compete economically with coal and other fuel sources. Figure 4-5 provides an overview of world coal consumption since 1980 with projections to 2010. Consumption is projected to rise moderately until 2000. After 2000, coal consumption is projected to increase dramatically to meet world energy needs.

Of the countries having coal reserves, three hold the majority of the world's reserves—China, the former Soviet Union, and the United States. Figure 4-6 provides a graphic representation of the world's coal energy reserves by selected countries. As can be seen from Figure 4-6, the former Soviet Union, China, and the United States have the largest shares of coal reserves.

World Coal Production

In the past, China has been a world leader in the production of coal, along with the United States and the former Soviet Union. China averaged about 1.2 billion short tons of coal production per year. Other major producers included Poland, India, Australia, and South Africa. Of the major coal producers, China, the United States, India, and Australia are expected to expand coal production until 2010. This expansion will be used to meet domestic and export needs. The major importing countries with lacking coal reserves include Japan, South Korea, Taiwan, Brazil, and several western European countries. Additionally, there are

World Coal Consumption with Projections to 2010

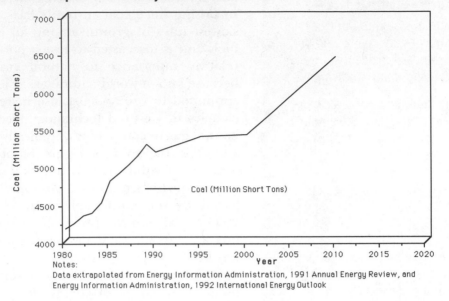

Notes:
Data extrapolated from Energy Information Administration, 1991 Annual Energy Review, and
Energy Information Administration, 1992 International Energy Outlook

FIGURE 4-5 This graph provides an overview of world coal consumption since 1980 with projections to 2010. While this consumption is forecasted to increase modestly until 2000, dramatic increases are expected after that year.

World Coal Reserves

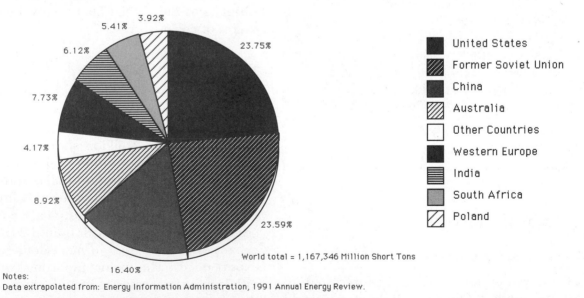

Notes:
Data extrapolated from: Energy Information Administration, 1991 Annual Energy Review.

FIGURE 4-6 World coal reserves are estimated at 1,167,346 million short tons. The largest shareholders of coal reserves include the United States, the former Soviet Union, and China.

predictions that a number of countries will experience a growing demand for coal as their energy reserves become depleted. These countries include Britain, Germany, Poland, and the Czech and Slovak Federal Republic.

Based upon the distribution of world coal reserves and the growing international need for energy, coal trade is projected to expand until 2010. Of the major producers, Australia, South Africa, and the United States are expected to account for over two-thirds of the traded coal by 2000.

United States' Coal Consumption

A significant amount of coal is used for the production of electrical energy within the United States. Its importance as an energy source is presumed to be even greater by the turn of the century. Contingent upon future world oil prices and economic growth, coal's share of the total United States energy production is estimated to rise from 31 percent in 1990 to a range between 35 and 38 percent by 2010.

Of the coal extracted, subbituminous and bituminous coal account for the largest percentage used, as illustrated in Figure 4-7. Anthracite and lignite account for the remainder of the coal produced. Figure 4-7 also includes projected estimates for coal production through 2010. Note that coal production in the United States is projected to increase in the coming years. However, the production of anthracite is expected to decline. As illustrated in Figure 4-8, the overall share of anthracite in the production of coal has steadily declined. In fact, the use of anthracite has been in a fluctuating state of decline since 1917 when a high of about 98 million short tons were produced. This decline has been a result of dwindling anthracite coal reserves, the intensified production of western coal, and greater use of large-surface mining operations that are more productive than underground mines. (The differences between surface and underground mining is presented later in this chapter.)

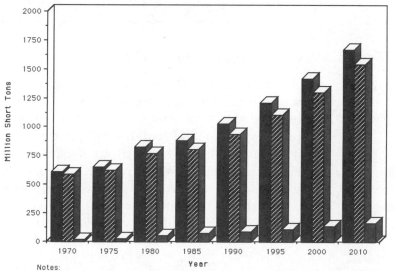

United States Coal Production with Projections to 2010

Million Short Tons

2000
1750
1500
1250
1000
750
500
250
0

1970   1975   1980   1985   1990   1995   2000   2010
Year

Notes:
Data extrapolated from: Energy Information Administration, 1991 Annual Energy Review.

FIGURE 4-7 United States' coal production has increased steadily since 1970 with bituminous and subbituminous coal making up the bulk of production.

■ Total Production

▨ Bituminous and Subbituminous

■ Lignite and Anthracite

Anthracite Coal Production in the United States with Projections to 2010

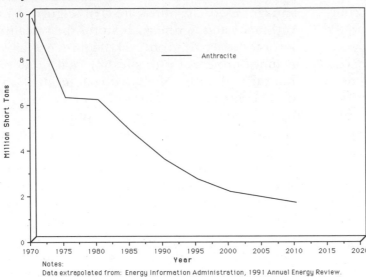

FIGURE 4-8 The use of anthracite has decreased steadily since 1970. However, this decline has not reduced the use of coal in providing energy for the United States.

Notes:
Data extrapolated from: Energy Information Administration, 1991 Annual Energy Review.

Doubling Times and Coal—Math Interface 4-2

The following problems help to illustrate the effects that doubling times (described in chapter 1) have on the total amount of coal available to a society.

1. Assume that there are 260,000 million tons of coal available in the United States. How many years would it take to consume all of this coal if each year 1,500 million tons of coal were consumed?

    Consumption at no growth would equal _____ years.

2. Coal production is expanding at an annual rate of 2 to 3 percent. If the coal demand increased by 3 percent each year, how many years would it take to deplete the United States'

reserves if this annual growth rate continued until the reserves were exhausted?

Reserves would be exhausted in _____ years.

Coal Reserves—East and West

To delineate better the availability of coal in the United States, reserves are divided into two regions: the West and the East. The eastern reserves include the coal that is located roughly from Oklahoma eastward, as illustrated in Figure 4-9. The eastern reserves are mostly bituminous coal. Western reserves are those found throughout the Rocky Mountains and westward to the Pacific Ocean. These reserves are mostly lignite and subbituminous coal.

Coal Distribution in the U.S.

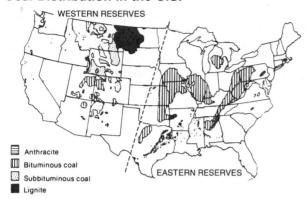

Legend:
- Anthracite
- Bituminous coal
- Subbituminous coal
- Lignite

FIGURE 4-9 Coal is found in many regions throughout the United States. Both the eastern and western regions are used to supply coal to our society.

Currently, more coal is mined east of the Mississippi River than in the West, but the West's total production share has increased almost every year since 1965. The main reason for this expansion in the use of western coal is a result of the Clean Air Act Amendments of 1990. As noted earlier, the low-sulfur coal that is concentrated in the West burns cleaner than eastern coal. Despite the demand for low-sulfur coal, projections do not indicate that western coal will exceed coal production in the East until sometime after 2010. This is because western coal has additional shipping costs associated with long hauling distances and the technical problems of converting existing bituminous coal-fired plants to burn western subbituminous coal.

Coal Use in Domestic Markets

The main consumers of coal in the United States are electric utilities, as illustrated in Figure 4-10. This sector should account for virtually all of the projected growth in domestic coal consumption through the year 2010. As illustrated in Figure 4-10, the electric utilities consumed 84 million short tons of coal in 1950 and have steadily enlarged this share of coal use each year. By 2010, the United States' electrical utilities could be consuming close to 1100 million short tons of coal each year. Other sectors using coal are projected to remain stable in the coming years. One other segment of the United States' coal industry that is anticipated to increase is that of coal exports.

United States' Coal Exports

During the next two decades, United States' coal exports are expected to rise substantially. Figure 4-11 illustrates the projected growth of coal exports with projections to 2010. Based upon data collected by the United States Energy Information Administration, coal industries could be exporting more than 230 million short tons of coal by 2010. This growth will be a result of three factors: declining coal production in Europe, an enlargement in electricity demand throughout Asia, and the limited potential of other countries to increase their exports after the year 2000. The majority of the estimated growth in United States' coal exports is expected to occur after 2000, as illustrated in Figure 4-11. At this time the United States and Australia are projected to compete internationally for the coal export market. However, the United States' coal industry is expected to have an advantage in the European marketplace due to lower freight costs.

> **Coal Exporting—Energy Issue 4-2**
> Should the United States capitalize on its coal reserves by exporting this resource to other countries, or should the United States conserve this resource to offset its impending energy crisis as petroleum reserves become exhausted? Consider the data presented in Figures 4-6 and 4-10. What are the benefits and consequences of exporting coal to other countries?

United States Coal Consumption by Sector with Projections to 2010

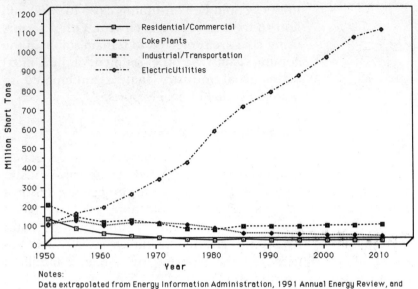

FIGURE 4-10 United States' coal consumption by sector is illustrated with projections to 2010. Note that the energy sector (electric utilities) is the fastest-growing sector of the four presented.

Notes:
Data extrapolated from Energy Information Administration, 1991 Annual Energy Review, and Energy Information Administration, 1992 International Energy Outlook

Projected United States Coal Exports to 2010

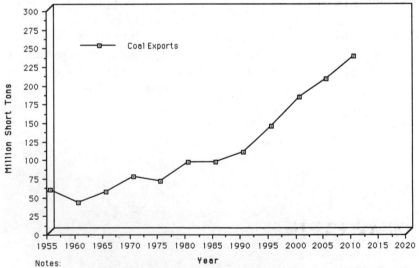

FIGURE 4-11 The United States is forecasted to become a major international exporter of coal in the future.

Notes:
Data extrapolated from Energy Information Administration, 1991 Annual Energy Review, and Energy Information Administration, 1992 International Energy Outlook

COAL HANDLING

Coal Mining Overview

Coal is found in seams, often called strata (stratum is the singular form of coal). Strata are defined as layers of coal usually about 20 to 50 feet thick. The coal strata occur at various depths and angles. Some are found so deep that they must be mined by using underground mining technology. Other strata are close enough to the surface that surface mining can be performed. Whether or not a coal seam is mined depends upon its thickness, depth, continuity, angle of formation relative to the surface, rank, and sulfur and ash content.

Once coal has been located, the first step (before mining begins) is to assess the quality and quantity of the coal. This process begins by taking coal samples. Coal core samples are obtained using a drilling rig; then the samples are brought to the surface for inspection. Following the coal's inspection, future mining operations can be determined and planned. These operations fall into two categories: surface (strip) mining or underground mining.

Surface (Strip) Mining

Strip or surface mining has several advantages over underground mining. These include higher production capacity in a workday, employment of a smaller workforce, recovery of nearly 100 percent of the coal seam, and fewer hazards than underground mining. Although surface mining techniques can be used in mountainous or flat terrain, these techniques are best utilized when the area to be mined is on relatively flat terrain and the largest equipment is used. This type of surface mining is known as area mining and usually occurs in strips one mile long and 100 feet wide. Other types of strip mining include con-tour mining (mining around a slope), mountaintop removal (a type of contour mining in which stripping proceeds all the way across the top of a mountain), and augering (a supplementary mining method used in conjunction with stripping that employs an augerlike drill to bore and remove coal).

The growth in surface mining has had a direct relation to the development of large excavating equipment that is used to strip away the soil and rock above the coal. The excavating equipment commonly used includes draglines, power shovels, and transport trucks. Each of these pieces of equipment is able to move tremendous amounts of coal, soil, or rock. Draglines, for example, are capable of taking more than 300 tons of material in a single pass. Figures 4-12 and 4-13 illustrate the use of draglines and power shovels (respectively) in the removal of coal.

FIGURE 4-12 Draglines, such as the one illustrated, can remove more than 300 tons of coal or overburden (soil on top of coal seam) in a single pass. (Courtesy of the American Petroleum Institute)

FIGURE 4-13 The power shovel shown loading the transport truck (in the foreground) can handle capacities in excess of 250 tons. (Courtesy of the Kerr-McGee Corporation)

Once strata of coal are located and scheduled to be strip-mined, the first step in the mining process is to remove the soil and the overburden. The *overburden* is defined as the material lying between the topsoil and the coal seam or stratum. Earth is removed in long strips by using draglines or power shovels. This overburden is collected behind the mine where it will be used later to restore the land surface. Once the coal strata are exposed, they are drilled and blasted into smaller pieces for shoveling. Large coal shovels and trucks then enter the surface mine to extract the coal.

Land Reclamation

After the coal strata have been removed, the overburden is placed back in the hole, or strip, and the topsoil is replaced. The topsoil is then planted and brought back to a usable condi-

tion. It should be noted that reclamation does not restore the land surface to its original shape. This is not possible since the removed coal seam might have been seventy feet thick and would call for huge amounts of fill for complete restoration of the land. Depending upon the size of the coal mine, the type of equipment used, and the depth of the coal seam, the land involved may not be productive (except for mining) for a period of 5 to 10 years. This is because the reestablishment of vegetative cover and animal species is a slow and deliberate process.

Underground Mining

When coal seams or strata are found too deep in the ground for surface mining, underground mining technologies are used. Three methods—conventional, continuous, and longwall mining—are commonly used. The differences are as follows:

1. Conventional mining uses a coal-cutting machine (analogous to a chain saw) to cut blocks of coal. Compressed-air drilling machines are used to drill holes for explosives or compressed air. The explosives or air blasts break the coal into small pieces for ease of loading and removal.
2. Continuous mining uses machines that cut the coal from a seam without using explosives. Figure 4-14 shows an example of a continuous mining machine. Once cut, the coal is placed on conveyor belts or trams and brought to the surface for further processing.
3. Longwall mining (see Figure 4-15) removes the coal and puts it on the conveyor belt in one operation. A cutting machine moves back and forth across a wide stratum of coal. Hydraulic jacks support the roof of the mine. As the machine

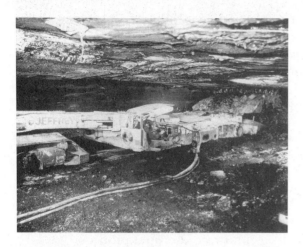

FIGURE 4-14 This continuous mining machine is used to remove coal from the underground mining operation. (Courtesy of Sun Company, Inc.)

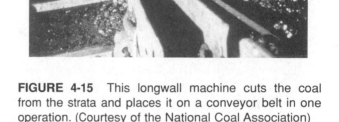

FIGURE 4-15 This longwall machine cuts the coal from the strata and places it on a conveyor belt in one operation. (Courtesy of the National Coal Association)

moves further into the stratum, the roof is allowed to collapse behind it. Because the underground shifts result in surface depressions (sink holes), some efforts to eliminate this type of mining are currently under way.

Mining Costs

In the eastern region about 86 percent of the coal is deep underground, while 14 percent is near the surface. The depth of the coal seam is one economic factor in determining whether the coal is to be surface-mined or mined underground. Underground mining is approximately three times as costly as strip mining.

Underground coal in the western region forms roughly 60 percent of the reserves, while surface coal constitutes the remaining 40 percent. Because of its shallow depth (less than 200-feet deep), some western coal reserves can be surface-mined. This surface mining substantially reduces the cost of the coal.

Coal Transportation

After the coal has been removed by either surface or underground mining, it is crushed into pieces about three-fourths of an inch in diameter. The coal is crushed for two reasons. The first is to free the coal from mineral impurities. The second is to reduce the coal to a marketable size. After crushing, the coal is thoroughly washed to remove impurities. The coal is then transported to the electrical generating power plant by train or barge. Trains with 100 cars or more bring the coal directly into the power plant facility where it is further crushed into a fine powder before being used.

Coal Slurry Transportation

Another transportation technique is *coal slurry*. This method was developed to help eliminate coal handling and to reduce transportation costs. In this method, coal is first crushed to a fine powder at the mine. After washing, it is mixed with water to form a slurry. The slurry can be pumped through pipelines to its destination, the power plant. Once the coal reaches the power plant, it is dewatered or dried for use. The coal slurry process eliminates many of the problems associated with coal handling that would otherwise be produced if transported by rail or barge.

The United States Office of Technology Assessment (OTA) has concluded that specific conditions dictate whether slurry, railroad, or barge transport is the least costly method of moving coal. These conditions include water availability, environmental effects of pipeline routes, location of pipelines and rights-of-way, and the cost to dewater or dry the coal.

COAL-FIRED ELECTRIC GENERATING PLANTS

The purpose of the coal-fired electric generating plant is to convert the chemical energy in coal to electrical energy. When the coal arrives at the plant, it is initially placed in storage. Plants typically have 60 to 90 days of storage to allow for rail or mining strikes, mechanical failures, and other unanticipated delays.

Plant Operation

The coal is transported into the plant by large conveyor belts. Figure 4-16 represents a schematic of a typical electrical generating plant that produces 750 megawatts of electrical power. In this figure, the numbers in parentheses on or next to the various components help to illustrate the operation of a coal-fired electric generating plant. The general operation of the plant is as follows:

1. Coal is brought into the plant via a conveyor belt from the coal storage area.
2. The coal is then crushed into a fine powdered mixture in the coal mill (1).
3. The coal powder is mixed with large quantities of air and sent to the boiler (2). The burn rate for this plant is 400 to 425 tons of coal for each hour of operation.

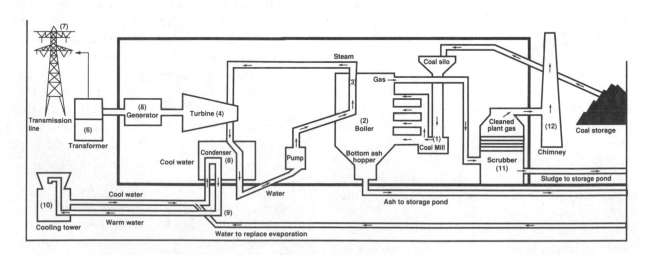

FIGURE 4-16   This is a schematic of a large coal-fired electric generating plant.

4. The air (6,000,000 pounds per hour) and coal are blown into the boiler at several levels, creating a large tower of fire inside the boiler (2). The walls of the boiler are made of tubing through which water flows. There are approximately 200 miles of tubing in the boiler.

5. The burning of the coal in the boiler heats the water until it is a high pressure steam (3). This steam is then sent through several stages of turbine rotors, which rotate the turbine (4). The steam turbine depicted in Figure 4-16 is capable of producing 1,000,000 horsepower. The turbines, which turn at 3600 rpm, convert the steam to torque used to rotate the generator (5). The tips of the turbine blades travel at speeds above 1220 mph.

6. The electricity produced by the generator is at 24,000 volts and is sent to a transforming station (6) where the voltage is increased to 345,000 volts for distribution to the power-grid system. Here the electricity is further distributed through transmission lines (7).

7. The exhaust from the boiler is sent through a *scrubber* to remove ash, sulfur, and other contaminants (11).

8. The exhaust is then directed to the chimney and released high into the air to dissipate the remaining pollutants (12).

9. Cooling towers (10) cool the water (used in the boiler [2]) in a loop (9) that circulates back to a condenser (8). Here, steam in the boiler turbine loop is condensed back to water to begin the cycle again.

New Coal-Fired Power Plants—Energy Issue 4-3

Contrast the economic, political, environmental, and social advantages and disadvantages of building a new coal-fired power plant to help meet expanding electrical demand in our society.

Fluidized Bed Combustion

There are other means of burning the coal in a power plant. One newer method is called *fluidized bed combustion*. This technology allows the combustion of coal in a more environmentally acceptable manner. In the electric utility industry, a major dilemma has been to design a power plant that can burn high-sulfur coal and still meet sulfur dioxide (SO_2) emission standards established by the federal government. Electric utilities have relied on the use of scrubbers to remove ash, sulfur, and other contaminants. This technology, however, is costly to maintain and operate and may not be suitable for high-sulfur coals. The use of a fluidized bed combustor eliminates some of these problems.

In this fluidized bed combustion system, crushed coal is mixed with limestone and burned while suspended on a "bed" of heated air. The calcium in the limestone combines with the sulfur in the coal to form calcium sulfate, a dry, solid waste product. This system not only eliminates the need for scrubbers, but the solid waste disposal problem is minimized. The process also uses lower combustion temperatures, resulting in reduced nitrogen oxide emissions. Fluidized bed combustion can also cleanly burn low-grade fuels—such as peat, tree bark, industrial and sewage sludge, oil shale, lignite, and even cow manure. Other types of combustion and generating processes are also being studied to make the combustion process more efficient and to reduce pollutants. Some include pressurized fluidized bed combustion, post combustion technology, limestone injection technology, combined cycles, and cogeneration.

ALTERNATIVE FUTURES

Coal Gasification and Liquefaction

The production of gaseous or liquid fuels from coal dates back well over 100 years, as the use of coal gas predates the use of natural gas. The first attempts to use coal gas occurred during the eighteenth century. It was at this time that coal gases were used in the production of coke for the steelmaking industry. Coal gases were also marketed for lighting streets and homes during the early nineteenth century. Germany, during World War II, demonstrated the ability of processing both liquid and gaseous fuels from coal. To distinguish between the two types of processes: coal *gasification* is defined as the process of converting coal to natural gas; coal *liquefaction* is the process of converting coal to liquid products. The products resulting from coal liquefaction are similar to oils or petroleum products.

Currently, there are at least eight processes used for coal gasification and liquefaction. While the processes are not necessarily new, industries have been experimenting with and improving them over the past 15 years. When the processes become economically advantageous, they will be ready for further development and implementation. Also, as energy prices continue to increase, so will incentives to develop coal gasification and liquefaction.

Coal Gasification Process

The processes for coal gasification are constantly improved. The research is aimed at allowing the various types of coal with distinct heating values, carbon contents, and impurities to be converted into gas or liquid fuels. Because of the different characteristics of the various types of coal (lignite, subbituminous coal, bituminous coal, and anthracite), the processes are still undergoing research and development.

Coal Gasification Reactor

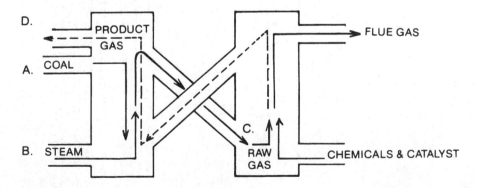

FIGURE 4-17 This is a simplified schematic of a coal gasification plant. Coal, steam, high pressures and temperatures, and a catalyst are needed to produce the gas.

While constantly changing, the basic technology of the coal gasification process is simple. Generally, coal, reacting with steam and hydrogen in the presence of pressure and heat, produces raw gases, usually of low to intermediate Btu quality. Figure 4-17 illustrates a simple coal gasification process.

In the coal gasification process, coal is fed into the high-pressure, high-temperature reactor at point A shown in Figure 4-17. Steam is then fed into the reactor at point B. The resulting product is termed raw gas, which is fed into the second stage at point C. Various chemicals and a catalyst are then added to remove hydrogen sulfide and carbon dioxide. A catalyst is a chemical that alters the rate of a chemical reaction and is itself unchanged by the process. The reactor used in the coal gasification process is where the coal contacts the catalyst under controlled conditions of heat and pressure. These chemicals also help to purify and upgrade the hydrogen sulfide, carbon dioxide, and methane gas that result from the process. This upgraded gas is sent out at point D as product gas that can be sent directly to the existing refinery processes.

The product gas must then be upgraded at the refinery by removing hydrogen sulfide and carbon dioxide. The Btu content of the methane must also be increased before it can be used commercially. Water, heat, hydrogen, and several catalysts are used to remove the carbon dioxide and hydrogen sulfide, and to purify and upgrade the methane—thus creating a higher Btu-content gas. These upgrading processes have become standardized within the industry.

Coal Gasification Costs

The handling of coal for gasification is costly. The conversion process accounts for only about one-third of the cost of a gasification plant. The major expense results from materials handling, gas treating, and other operations.

The cost of handling and transporting coal is only one disadvantage to utilizing coal as an energy resource. Coal also leaves about 10 percent ash, requiring disposal. One final dilemma with coal is caking. *Caking* is a tendency of coal, particularly eastern coal, to fuse under high pressure and at temperatures greater than 2,000° F. This causes plugging of the reactor vessels and disruption of the gasification system. The process of reducing the caking is called devolatilization (*devolatilization* is the removal of volatile material in a substance). When devolatilization is used in gasification, the process normally occurs within the first stage of the reactor. Devolatilization also increases the processing cost.

Coal Gasification Yields

The first significant coal gasification technology was developed in Germany. The fuel produced was initially called **town gas** and, subsequently, natural gas. Many industries are currently involved in research and development activities dealing with coal gasification. Several industries are working directly with the U.S. Department of Energy to develop coal gasification and liquefaction processes, both experimentally and for demonstration purposes.

In the coal gasification process, three combustible gases—carbon monoxide, hydrogen, and methane—are produced. Carbon monoxide and hydrogen have roughly the same heating value. Methane has approximately 60 percent more heating value than hydrogen or carbon monoxide. It is, therefore, similar to natural gas. Carbon dioxide, nitrous oxide (oxidies of nitrogen), and hydrogen sulfide gases are produced in the coal gasification process, also. These are noncombustible gases that have no current use but do have an effect on people and the environment. Carbon dioxide and nitrous oxide are considered major greenhouse gases, and hydrogen sulfide is poisonous.

To illustrate the quantity of gas that can be produced from coal, the U.S. Department of Energy worked with an Illinois company to convert coal into natural gas. The project was designed to convert 2,200 tons of coal per day into 18 million cubic feet of natural gas. In addition, 100,800 gallons per day of synthetic crude oil were produced in the conversion process. A second project was designed to convert 2,700 tons of high-sulfur coal into 22 million cubic feet of *syngas* (synthetic gas), and an additional 16,830 gallons of *synoil* (synthetic oil) were produced daily.

Coal Liquefaction

The technology of coal liquefaction is very similar to that of coal gasification. The final product is a synthetic oil that can be processed by conventional refining technologies. Most liquefaction processes under development involve various approaches to reacting hydrogen with coal to form liquid fuels. One ton of coal yields about 100 gallons of synthetic oil in a typical coal liquefaction process.

Coal Liquefaction Process

The liquefaction process involves several stages. As shown in Figure 4-18, hydrogen, coal, and a slurry of mixed oils are preheated and sent to the reactor. In the reactor the mixture is subjected to a catalyst, temperatures of 850° F and pressures of 2,000 to 4,000 lbs per square inch. Under these circumstances, oil and gas

are produced and separated to make the desired liquid fuel. Some low-grade oil and hydrogen gases are recycled back through the process. The higher-grade oil is sent to the refinery. Hydrogen sulfide, ammonia, ash, and other solid residues are also removed.

Whatever technology emerges as successful, there appears to be great promise for the use of coal as a fuel alternative to petroleum. The social benefits from the use of coal are overwhelming. Coal, which is currently the most abundant fossil fuel in the United States, is supplying much of the energy needed to maintain our current standard of living. In addition, the huge quantity of coal available in the U.S. has the potential to maintain some level of economic and energy security as the United States draws down its reserves of oil and natural gas. However, this economic and energy security do come with a social and environmental cost.

*Coal Gasification/Liquefaction—
Energy Issue 4-4*

Explain the purposes of developing coal gasification and liquefaction processes. What are the social and environmental costs of implementing this technology?

SOCIAL AND ENVIRONMENTAL COSTS OF USING COAL

The social and environmental costs of using coal are numerous. While some of these costs are slight, others present a real danger to humans, plants, and animals. The various social and environmental costs presented here should be addressed and corrective action taken in order for coal to be utilized in an environmentally safe and socially responsible manner.

Coal Liquefaction Process

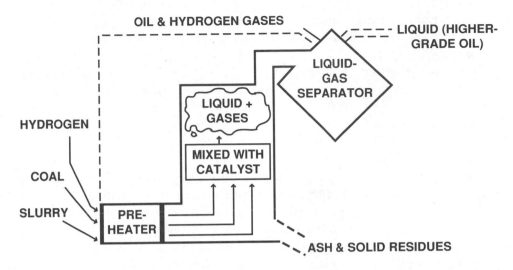

FIGURE 4-18   Coal can be converted to liquid petroleum by adding a hydrogen catalyst and thermal energy as in this simplified schematic of a coal liquefaction plant gas.

Social and Environmental Costs of Mining

The social and environmental costs of surface mining include erosion, topsoil destruction, and habitat loss. Where surface mining techniques are used in dry climates, reclamation may not completely restore the vegetation cover. This makes the area barren and unsuitable for future use. Additional problems resulting from inadequate reclamation include the buildup of silt in downstream waterways (which can cause ecological disruption) and acidification of water supplies (which is toxic to plants and fish).

The social and environmental costs in underground mining include danger to miners from accidents, land use, fires (both underground and in waste piles outside the mine), and acid mine drainage. Acid mine drainage occurs when water flows in and out of the mine as it comes in contact with the coal. This results in a dilute sulfuric acid that is detrimental to

aquatic life. This acidity also makes the water unusable for industrial or municipal water supplies. (The same process occurs with surface mining.) One added social cost in both underground and surface mining is the long-term health of miners. Those who breathe coal dust for long periods of time are prone to "coal worker's pneumoconiosis" or "black lung." This disease does not allow the lungs to clean themselves and leads to a form of emphysema or other respiratory diseases.

Social and Environmental Costs of Burning Coal

There are several social and environmental costs of converting coal into energy. These include resource depletion, global warming, and acid rain. Resource depletion is a very serious concern when dealing with nonrenewable fuels. Although coal may be the most plentiful nonrenewable energy resource on earth, every ton of coal burned is lost for use by future generations. This raises numerous questions as to our responsibility to future generations and their access to sources of energy.

Global warming, as presented in chapter 3, results when certain atmospheric gases (i.e., carbon dioxide, methane, chlorofluorocarbons, nitrous oxide, and ozone) trap solar rays that would otherwise be radiated back into space. The burning of coal by all energy-using sectors contributes to global warming by the release of carbon dioxide, methane, and oxides of nitrogen. Additionally, the conversion processes of coal gasification and liquefaction contribute to this problem as well. Note that the United States is one of the world leaders in the emission of greenhouse gases. In 1990, the United States released more than 1 billion metric tons of carbon dioxide, 15 million tons of nitrogen oxides, and 1540 million pounds of chlorofluorocarbons. While the amount of emissions fluctuate each year, no significant progress has been made in their elimination.

The use of coal results in *acid rain*, which is produced when sulfur dioxide from the burning of coal rises into the air. (The United States emits in excess of 20 million tons per year of sulfur dioxide with three-quarters coming from the burning of fossil fuels by electric utilities.) The sulfur then mixes with airborne moisture to produce sulfuric acid. When rain, snow, or fog is present, this sulfuric acid falls to the ground. The augmented amount of acid (besides damaging buildings, monuments, forests, crops, and so on) raises the acidic levels of lakes and streams. The increased acidity tends to kill certain aquatic organisms and, in turn, disrupts biological food chains. This results in food shortages causing fish and other aquatic species to die from starvation.

To help reduce the amount of acid produced by a power plant, the exhaust is washed with an alkaline solution. Quite often lime is mixed with water and sprayed into the exhaust to neutralize the sulfur. Some power companies also use helicopters to spray lime into surrounding lakes to neutralize the water to an acceptable level. Liming by this method is considered by some companies to be a cost-effective means to control the acidic solutions of lakes. The result is protection of the fish population. However, the problem of producing acid rain at the power plant is still not solved. To solve the problem, lower sulfur levels in coal, less coal usage by society, or stricter power plant standards are needed. Note that coal-burning power plants in the Ohio River Valley and lower Midwest contribute to acidification of lakes throughout the Northeast and Canada. In fact, an estimated 50 percent of the acid rain in Canada is a result of sulfur dioxide pollution in the United States.

Finally, the byproducts of sulfur dioxide (known as sulfates) are recognized as major contributors to atmospheric haze (i.e., smog that is a result of ozone pollution from the burning of hydrocarbon fuels) and airborne pollution. This pollution can lead to bronchitis in children, decreased breathing capabilities in adults, and life-threatening conditions for those with respiratory problems who are exposed to such pollution for extended periods of time. Scientific data has shown that some 86 million people in the United States are subjected yearly to pollution levels that exceed at least one national air quality standard.

Social and Environmental Costs of Liquefaction and Gasification

The techniques of gasification and liquefaction require large amounts of water in their processing of coal. Many areas of the United States, particularly western regions, are in short supply of water. The use of liquefaction and gasification plants further reduces water supplies in these areas. Coal conversion processes also release molecules known as polycyclic aromatic hydrocarbons. These molecules are carcinogenic, and evidence indicates that there is a risk for workers in the coal conversion processes (e.g., electric generating, coke, gasification, and liquefaction plants). One final social concern regarding liquefaction and gasification is that of cost. These technologies are expensive. Because they are still, experimental in many cases, the fuels currently processed are more costly than the fuels that they are intended to replace. As traditional fuel sources experience drawdown, how costly will these replacement fuels remain? This answer will have to be formulated in light of the numerous technical and political problems of resource depletion.

Environment and Electric Power Plants— Energy Issue 4-5

What consequences would result if all coal-fired electric generating power plants had to become efficient and less environmentally damaging? Consider economic, environmental, and technological aspects of the change.

POINT/COUNTERPOINT 4-1

TOPIC

Coal Use—The Future

Theme: The potential for increasing the use of coal in the future is likely to become a reality. There is certainly an abundance of resources both in the United States and internationally. This discussion deals with identifying the effects of coal use on society if it is decided that this energy resource should be significantly increased to meet the demand of electrical energy in the future. After further research of the social and environmental consequences (both positive and negative) of using coal, attempt to answer the following questions in a classroom discussion.

1. What are the social conditions causing the expanding demand for electricity and coal use? Are these social conditions necessary for basic human survival? Identify those social conditions that should be maximized or minimized based on a philosophy of conserving energy resources.

2. What steps could be taken to reduce the need for electricity, and how could the public be convinced that electrical energy demand should decrease to counter the negative consequences of resource depletion and pollution?

3. Assuming oil and natural gas have been depleted in the United States, what energy sources would likely be used? What overall consequences could be expected to each of the energy-use sectors, to society, and to the environment? How might society be transformed as oil and gas reserves were depleted?

4. As the traditional energy sources of oil and natural gas become depleted worldwide, more capital and technology must be utilized to transform other energy sources into a usable form (such as via coal gasification and liquefaction). Thus, energy is coming less from a raw material, which is either dug or pumped out of the ground and used, and more from a manufactured product. The addition of more capital and technology to process fuels will have a variety of effects on the world economy. Select a developing country with coal reserves and identify the consequences of redirecting capital and technology to meet the country's energy needs by using coal gasification and liquefaction. What sectors of society would be hurt the most? Identify the prospects for reducing energy inequities between the rich and poor. What international tensions could be expected from competition to develop more efficient energy conversion technologies and to extract the remaining oil and gas reserves?

SUMMARY & REVIEW

Summary/Review Statements

1. Coal is the most abundant of nonrenewable energy resources on earth.
2. The United States has approximately 23 percent of the world's coal reserves.
3. The four types of coal found today in the United States are lignite, subbituminous coal, bituminous coal, and anthracite.
4. Three major impurities in coal are moisture, ash, and sulfur.
5. As the carbon content of coal rises, the heating value generally increases.
6. Lignite has the lowest heating value, and bituminous coal has the highest heating value.
7. The least abundant coal is anthracite.
8. Both underground and surface mining are used to extract coal strata from the earth.
9. Coal can be transported by rail (train), by barge, by truck (out of the mine), and by coal-slurry pipeline.
10. One major disadvantage of using coal is environmental pollution.
11. The coal slurry method of transportation eliminates the need for traditional vehicles for coal transportation—such as trains, barges, and so on.
12. Some of the more important technological processes in an electrical generating power plant include coal storing, milling, conveying, using combustion in a boiler, scrubbing, and electricity generating.
13. Fluidized bed combustion is now being used to help reduce acidic exhaust in coal-fired power plants.
14. Coal gasification is a process that takes coal and converts it into a product similar to natural gas.
15. Coal liquefaction is a process that takes coal and converts it into crude oil, which can then be refined.

Discussion Questions

1. Compare and contrast the environmental disadvantages of underground versus surface mining. Consider the costs and the relationships to the environment. Which type should be expanded in the future and why?
2. What are some advantages and disadvantages of using lignite? Relate the answers to the location of the resource, the location of the applications, and the future use of coal.
3. What effects could the increased use of coal gasification and liquefaction have on future energy resources? Consider both short-term (10 to 15 years) and long-term (100 years) effects.
4. Will the affluent of the world be willing to reduce their demand and consumption of energy? What are the social and environmental costs of not reducing demand and consumption? What alternatives are there to this reduction?
5. High energy prices are often a good incentive for energy conservation and for technological innovation (new products, goods, or services). What are the negative consequences of high energy prices? What sections of society are penalized the most? Identify strategies for equalizing these inequities by helping those most affected.

chapter 5

Petroleum Resources

PURPOSES

Goals of Chapter Five

The goal of this chapter is to examine the use of petroleum. Petroleum resources provide the largest portion of all fossil or finite fuels used in the United States. They are a significant part of the total energy being supplied to our society.

This chapter provides an overview of petroleum reserves, production rates, and the technologies associated with the petroleum industry. These technologies are then evaluated on the basis of the social, environmental, and economic implications. At the completion of this chapter, you will be able to

1. Describe how petroleum energy resources are used.
2. Analyze the technologies associated with petroleum exploration.
3. Analyze the technologies associated with the production of petroleum.
4. Analyze the technologies associated with the transportation of petroleum.
5. Identify the refining processes associated with crude oil.
6. Identify the chemistry and terminology of petroleum products.
7. Evaluate oil shale as an alternative energy resource.

8. Analyze the environmental and social costs of petroleum use.

Terms to Know

By studying this chapter, the following terms can be identified, defined and used:

Anticline
Barrel of oil
Bubble cap
Camphene
Catalyst
Choke
Cracking
Crude oil
Development well
Discovery
Dry hole
Enhanced recovery
Existing reserves
Fault
Fire flooding
Fractional distillation
Future reserves
Gathering lines
Gravimeter
Greek fire
Impermeable rock
Improved recovery
Injection well

In situ combustion
Jack-up drilling rig
Kerogen
Magnetometer
Motor octane
Oil pools
Oil shale
Older petroleum
Outer continental shelf
Oxygenated fuels
Petroleum
Proven reserves
Reformulated gasoline
Research octane
Retorting
Road octane
Rock oil
Secondary recovery
Seismograph
Semisubmersible rig
Sour crudes
Steam stimulation
Strategic Petroleum Reserve
Stratigraphic trap
Stripper wells
Sweet crudes
Technically recoverable resources
Undiscovered resources
Wildcatter
Wildcat well
Young petroleum

INTRODUCTION TO PETROLEUM

Petroleum Use in Antiquity

Like other minerals, petroleum has been used since antiquity. Among the people first known to use petroleum were the Chinese almost 3000 years ago. They used this resource as a fuel for both heating and lighting. The Babylonian Empire (2500–540 B.C.) used oil for lighting and asphalt. The asphalt served a variety of purposes including waterproofing the walls of Jericho. Petroleum was also known throughout much of Greece by the year 500 (e.g., an oil well was dug on an island in the Ionian sea). The Greeks used oil for lighting, as well as in weaponry. The weapon, known as *Greek fire*, was invented by a Syrian refugee, Kallinikos, in 673. Greek fire was a petroleum-based incendiary that was used to burn enemy ships and siege machinery.

While other uses of petroleum can be illustrated, note that the oil industry is comparatively young. It was not until the 1850s that petroleum became commercially exploited on a large scale.

Historic Use of Petroleum in the United States

The commercial exploitation of petroleum in the United States was a result of many separate social and technical developments that came together in such a fashion that they transformed society. From the commercialization of oil came the arrival of the petroleum industry and its related products, inexpensive fuel for heating and lighting, and new forms of transportation. However, this commercialization could not have occurred without the social and technical needs for inexpensive fuels for lighting and lubricants to replace the expensive animal and vegetable oils used in the nineteenth century.

With the development of modern civilization, the growth of industrialization and an increasing population led to a demand for artificial lighting beyond that of a wick dipped into animal grease or vegetable fat. The upper classes had long relied on the oil extracted from the blubber of sperm whales for lighting. As the demand for this oil grew, the whale population was severely reduced. By the late 1800s, whalers were traveling as far as the North Pacific to obtain whales; and, by 1840 the whales off the coast of Greenland were hunted almost to depletion. This resulted in rising prices for whale oil and the urgent need for a substitute fuel source for lighting.

One of the fuels developed and made available commercially for illumination was camphene. *Camphene* is a derivative of turpentine that produces an excellent flame for lighting but has the drawback of being very inflammable—it had a tendency to explode when used. Other fuel sources for lighting included town gas distilled from coal and kerosene (also known as coal oil because it was first distilled from coal). Both sources were expensive. Kerosene also had two other problems that limited its large scale use. First, there was no large supply available; and second, there was no oil lamp that could burn kerosene cleanly (the kerosene lamp had not been developed by the 1850s).

The search for an inexpensive fuel for lighting and lubrication led a group of investors, the Pennsylvania Rock Oil Company, to exploit a resource known as *rock oil* (i.e., petroleum) in Titusville, Pennsylvania. *Rock oil* was a term used to distinguish petroleum from other vegetable oils and animal fats. Its primary use prior to 1859 was as a folk medicine. Physicians and merchants sold the rock oil under the name of Seneca Oil. This artificial medicine was believed to relieve suffering from headaches, toothaches, upset stomachs, and rheumatism. Seneca Oil was also used as a topical ointment for animals.

In 1855, investors led by George Bissell received a scientific report that outlined the value of rock oil. This report illustrated that rock oil had the potential to be distilled into many fractions of carbon and hydrogen. One of these fractions was suitable for a high-quality illuminating oil. In fact, in a simple refining process, 70 percent of the crude petroleum could be converted to illuminating oil for only 4 to 10 cents per gallon.

By 1859, the Pennsylvania Rock Oil Company had secured sufficient capital and had devised a method of obtaining the oil from the ground. The method used by Bissell and his associates was to drill for the oil, as was commonly done in boring for salt. On August 27, 1859, the Pennsylvania Rock Oil Company drilling rig struck oil at a depth of 69.5 feet at Titusville, Pennsylvania. Within ten years, the production of oil expanded from 500,000 to 4,215,000 barrels, and the petroleum industry developed nationally.

Unit of Measurement—Historic

The term *barrel* dates from this early period of production when oil was shipped in wooden barrels from the well to the refinery. One *barrel of oil* contained 42 gallons of oil in this system of measure. The 42-gallon barrel remains the basic unit of measurement for the oil industry today. However, the vast majority of oil is no longer shipped in barrels. (Shipping methods will be presented later in this chapter.)

In addition to the barrel, oil is often referenced by the ton. For example, about 6.6 barrels of crude oil equals one ton. Often, tons of oil are used when comparing the number of Btu's in oil. There are nearly 37,000,000 Btu's in one ton of oil. Also, there are about 5,600,000 Btu's in a barrel of oil.

PETROLEUM IDENTIFICATION AND CHARACTERISTICS

Origin of Petroleum

Like coal, petroleum is a fossil fuel that has been produced by the decomposition of organic materials. Petroleum deposits are usually found in sedimentary basins. Although the exact mechanism of their formation is still a subject of scientific research, oil deposits are believed to represent the remains of microorganisms that inhabited the ocean floors. These organisms were probably part of a prephotosynthetic past when free oxygen was absent from the earth. The microorganisms obtained their needed oxygen by extracting it from organic molecules. Through time, other plants and animals lived and died in the shallow seas and coastal waters where light could penetrate the water to support photosynthesis. As these plants and animals died, their corpses settled into the bottom mud of these waters where they decayed. The result was a residue of hydrocarbon molecules (compounds of hydrogen and carbon) left behind. As these hydrocarbon molecules were buried by the continual accumulation of sediments and other decaying plant and animal compounds, they were subjected to high temperatures and pressures. Through heat, pressure, and time, the hydrocarbon molecules were cracked (a process where further molecular rearrangement occurs) into lighter molecules. These lighter molecules formed various combinations of fuel oils, gasoline, and natural gases.

Though the oldest known petroleum deposits may have been formed about a half billion years ago, most of the commercially available deposits are thought to be less than half that old. Today, petroleum is still found in the early stages of formation around the world.

Studies of these early-stage bottom sediments show that to achieve the first recognizable states of conversion of organic sediments into oil requires some 3000 to 9000 years.

Since the majority of all petroleum was formed on the seafloor, it contains a high proportion of seawater (in which droplets of oil are dispersed). With time these sediments are buried under more layers of sediment and organic matter. As the layers above the oil become more consolidated, the petroleum moves upward because of the movements of the earth above. The petroleum eventually separates from the water, by virtue of its lower density or greater viscosity. As sea is replaced by land, the oil remains in the earth below. The oil moves from the rocks in which it was condensed and rises through porous formations until it is blocked by impervious rock. If blocked by impervious rock, it collects in pools. These *oil pools* are surrounded by rock that holds the oil in its pores until released by movement of the earth, by erosion, or by oil drilling.

Characteristics of Petroleum

The chemical and physical properties of *crude oil* are extremely varied. Chemically, *petroleum* is a mixture consisting mostly of hydrocarbons (i.e., hydrogen and carbon molecules). Oil may contain amounts of other compounds that include elements of oxygen, nitrogen, and sulfur. Aside from these major constituents, there may also be traces of vanadium, nickel, and smaller traces of chlorine, arsenic, and lead. Note that the chemical composition of oil differs depending on its source and the geological conditions under which it was formed.

The constituents of petroleum range from gases to waxy and tarlike compounds. Generally, *young petroleum* tends to have hydrocarbons with high molecular weights while *older petroleum* has a larger percentage of lower molecular weight (lighter) hydrocarbons. A further analysis of the characteristics of petroleum is included later in this chapter under the heading of Refining Petroleum.

PETROLEUM AVAILABILITY

World Petroleum Consumption

There are many estimates of the total world oil supply. These estimates often examine the history of oil production and then extrapolate into the future to estimate the quantity of the resource left to be used. One that is derived by M. King Hubbert predicts that oil production will continue to increase until about the year 2000, at which time one-half of the total world oil will have been consumed. Figure 5-1 illustrates this theory. The time period normally ranges from about 1900 to around 2100, the center or peak of the curve being about the year 2000. Updates of this curve are continually made showing the peaks of the curve being reached earlier or later than 2000. The area under the curve shows total world oil supplies. The curve shows that if the petroleum need is greater than the supply of oil after the year 2000, other expensive hydrocarbon resources or alternative resources will have to be developed to compensate for the decline in crude-oil production. Synthetics, unknown reserves, various renewable resources, and heavy tars refined into oil may be exploited to fill this anticipated decline in crude-oil production.

The need for oil worldwide should continue to grow in response to economic growth. Total world consumption of oil is anticipated to grow by a little more than 1 percent per year through 2010. This is less than the projected annual growth rate in the gross domestic product (GDP) of most countries (currently at almost 3 percent per year). This oil consumption rate of growth is also considerably less than during the time span between 1985 and 1990. During that period, oil consumption increased by approximately 2 percent per year. Figure 5-2 provides a graphic representation of world petroleum consumption. In this graph are projections to 2010. From the graph it can be seen that the consumption of oil continues to increase. Growth will continue at about 1.35 percent per year until 2010.

The basic reason for the continued growth in oil consumption is that, as world economies grow and energy demand increases to meet this need, oil is forecasted to remain the fuel of choice for many users. This foreseen growth of petroleum use is presumed because neither natural gas nor renewable fuels are projected to displace oil's dominant position in meeting the needs of the transportation sector. However, the increase in world oil consumption will be offset by improvements in energy efficiency, conservation, and use of other fuels for some applications. These applications include the use of natural gas for electricity generation and as a fuel source in the transportation sector. Other applications include the continued development of nondepletable and renewable energy sources for electrical generation.

The demand for oil will continue to be most rapid in the developing nations of Asia and the Middle East. From 1985 to 1990, consumption of oil in the Middle East increased 3.7 percent, and consumption increased 5.2 percent in the Far East and Oceania. This growth rate is well above the growth in total world oil consumption. Although the Far East is predicted to reduce its energy consumption rate, it is still expected to be twice as high as that of the rest of the world.

M. King Hubbert's Estimate of World Petroleum Reserves

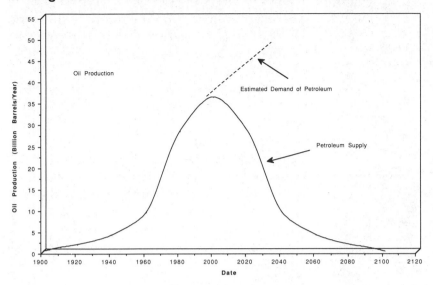

FIGURE 5-1 M. King Hubbert predicts that oil production will continue to increase until about the year 2000, at which time one-half of the total world oil will (by his calculations) have been consumed. The curve illustrates that if petroleum demand is greater than supply after the year 2000, other resources will have to be used to compensate for the decline in crude-oil production.

World Oil Consumption 1989 with Projections to 2010

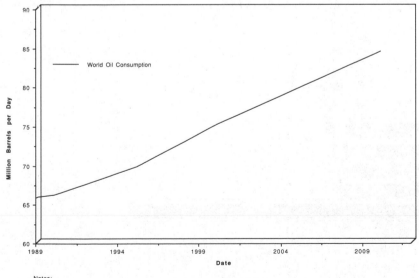

Notes:
Data extrapolated from Energy Information Administration, 1992 International Energy Outlook.

FIGURE 5-2 The demand for oil is projected to increase by 1.35 percent per year until 2010.

World Petroleum Production

The major influence on the future of the world oil market is dependent upon the countries holding the largest reserves. The Organization of Petroleum Exporting Countries (OPEC) has three-fourths of the world's oil reserves. The countries collectively part of OPEC include Algeria, Ecuador, Gabon, Indonesia, Iran, Iraq, Kuwait, Libya, Nigeria, Qatar, Saudi Arabia, the United Arab Emirates, and Venezuela. Figure 5-3 illustrates the reserves held by these countries. As can be seen, these countries have almost 1 trillion barrels of oil and have a high probability of more still to be discovered. Within OPEC, Saudi Arabia has the highest reserves (exceeding 245 billion barrels of oil), followed by Iraq, the United Arab Emirates, Kuwait, and Iran, with about 100 billion barrels each.

OPEC Oil Reserves

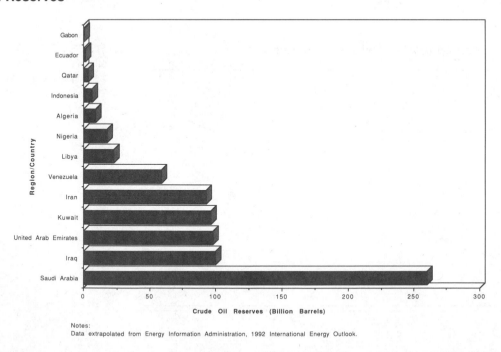

Notes:
Data extrapolated from Energy Information Administration, 1992 International Energy Outlook.

FIGURE 5-3 The Organization of Petroleum Exporting Countries (OPEC) has three-fourths of the world's oil reserves. Together, these countries have almost a trillion barrels of oil.

As a result of a lack of surplus in oil production outside of OPEC, price increases will likely occur to balance the demand and supply of petroleum. This will become true particularly as OPEC reserves begin to stabilize in the next 4 to 15 years. Numerous factors will influence the decline in the rate of increase in world oil production. These factors include changes in demand resulting from global economic growth, conservation, fuel substitution, national and international energy policies, and environmental concerns. Also, international political events (such as the relationships among OPEC and non-OPEC countries and the structural changes occurring in centrally planned economies) will make predicting the future of oil availability uncertain at best. This difficulty in making precise predictions should be further evidenced by reviewing Figures 5-4 and 5-5.

As can be seen from Figure 5-4, petroleum production and consumption is projected to increase through 2010. The consumption of oil and production levels are very close through the year 1995; and by 2010, world oil consumption could outpace production. In Figure 5-5, note that the countries of the Middle East hold approximately 65 percent of the world's total reserves of petroleum. The greatest share of these reserves is additionally held by OPEC. When one considers this imbalance of resources held by the countries of the Middle East (and especially OPEC) and the rate of growth in petroleum consumption, it should be evident that making precise predictions of the future of oil production is difficult at best. Political tensions in the Middle East could easily create another disruption of petroleum production as occurred during the Persian Gulf War.

Although the likelihood of another Persian Gulf war has been decreased by the reduction of Iraq's military strength, clearly political tensions in the region have not dissipated; and, the dependence on Persian Gulf oil supplies is increasing. Note further that the petroleum reserves in the Persian Gulf are among the least expensive to produce. Production costs of

World Oil Consumption and Production with Projections through 2010

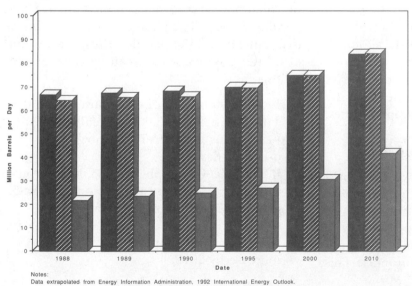

FIGURE 5-4 The world consumption of petroleum reaches world production levels by 1995.

Notes:
Data extrapolated from Energy Information Administration, 1992 International Energy Outlook.

World Crude Oil Reserves

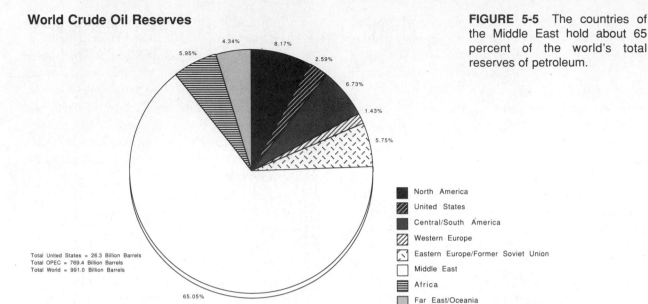

FIGURE 5-5 The countries of the Middle East hold about 65 percent of the world's total reserves of petroleum.

4.34%
5.95%
8.17%
2.59%
6.73%
1.43%
5.75%

- ■ North America
- ▨ United States
- ■ Central/South America
- ▨ Western Europe
- ▨ Eastern Europe/Former Soviet Union
- □ Middle East
- ▤ Africa
- ▨ Far East/Oceania

Total United States = 26.3 Billion Barrels
Total OPEC = 769.4 Billion Barrels
Total World = 991.0 Billion Barrels

65.05%

Notes:
Data extrapolated from Energy Information Administration, 1992 International Energy Outlook.

Persian Gulf crude oil are only a few dollars per barrel. In 1989, OPEC supplied almost 36 percent of the world's oil. By 2010, this dependence could increase to over 48 percent. This increase in dependence must be addressed if economic and world security is to be ensured for future generations. Note also from Figure 5-5 that the United States has only a little over 2 percent of the world's crude petroleum reserves. This percentage is equal to about 26 billion barrels of oil. While this may seem to be a huge resource (and surely more oil discoveries are predicted), it will not meet the needs of the country based upon present consumption rates.

United States' Petroleum Consumption

It is clear that the United States is the world leader in national petroleum consumption. Citizens in the United States use more oil per person than any other group of people in the world. Figure 5-6 illustrates oil consumption of various countries and types of economies. This graph also provides information about the total world consumption of oil for the years 1990 (actual) and 2010 (projected). An interesting calculation can be made from the data in Figure 5-6 to show how much petroleum people use in the United States. The data in Figure 5-6 indicates that 17.5 million barrels of petroleum were consumed in the United States per day in 1990. The United States also is home to about 249 million people. Thus, the amount of oil per day per person can be calculated as follows:

- 17,500,000 barrels of oil are used each day
- This is equal to 735,000,000 gallons of oil (42 gallons per barrel × number of barrels used per day)
- If 249,000,000 people use 735,000,000 gallons each day, then each person used 2.95 gallons of petroleum each day in 1990 (735,000,000 / 249,000,000). Note that this petroleum consumption is expected to increase by 2010.

Over the next 20 years, the United States' need for petroleum products is forecasted to increase by between 0.6 and 1.4 percent per year. This would place the United States requirements for oil between 17.9 and 19.5 million barrels per day in 2000 and between 19.3 and 22.4 million barrels per day by 2010. Figure 5-7 illustrates this foreseen increase (using an average of 0.9 percent increase per year) in United States' consumption.

World Oil Consumption (Million Barrels per Day)

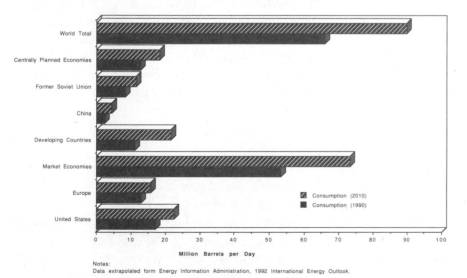

Notes:
Data extrapolated form Energy Information Administration, 1992 International Energy Outlook.

FIGURE 5-6 This figure illustrates oil consumption of specific countries and types of economies. It also provides information about the total world consumption of oil for the years 1990 (actual) and 2010 (projected).

While the demand for petroleum has generally increased since 1950, United States' petroleum production has been in a state of decline since 1988 (see Figure 5-7). This decline is a function of many factors including the aging fields of oil (older fields produce less oil) and the economics of petroleum production. In the chart, the decline in petroleum production is presented under the heading of United States Petroleum Production. This annual rate of decline is approximately 0.8 percent. During the 1950s and 1960s, production capacity of petroleum products often exceeded demand. However, by the 1970s the need for oil had increased while the average productivity of wells in the United States had begun to decline. Figure 5-7 also provides useful information regarding petroleum consumption during the 1980s with projections to the year 2010. As can be seen by the graphic representation, since 1983 the consumption of petroleum products has continued to increase. This increase has resulted partially from lower crude-oil prices. In 1990 and 1991, the warm winters and a stagnant economy reduced petroleum consumption to approximately 17 million barrels per day for 1991. However, projections for the future show that the consumption of petroleum products is anticipated to increase by an annual rate of 0.9 percent.

Figure 5-7 illustrates that the growth in petroleum consumption, coupled with the decline of United States' petroleum production, will lead to the expansion of oil imports in the future. The total imports are predicted to increase to between 10.2 and 15.4 million barrels per day by 2010. At this rate of increase, oil imports could account for 53 to 69 percent of the total United States' petroleum demand. Thus, the United States would use about 25 percent of the anticipated world total of petroleum consumed in 2010.

United States' Petroleum Production

The decline of United States' oil production is a result of aging fields (i.e., extraction of the resource) and less drilling occurring by major U.S. oil companies. This reduction in drilling in the United States has been offset by investments by U.S. oil companies in foreign areas. For the first time in the history of oil production, U.S. companies are now spending more in foreign countries than in the United States. The reasons for this shift in foreign interests are twofold. First, the United States has been well-explored with few chances of finding another large field without major negative effects on environmentally sensitive areas. Second, production costs are higher than elsewhere because of the size and age of the oil fields in the United States. Additionally, the environmental restrictions imposed on offshore areas and in Alaska are expected to reduce any large-scale future potential production. Figure 5-8 illustrates the production of crude oil in the United States and Alaska. As can be seen in this figure, most crude oil collected came from the lower 48 states and onshore wells. Note also that Alaskan oil fields are expected to continue to decline until 2005. Sometime after this date, supplies from the Point McIntyre Field (discovered in 1989) are projected to help supplement United States' production. A further presentation of United States' petroleum supplies is presented later in this chapter in the section entitled Supplies of Petroleum.

Dependence on Foreign Crude Oil— Energy Issue 5-3

Identify the social consequences and environmental costs or benefits from an increasing dependence on foreign crude oil.

United States' Petroleum Production, Consumption, and Imports 1950 with Projections to 2010

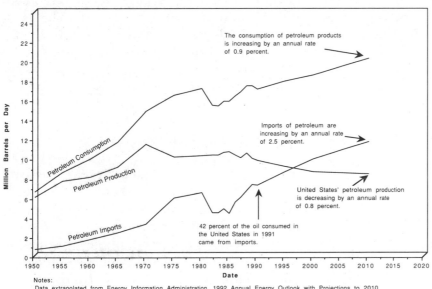

Notes:
Data extrapolated from Energy Information Administration, 1992 Annual Energy Outlook with Projections to 2010 and Energy Information Administration, 1992 International Energy Outlook.

FIGURE 5-7 The annual rate of petroleum production in the United States has been in a state of decline since 1988. However, the demand for petroleum products has continued to increase since 1983. Imports of petroleum are expected to increase by an annual rate of 2.5 percent over the next 20 years.

United States' Crude Oil Production with Projections to 2010

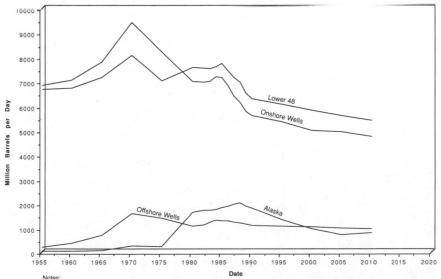

Notes:
Data extrapolated from Energy Information Administration, 1992 Annual Energy Outlook with Projections to 2010 and Energy Information Administration, 1991 Annual Energy Review.

FIGURE 5-8 Most crude oil collected in the United States comes from the lower 48 states and onshore wells. Note how crude oil production in the United States has been in a general state of decline.

The amount of imported petroleum used by the United States changes constantly. In 1973, at the beginning of the oil embargo, the United States imported about 34 percent of its petroleum. Concerned about the increasing use of imported oil, the federal government established a goal of becoming oil independent. By 1985, United States dependency on foreign oil had dropped to about 27 percent. In the meantime, the price of crude oil had risen to an average world price of about $32.00 or more per barrel.

Strategic Petroleum Reserve

In response to increasing oil prices and with the goal of becoming petroleum independent, the United States government established the *Strategic Petroleum Reserve*. This reserve was to be used in the event of supply disruptions from exporting oil countries. In 1977, the Strategic Petroleum Reserve began storing crude oil, and by the end of 1989, it contained about 580 million barrels of oil. Sales to oil companies of this petroleum first occurred during the 1990 Iraqi invasion of Kuwait.

The adequacy of the Strategic Petroleum Reserve is often measured in the number of days the reserve can supply petroleum to consumers without the use of imports. This is an important characteristic of the reserve should a supply interruption occur. Through 1985 the energy security of the country increased every year due to additions in the reserve. However, in 1986 the reserve declined for the first time—from 115 days in 1985 to 94 days in 1986. This data is illustrated in Figure 5-9.

As world petroleum prices began to decrease (as low as $12.00 to $15.00 per barrel from 1987 to 1989), oil companies bought more petroleum from foreign suppliers. At lower prices, it is less costly to buy foreign oil than to drill for domestic sources. During the early 1990s world crude oil prices continued to fall, resulting in an increased United States' dependency upon foreign oil. The United States clearly has not fulfilled its goal of becoming oil independent. In 1991, for example, 42 percent of the petroleum used in the United States came from imports.

Strategic Petroleum Reserve, 1977 – 1991

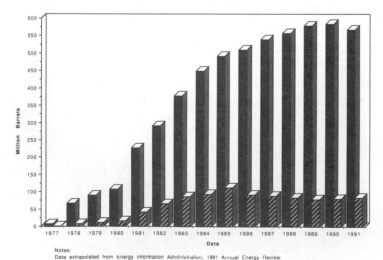

Notes:
Data extrapolated from Energy Information Administration, 1991 Annual Energy Review

FIGURE 5-9 The United States government established the Strategic Petroleum Reserve in an attempt to become energy independent. This figure presents data (through 1991) on the number of days the reserve can supply petroleum to consumers without the use of imports and barrels of oil contained in the reserve.

UNITED STATES' PETROLEUM DEMAND BY SECTORS

Petroleum Demand in the Transportation Sector

The transportation sector is projected to remain the leading end user of petroleum products through the turn of the century. Figure 5-10 provides a graphical representation of the energy demanded in various sectors for the year 1990 and projected needs in 2010. This data is presented in million barrels per day. In 1990 the transportation sector consumed about 11 million barrels of oil per day. This was about 64 percent of all the oil used in the United States. By 2010, this sector is estimated to consume approximately 13 million barrels per day. This would represent 66 percent of all the petroleum estimated to be used in the United States. While this percentage appears high, it is fairly accurate. Remember that the transportation sector currently consumes approximately 23 percent of all the energy consumed within the United States. (If necessary, review Figure 2-17 in chapter 2.) Gasoline converted from crude oil is anticipated to reach levels of 7.9 to 9.1 million barrels per day by 2010. (In 1990, approximately 7 million barrels of oil per day were converted to gasoline.) The use of nontraditional fuels in the transportation sector (e.g., natural gas, alcohol, and electricity) is not expected to affect the consumption of gasoline until sometime after 2000. Nonetheless, these nontraditional fuels are not expected to satisfy more than 1 percent of the energy demanded by the transportation sector in 2010. The anticipated increase in the price of gasoline is somewhere between $1.30 and $1.74 per gallon by 2010. Diesel fuel would increase in price between $1.27 and $1.69 per gallon. These prices include state and federal taxes, but do not include local taxes.

Diesel fuels are also changing to reduce the sulfur content to about one-fifth the level that is permitted in other distillate fuels, such as home heating oil and the fuel used in stationary generators. The change is expected to occur in diesel fuels for on-highway use. These fuels make up about 45 percent of all distillate fuel oil used in the United States.

*Beginning in 1995, gasoline sold in the areas doing the poorest job of satisfying the federal standards for ground-level ozone (a cause of smog) will be required to sell **reformulated gasolines**. Reformulated gasolines (RFG) are gasolines that are blended with oxygen and other compounds. These gasolines will be used to reduce ozone-producing compounds and to help reduce toxic emissions by 15 percent from 1990 levels. The areas required to use reformulated gasoline account for 25 percent of all the United States' gasoline use. The areas include Baltimore, Chicago, Hartford, Houston, Los Angeles, Milwaukee, New York, Philadelphia, and San Diego.*

The new regulations implemented by the federal government will make it difficult for the leading petroleum products to be interchangeable from one season to another or across geographical areas. Although these changes are needed to better the air quality in the United States, they do not come without cost. Projections indicate that by 2000 half of the gasoline in this country will be reformulated. Oxygenated gasolines are expected to cost an added 2 cents per gallon, while reformulated blends will be approximately 8 cents higher per gallon.

After reviewing the foregoing information and Figure 2-13 (in chapter 2), provide some alternatives to the plans implemented by the federal government. In providing these alternatives, attempt to reconcile the needs of people to transport themselves, goods, and the services that they provide with the perceived rights of an individual to own and drive an automobile. Also, attempt to answer the following questions:

1. *How can the urban infrastructure be redesigned to reduce dependence on the automobile?*

2. *What tax incentives could the federal government introduce to help reduce dependence on oil by the transportation sector?*

3. *How can individuals be encouraged to carpool, bike, use mass transit systems, or walk instead of using an automobile for short trips?*

4. *What are some long-term environmental consequences of the United States' dependence on petroleum energy sources for the transportation sector?*

Petroleum Demand in the Industrial Sector

The industrial sector is the second largest user of petroleum based products. (Review Figure 5-10.) This sector, which is a heavy user of petrochemical feedstocks, accounts for more than 25 percent of the United States' oil demand. This sector is predicted to increase its use by an annual rate of about 1 percent. Although the consumption of liquefied petroleum gas and petrochemical feedstocks continues to increase, the largest share of petroleum products consumed by the industrial sector was in the form of miscellaneous products. These products consist of asphalt, road oil, petroleum coke, and lubricants. Liquefied petroleum gas consumption is expected to increase at a rate of 1.9 to 2.8 percent annually. The use of petrochemical feedstocks is also forecasted to increase at an annual rate of 1.6 to 2.2 percent until 2010.

Petroleum Demand by Various Sectors (1990 and Projected Use in 2010)

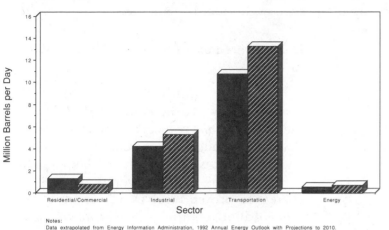

Notes:
Data extrapolated from Energy Information Administration, 1992 Annual Energy Outlook with Projections to 2010.

FIGURE 5-10 This figure illustrates the petroleum demand (for 1990 and projected for 2010) by various sectors in the United States. The largest consumer of petroleum products is the transportation sector followed by the industrial, residential/commercial, and electric utilities sectors.

■ Million Barrels per Day (1990)

▨ Million Barrels per Day (2010)

Petroleum Demand in the Residential and Commercial Sectors

Currently, the residential and commercial sectors account for about 8 percent of the petroleum demand in the United States. The use of petroleum products in the residential and commercial sectors has decreased over the past several years. This trend is expected to continue into the next century. As can be seen in Figure 5-10, the residential and commercial sector will consume less than 1 million barrels per day of petroleum by 2010. This would be approximately 4 percent of all the petroleum demanded by United States' consumers. The reasons for these decreases is a result of the residential and commercial sectors switching to other heating fuels, predominately natural gas. This switch to natural gas is because it is cheaper and more efficient.

Petroleum Demand in the Energy Sector

The use of petroleum in the energy sector (i.e., electric utilities) has decreased during the past several years. This decrease was a result of many utility companies switching from oil to coal or natural gas. However, the use of petroleum in the energy sector is showing signs of increased use. (Review Figure 5-10.) This increase is expected to be between 1.2 to 4.7 percent annually, which is a reflection of residual fuel oil used by electrical generating plants becoming more economically competitive with other fossil fuel prices. By 2010, the energy sector is projected to demand approximately 3 to 4 percent of the petroleum used in the United States.

SUPPLIES OF PETROLEUM

Proven, Undiscovered, and Technically Recoverable Resources

When presenting information on the types of oil reserves available, it is helpful to classify the reserves into the following categories. *Proven reserves* are the estimated quantities that geological and engineering data show with reasonable certainty to be recoverable in future years.

Undiscovered resources are those found outside known gas and oil fields in which the presence of resources has been confirmed by exploratory drilling. *Technically recoverable resources* are those reserves that should be able to produce crude oil but may not be economically profitable. These three terms are used to distinguish between existing reserves and projected reserves of oil. Awareness of this distinction is helpful for understanding the estimates published in energy literature. A further analysis of these terms follows.

Proven Reserves

Proven reserves are those supplies that have been located and are known to be recoverable with existing facilities, present technology, and at current costs and price levels. Resulting from price changes, proven reserves are recalculated regularly for each oil field. Proven reserves can be decreased by production increases, abandonments, low oil-well performance, and price reductions per barrel for crude oil. Of course if world oil prices increase or new technology is introduced, oil that was once considered nonrecoverable may be included among proven reserves.

United States' proven oil reserves are often termed *existing reserves*. Many United States' petroleum companies and governmental agencies predict that at current rates of consumption, existing oil reserves will quickly decline over the next 15-year period. However, some scientific estimates suggest that United States' existing oil reserves will drop significantly beginning within the next 5 to 10 years. For example, estimates by the Worldwatch Institute indicate that existing oil reserves in North America will be depleted within the next 8 years (based on current production and consumption rates). However, as exploration and

drilling technology continue to improve and more future oil reserves are located, estimates such as this may be extended.

Undiscovered and Technically Recoverable Reserves

Undiscovered and technically recoverable resources may also be counted on for use in the future. *Future reserves* consist of all reserves that, in the judgment of geological engineers and economists, are likely to be added to proven reserves. Undiscovered reserves may change to proven reserves as a result of additional drilling depths or extensions of existing oil fields. Technically, recoverable resources may also be counted on later as increases in oil prices make otherwise nonrecoverable oil economically recoverable.

NEW DISCOVERIES

Price as a Controlling Factor in Technological Change and Exploration

The number of new discoveries of oil in the United States is controlled by price. Figure 5-11 illustrates how the world oil prices change over a period of years (since 1979). Figure 5-11 also illustrates the average price of petroleum in the United States and the number of wells drilled. As world petroleum prices increase, United States' oil companies typically find more oil. This is accomplished by drilling more wells, improving drilling technology, and drilling deeper. For example, oil companies can now drill more than 25,000–feet deep, thus finding increased amounts of oil. With advancing technology, it may be possible to find technically recoverable resources into the twenty-first century. However, the price to acquire the new oil may make petroleum use cost prohibitive for certain groups. As can be seen from the data in Figure 5-11, United States' consumers have generally paid a lower cost for crude petroleum products than has the rest of the world. Eventually, however, United States' consumers are expected to pay a greater share of the costs of this resource. Note also that the annual growth rate is expected to be 2.2 percent for both world and United States' crude petroleum prices.

In Figure 5-11, it is also possible to see the relationship of crude oil prices in the United States to the number of wells drilled. During the 1950s and 1960s, production of oil wells in the United States exceeded demand. As shown in Figure 5-11, even as prices remained level during this time, more wells were drilled to extract this resource. By 1970 to 1987, the wells drilled tended to reflect the oil prices (e.g., increasing or decreasing) within the United States.

Note also that the majority of existing known oil resources in the United States have been exploited. If consumers continue to demand petroleum at ever increasing rates, it may be necessary to depend upon resources in wildlife preserves, natural parks, and other areas that have been designated as natural areas.

Outer Continental Shelf

Large quantities of oil have been found on the outer continental shelf. The *outer continental shelf* is an area surrounding the North American continent, as shown in Figure 5-12. This shallow area extends from the shore as far out as 100 to 800 miles. Because of the large amount of oil discovered on the outer continental shelf, increasing quantities of offshore oil is taken from this area. The United States has developed this resource over the past 20 years—causing proven reserves to increase greatly.

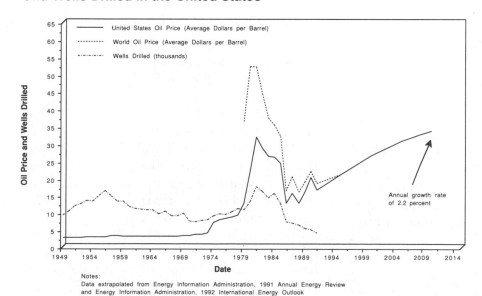

Petroleum Oil Prices (United States and World, 1949 – 2010) and Wells Drilled in the United States

FIGURE 5-11 This figure illustrates the average price of petroleum in the United States and the number of wells drilled with projections to 2010.

U.S. Marine Producing and Potential Petroleum Areas

- ① Beaufort Sea
- ② Bristol Bay
- ③ Gulf of Alaska
- ④ So. California
- ⑤ N.W. Gulf of Mexico
- ⑥ N.E. Gulf of Mexico
- ⑦ Blake Plateau
- ⑧ Baltimore Canyon
- ⑨ Georges Bank

Outer Continental
Shelf Area

FIGURE 5-12 The outer continental shelf surrounding North America will continue to provide much of our existing crude oil reserves in the future.

Drilling in the Outer Continental Shelf— Energy Issue 5-6

Based upon your knowledge of the first several chapters, identify at least three disadvantages and three advantages of drilling for oil in the outer continental shelf.

Approximately half of the estimated undiscovered oil resources in North America are found on the outer continental shelf. Table 5-1 provides an overview of these estimated undiscovered recoverable crude oil resources. As can be seen in Table 5-1, the reserves have been reported using two estimates (i.e., low and high) and a mean. The low value represents a 95 percent probability (a 19 in 20 chance) that there is at least this amount of petroleum that can be recovered in the future. The high value represents a 5 percent probability (1 in 20 chance) there is at least this amount to be recovered in the future. (The totals for the low and high values have not been obtained by arithmetic summation. They are derived by statistical methods.)

Table 5-1
Estimated Undiscovered Recoverable Crude-Oil Resources (billion barrels)

Region	Mean	Low	High
Onshore and State Waters	**33.3**	**19.6**	**51.9**
Alaska	13.2	3.6	31.3
Pacific Coast	3.5	1.5	6.6
Colorado Plateau and Basin and Range	1.5	0.5	3.4
Rocky Mountains and Northern Great Plains	4.5	2.7	6.9
West Texas and Eastern New Mexico	2.6	1.5	4.0
Gulf Coast	4.2	2.4	6.7
Mid-Continent	1.9	1.2	2.7
Eastern Interior	1.8	1.3	2.4
Atlantic Coast	0.2	0.1	0.5
Federal Offshore	**16.1**	**9.2**	**25.6**
Alaska	3.4	0.6	9.4
Pacific Coast	3.4	0.9	8.3
Gulf of Mexico	8.6	4.9	13.6
Atlantic Coast	0.7	0.1	2.3
United States Total	**49.4**	**33.3**	**69.9**

Notes:
Data extrapolated from the Energy Information Administration, *1991 Annual Energy Review.*

Another region that is part of the outer continental shelf is the Arctic slope. This area is known to have great possibilities for the discovery of oil. The region is represented by the Prudhoe Bay field and the Point McIntyre field in Alaska. Moreover, the Baltimore Canyon, Blake Plateau troughs, and the Georges Bank on the Atlantic outer continental shelf contain large oil reserves. These regions, however, contain less than the reserves estimated to exist in offshore areas of the Gulf of Mexico. Federal offshore reserves are estimated to range between 9.2 and 25.6 billion barrels of oil. (Review Table 5-1.)

The remainder of the reserves are located onshore. As can be seen in Table 5-1, most of these reserves are located west of the Mississippi River. Onshore reserves are estimated to range between 19.6 and 51.9 billion barrels of crude oil. The total estimated undiscovered recoverable crude oil reserves of the United States range between 33.3 and 69.9 billion barrels of crude oil. Although these reserves may appear to be vast, based on current and projected consumption rates, these limited supplies could create changes in petroleum use within the next decade. Such changes may include an increased emphasis on conservation, use of imported oil, and substitution of petroleum with renewable or synthetic oils.

Synthetic Oils

The use of synthetic (artificial) oils produced from coal or other solid fossil fuels (i.e., oil shale, which will be discussed later in this chapter) will only begin after coal liquefaction has been improved. As presented in chapter 4, liquefaction is a process in which coal is processed into a crude oil that can be further refined. Such processes are still considered a "parasitic" technology since they consume almost as much energy in process as they deliver in output. Although synthetic liquids are a minor contributor, they should nevertheless add to the total petroleum supply as liquefaction processes improve.

EXPLORATION OF PETROLEUM

Discovery of Petroleum Formations

The exploration for petroleum uses a distinct technology. *Petroleum exploration* is defined as the technical processes used to find future reserves of petroleum. Exploration is primarily concerned with geologists searching for and locating possible petroleum formations. Geologists use the scientific study of the origin, history, and structure of the earth to locate these formations.

Early oil explorers used guesswork and played hunches. Today's geologists work from a knowledge of the chemistry of crude oil, the shape of its formation, and the kinds of areas where oil is frequently found.

Formation of Petroleum

Formations containing oil may lie under mountains, deserts, marshes, or seas. The oil bed may be one to two miles below the surface of the earth, or deeper than 25,000 feet. As earlier noted, these hydrocarbon formations were created from decaying organic matter (aquatic plants and animals that lived and died in and around slowly deposited sea-bottom sediment). Over hundreds of millions of years, great masses of this sediment were slowly and continuously turned into hydrocarbon fuels. In fact, much of the sediment washed and scraped into the surrounding seas, forming the outer continental shelf. Within these formations, natural gas and oil are both formed by geological processes.

Types of Petroleum Formations—Anticlines, Faults, and Stratigraphic Traps

Three types of land formations are known to hold crude oil reserves. As Figure 5-13 shows, oil may be trapped in an *anticline*, an upward fold or arch of rock layer sloping in opposite directions from a crest. Within this formation, the oil lies in small spaces between the grains of permeable rock. The oil may also settle near the top of the bulge or anticline in higher concentrations of hydrocarbons.

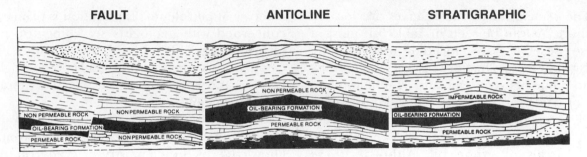

FAULT **ANTICLINE** **STRATIGRAPHIC**

FIGURE 5-13 Shown above are the three types of land formations known to hold crude oil.

Petroleum may also gather at a *fault* or place where the rock layers crack and slip from one another. This type of formation is also illustrated in Figure 5-13. In this case, nonporous or *impermeable rock* stops the oil flowing from the permeable rock. These faults or slips are caused by the internal forces of the earth shifting over geological areas.

A third type of petroleum formation is called a *stratigraphic trap*, as illustrated in Figure 5-13. The permeable rock layers bearing oil gradually taper off under nonporous or impermeable layers of rock. This type of formation is difficult to find because the rock tapers off gradually.

Locating Petroleum

Petroleum formations can be located with four methods. Each of these methods detects the rock formations that may contain oil or natural gas. These methods involve the use of the seismograph, the magnetometer, the gravimeter, and/or core samples.

Seismographic Studies of Rock Formation

A *seismograph* is a device that measures small shock waves sent down into the earth. As these shock waves hit rocks of different densities (permeable or impermeable), they bounce back to the earth's surface. Figure 5-14 illustrates how oil formations are recorded by a seis-

mographic system. Seismography is probably the most accurate current method of locating oil reserves. In the use of a seismograph, many recording wires with sensors are placed over a large area of the earth. A shock wave is then sent into the ground. The shock waves may be generated by a small dynamite blast, thumping the ground with a heavy weight, or (at sea) by releasing underwater intermittent blasts of compressed air. By measuring the time it takes for the shock waves to return, various rock formations and their depth can be detected and charted by geologists using computers.

Seismograph

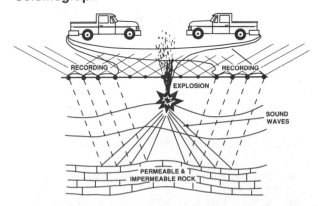

FIGURE 5-14 The seismograph uses shock waves sent into the ground and reflected back to sensors to locate oil reserves.

Magnetometer Studies of Rock Formation

Oil explorers also use an instrument known as the *magnetometer*. The instrument relies on the presence of iron in the rock formation. Some rocks contain more iron than others. Iron content is a function of their permeability. The more iron a rock formation contains, the more magnetic the rock. By measuring the force of the magnetism at the earth's surface, a general measure of the magnetism of the rock formation below can be determined.

Gravimeter Studies of Rock Formation

A third method of petroleum location requires the use of a *gravimeter*. The force of gravity varies from location to location on the surface of the earth. This difference in gravitational force is dependent upon the rock formation beneath the surface. By measuring the forces of gravity at the earth's surface, a pattern can be ascertained for detecting various underground rock formations.

Core Samples to Determine Rock Formation

In addition to the foregoing methods, core samples can be taken to locate petroleum and natural gas. When taking core samples, holes are drilled with hollow tubes at a likely site. These core sample holes may be drilled to a depth of 2 to 5 miles to obtain a specimen of the rock formations below the surface. A core sample is then removed from the hollow tube and examined for traces of oil and natural gas. Core samples can also help determine the ages of the various strata or layers of rock. Such core samples are generally taken about 20 feet in length. Figure 5-15 shows a typical core drilling rig. Often these rigs are portable so that they can be moved to different locations where petroleum might be found.

FIGURE 5-15 This drilling rig removes core samples for evaluation to determine the location of oil under the ground. (Courtesy of Chevron Corporation)

Isometric Mapping

On the basis of the exploration technologies used, various mapping techniques are applied to view an oil reserve. One such method is called *isometric mapping*. Drawing upon various computer programs, data taken from seismographic exploration, reservoir description, core analysis, and interpretation, the oil reservoir can be represented on computerized three-dimensional maps.

Petroleum Drilling Odds

Today, the United States is a leader in drilling technology. Because of the improved exploration technology, growing numbers of oil deposits are found. From early oil explorations up until about 1970, the odds of finding oil were about 1 in 10. This meant that the driller of an exploratory well would find petroleum or natural gas in about 1 of 10 wells drilled. Today, because of improved exploration technology, almost every exploratory well that is drilled becomes commercially successful.

PRODUCTION PROCESSES

Drilling and Removing Crude Oil

The process of production in the petroleum industry involves the use of technologies to drill and remove petroleum from oil formations. These technologies include preparation of the site, drilling, and pumping.

Site Preparation

Before oil drilling can be accomplished, the land must be cleared and graded. Access roads must be constructed to bring in heavy equipment. The area (whether wooded, swampy, sandy, or rocky) will determine the extent of economic involvement needed for this preliminary work.

Once the area is cleared and made accessible, provisions are made for a constant supply of water and power used for both production processes and worker convenience. In some cases, water must be brought long distances. This will, of course, affect the economics of drilling. Despite its size, most oil-well drilling equipment is ordinarily built at a location away from the site and transported to the exploration area. Figure 5-16 illustrates a typical onshore drilling rig. At offshore sites, the drilling equipment is floated to the site by various watercraft technologies specially designed

FIGURE 5-16 This figure illustrates an onshore rotary drilling rig. (Courtesy of Amoco Corporation)

to carry (i.e., float) the drilling equipment to the offshore site for assembly.

Types of Drilling—Cable and Rotary

There are two basic types of oil drilling equipment in use today: the cable-tool drill and the rotary drill. Cable-tool drilling is the older of the two drilling styles. This process employs equipment that raises and drops a bit and stem on the end of a cable. The heavy bit pounds its way into the rock by pulverizing it. The tool bit is periodically removed and the hole is flushed with water. This flushing removes the rock particles in a slurry form. When the hole is deep enough, it is steel-lined with a casing pipe to prevent underground water seepage. Today, very few cable-tool drills are used.

The most common method of drilling for crude oil is with rotary drilling methods. A rotary rig drills a hole by rotating a column of hollow steel pipes. A drilling bit is placed on the bottom of each pipe. The bit, as shown in Figure 5-17, carves, chisels, or grinds a hole into the earth. As the hole deepens, the crew adds new lengths of drill pipe to enable them to go deeper.

Bits for rotary drills range in size from less than 4 inches in diameter to more than 24 inches. They are constructed of high-quality steel and have several rotating cones covered with tungsten carbide teeth to grind through the rock. Long-tooth bits are used for hard rock formations.

When a bit becomes dull, the entire length of drill pipe must be removed, disassembled, and stacked near the oil rig. In an extreme situation, a bit may be worn smooth in one day and after drilling only a few meters.

Drilling Operation

To learn how a rotary drill works, refer to Figure 5-18. This figure illustrates the main operating features of a rotary drilling rig. At the top is the crown block, which acts as a support for a block and tackle. These and the traveling block are used to raise and lower the drill and pipe into the hole. The rotary table turns to connect the lengths of pipe and also turns the drill into the ground.

A mixture of water, clay, and chemical additives, known as *mud*, are pumped under pressure down through the drill pipe. The mud

then enters through a nozzle in the bit and returns to the surface. This constantly circulating mud cools and cleans the bit while transporting the cut rock and other materials to the surface.

The mud is also used to help control the pressure of any natural gas, oil, or water encountered while drilling. If, during drilling, the bit should encounter gas under pressure, the pressure would threaten to blow out the well. This pressure might damage the equipment (causing a safety hazard) or create an oil spill from a

Rotary Drilling Rig

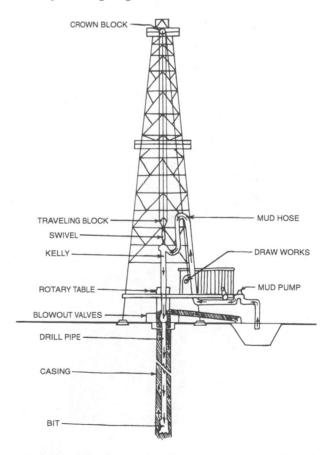

FIGURE 5-18 This figure illustrates the major parts of an onshore rotary drilling rig.

FIGURE 5-17 This rotary drilling bit is used to drill a hole through the earth and into the oil reservoir. (Courtesy of Dresser Industries)

blowout. The mud thus acts as a fluid "cork" to keep the gas and oil under control.

The casing, a heavy steel pipe that lines the well hole, is applied as the well is drilled. Figure 5-19 shows several oil-rig workers installing a casing pipe. Casing pipes near the surface are cemented into place to guide the drill. The surface casings protect freshwater supplies from contamination during drilling operations. Lower casing pipes are installed to keep loose dirt, rock, water, and other material out of the well in order to protect the reservoir. They are also used as a channel to bring oil and natural gas to the surface.

Pumping Units

Once the oil has been discovered, chemically analyzed, and accepted as commercially advantageous, the well is prepared so that the crude oil can be removed. The pumping pres-sure must be regulated to take advantage of the underground pressure. The oil well must not flow so fast as to waste the underground pressure. This regulating effect should yield enough oil to pay for the cost of drilling and allow for a profit. A pumping rig, as shown in Figure 5-20, may also be placed on the well to increase its flow. Ordinarily, this pump is not used until most of the oil has been removed from the well or until the pressure is no longer adequate to allow the crude oil to reach the surface.

Offshore Drilling

Offshore drilling refers to drilling for oil beyond the shoreline, either in coastal waters (i.e., outer continental shelf), far out at sea, or in state waters. The first offshore drilling was done by extending oil fields to the near shore-line. If the oil formation was thought to run under the water, the well was drilled by slanting it outward from the shore. This allowed the drillers to exploit this resource without signifi-

FIGURE 5-19 These "roughnecks" (trade slang for oil-rig worker) are installing the casing pipe of an onshore rotary drilling rig. (Courtesy of Chevron Corporation)

FIGURE 5-20 This oil pumping unit is used to pump oil out of the ground after the drilling rig has been removed. These pumping units are used if there is not enough natural pressure to remove the oil. (Courtesy of Marathon Oil Company)

cant new technology. Later, wells were drilled from piers, artificial islands, and finally, the platform oil drilling rig.

Types of Offshore Drilling

Several types of drilling rigs are used for offshore operations. The first, called a *jack-up drilling rig* (one that is jacked-up above the water surface), is a self-contained platform large enough to provide living quarters for a crew and storage for supplies. Legs that hold drilling equipment extend down to the ocean floor. If petroleum is found to be commercially marketable, the platform is converted to a production system. One example of this type of drilling rig is shown in Figure 5-21.

A second type of offshore rig uses a small, stationary platform that supports the rig. It is serviced by a floating vessel used to house supplies, provisions, and living quarters for the drilling crew.

A third type of offshore rig consists of a platform and a submersible barge towed to the drilling site. The barge is sunk to the bottom of the sea in order to bring the platform down to surface level.

A fourth type of rig is a mobile, self-elevating unit in which steel structures and legs can be jacked to platform level for towing. On location, the legs are lowered to the sea bottom.

A fifth type of offshore rig, the *semisubmersible rig*, has flotation sections under the surface of the water to keep the platform afloat. Figure 5-22 illustrates this type of drilling structure. The unit consists of a self-propelled drilling ship equipped with anchors, cables, and winches to hold it securely above the drilling site. Positioning is aided by propeller-driven thrusters. These vessels are designed for operation in deeper waters. If the petroleum reservoir contains large quantities

FIGURE 5-21 This is called a jack-up offshore rotary drilling rig. (Courtesy of Chevron Corporation)

FIGURE 5-22 This is an example of a semisubmersible rotary drilling rig. Flotations under the water keep the drilling rig afloat. (Courtesy of Apache Corporation)

of oil, several oil rigs, platforms, crew quarters, and so on, can be placed together and connected by walkways—as shown in Figure 5-23.

FIGURE 5-23 When the oil reservoir under a semisubmersible rig is very large, several platforms, crew quarters, and drilling units can be placed together. (Courtesy of Maxus Energy Corporation)

Petroleum Drilling Terminology

Many terms associated with the drilling of oil wells are used throughout the industry and in reference material. Several such terms are defined:

1. *Choke*–A choke restricts the size of the opening through which a well produces oil. The choke controls the flow of oil and/or gas in order to maintain the maximum efficient rate of flow. Chokes vary in size from about $\frac{1}{16}$ of an inch to 1 inch in diameter.

2. *Development well*–A development well is drilled in an area where oil or gas has previously been located. After a field has been discovered and the underground reservoir is roughly outlined by several exploratory wells, development wells are drilled to remove the oil or natural gas.

3. *Discovery*–A discovery occurs when an exploratory well has located a new deposit of oil or gas. A discovery well may open a new field where no oil or gas had earlier been found. It can also locate a previously unknown underground reservoir lying far beneath known producing areas. Not all discoveries are profitable. Many reservoirs are not large enough to make it financially worthwhile to extract the oil or natural gas within the well.

4. *Dry hole*–A dry hole is a well that either yields no oil or natural gas, or yields too little of these resources to make it financially worthwhile to extract. Dry holes are sometimes called *dusters*.

5. *Enhanced recovery*–Enhanced recovery, also called *secondary* (or *improved*) *recovery,* is a method used to increase the flow of crude oil from underground reservoirs.

6. *Injection well*–An injection well is used to pump water, artificial gas, or chemicals into the underground reservoir of a producing field. The object is to maintain the pressure needed to drive crude oil and natural gas to the surface or to sweep more crude oil from the reservoir.

7. *Wildcat well*–A wildcat is an exploratory well drilled in an area where there has been no previous oil production. According to oil lore, the term was first applied to a well drilled in country so remote that the only inhabitants were wildcats. Hence, the driller became known as a *wildcatter.*

8. *Stripper well*–A stripper well is an oil well that is so diminished that it is considered marginal in output (less than 10 percent of its original output per day). Marginal or stripper wells account for about 13 to 14 percent of domestic oil production.

ENHANCED RECOVERY METHODS

As indicated earlier, oil and natural gas are under natural pressure when discovered. However, as oil and gas are removed from the reservoir, the natural pressure decreases. Several enhanced recovery methods are used to either maintain the pressure or to remove more oil from an otherwise exhausted well. The most popular methods include steam stimulation and in situ combustion or fire flooding. Note that when using in situ combustion or steam stimulation, a portion of the resource is used by creating the steam or burning the resource to aid in the recovery process.

Steam Stimulation

Steam stimulation, shown in Figure 5-24, is widely used by the petroleum industry. This method of recovery pumps steam into injection wells to thin out and pressurize the oil formation. This process keeps the oil moving into the producing wells.

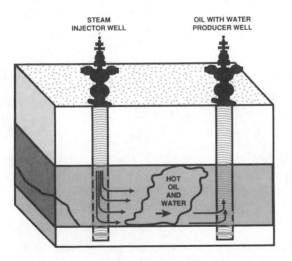

STEAM INJECTOR WELL

OIL WITH WATER PRODUCER WELL

HOT OIL AND WATER

FIGURE 5-24 Steam can be injected into an oil reservoir to increase the pressure and force more oil out of the producing well. This is one enhanced recovery method.

In Situ Combustion or Fire Flooding

A more exotic method of enhanced recovery is called fire flooding. With *in situ combustion* or *fire flooding*, thermal energy or heat from a slow-burning underground heat front causes the oil to thin out. This has a tendency to increase the pressure in the reserve, pushing it to the producing well. If commercial quantities of recoverable oil or gas remain in a reservoir after the use of a single secondary recovery method, additional secondary recovery techniques are used in various combinations to remove the petroleum.

> **Enhanced Oil Recovery—Energy Issue 5-7**
>
> Enhanced oil recovery methods have allowed more domestic crude oil reserves to be extracted. After researching these extraction methods in more detail, identify the economic and environmental costs of these recovery methods.

TRANSPORTATION OF CRUDE OIL AND PETROLEUM PRODUCTS

Overview of Transportation from Well to Refining Site

Transportation technology is involved in the transfer of crude oil from the well to the refining site. More than 15 million barrels of petroleum are moved each day to refineries within the United States. As shown in Figure 5-25, oil can be transported by pipeline, highway trucks, barges, railroad tank cars, or by tankers. The type of transportation depends upon various factors including distance to be traveled, amount of oil to be transported, costs of transportation, and the type of land or water the crude oil must be transported over.

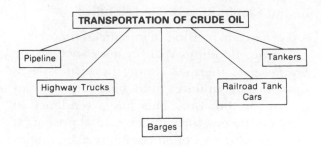

FIGURE 5-25 These are the major methods used to transport oil throughout the United States and around the world.

Pipeline

Domestic oil transportation by pipeline is the most extensively used method. There are more than 1.2 million miles of pipeline within the United States carrying petroleum to and from refineries. The use of pipelines accounts for approximately three-fourths of domestic crude oil movement. Figure 5-26 shows a typical oil pipeline under construction before being placed underground.

Modern pipelines vary from 2 to 48 inches in diameter. The larger pipelines carry greater quantities of oil to and from the refineries to storage tanks. The small pipelines, usually called *gathering lines*, carry the petroleum from the storage tank to truck lines or to other pipelines.

The pipeline has become one of the most economical transportation methods. The American Petroleum Institute estimated that it costs about one cent to ship one gallon of oil by a large-diameter pipeline from Texas to New York.

Pipeline Installation

The procedure for building a pipeline seems simple; but time, cost, and environmental effects must be considered. First, to secure land right-of-way, property owners must sign easements. Bulldozers and trucks must then clear a 50 foot right-of-way. A ditching machine must cut a ditch in the ground to contain the pipeline, as shown in Figure 5-27. The pipes are then welded together and tested for leaks with X-ray equipment. The pipe is insulated so that the oil is kept at a constant temperature. If it gets too cold, its viscosity (or thickness) may change, causing it to be more difficult to pump.

Insulation (commonly composed of asphalt, glass fiber, and felt) also helps to prevent any oil spill damage. The pipe is then placed in the ditch and backfilled to provide firm support. The pipeline may also be further protected from corrosion by the application of a low external voltage.

After these measures are taken, the land is graded and seeded to stop erosion. Petroleum is pumped through the pipeline at 0.7 to 1.1-5 mph depending upon the size and distance of the pipeline. Pumps are spaced about 30 to 60 miles apart, depending on terrain, type of oil, and size of the pipeline. These pumps help to move the oil to its needed destination.

FIGURE 5-26 One inexpensive method used to transport oil is by pipeline. (Courtesy of the American Petroleum Institute)

FIGURE 5-27 This machine is digging the trench in which the oil pipeline will be placed. (Courtesy of Equity Oil Company)

Using pipeline transportation, various types of oil may flow simultaneously through the pipeline. As many as 30 different shipments may follow one another in close succession. Shipping petroleum by this method helps to reduce consumer cost of the end product.

Tanker and Barge Transportation

Crude-oil water transportation involves both tankers and barges. Super tankers are special vessels whose hulls are divided into compartments for carrying liquid cargoes. Figure 5-28 shows a typical oil tanker. Barges are smaller vessels, but they are only capable of carrying oil to inland ports, such as the Great Lakes and some 28,000 miles of rivers and canals where tankers ordinarily cannot go.

The largest modern tankers are longer than 1000 feet and can carry 1.5 million to 2.75 million barrels of crude oil or petroleum products

at speeds of 15 to 17 knots (17.25 to 19.55 mph). Barges carry a smaller amount of oil, about 26,000 barrels. The United States currently has more than 350 tankers, and worldwide there are about 4,000 tankers.

Transportation by Truck

The truck mode of transportation represents about 29 percent of all oil moved within the United States. Tank trucks pick up products at refineries, terminals, and bulk plants and carry them to farms, factories, service stations, and homes. The trucks have an average capacity of between 6,800 and 13,000 gallons. These trucks are usually constructed of aluminum alloys, stainless steel, and reinforced fiberglass.

There is a reduced chance of a major oil spill with the transportation of petroleum by tank trucks. This is because of the smaller quantity carried as compared with that transported by ocean tankers. However, when an accident does occur, it is usually on a highway, thus, endangering the lives of other drivers. Transporting oil by trucks is more costly than using other forms of transportation.

FIGURE 5-28 Supertankers are used to carry large amounts of oil across oceans. (Courtesy of Texaco Incorporated)

Transportation by Railroad Tank Cars

Railroad tank cars have less capacity than tankers and some trucks. Nevertheless, this transport method moves a significant amount of oil. Transporting crude oil and petroleum products by railroad tank cars is less costly than using trucks. However, it is more expensive than using tankers and barges. Generally, the smaller the volume being transported, the higher the cost for transportation.

Transportation of Crude Oil—Energy Issue 5-8

What are the advantages of having numerous methods of shipping crude oil and petroleum products? Why is it not possible to use only one or two forms of petroleum transportation?

The Cost of Transporting Oil—Math Interface 5-2

The transportation of petroleum can be costly. Often the exact type of transportation is dependent upon the energy requirements to transport the oil. The following table illustrates the energy requirements to transport oil by different means. For example, the mileage when moving a ton of oil by pipeline is 275 miles per gallon of fuel. Additionally, it takes 450 Btu's to move a ton of oil one mile by pipeline. Using the table below and information from chapters 1–4, solve the following sequential oil transportation problems:

Mode of Transportation	Mileage (ton-miles/gal)	Energy Consumption (Btu/ton-mile)
Oil Pipeline	275	450
Railroad	185	670
Waterway	182	680
Truck	44	2,800
Airplane	3	42,000
(Data taken from Appendix K.)		

1. Twenty-five million barrels of oil must be shipped 800 miles. Shipping can be by pipeline or by waterway. How many tons of oil are there in 25 million barrels of oil if each gallon weighs 7.8 pounds?

   Tons of oil equal

   _____.

2. How much energy in Btu's would be needed to transport all this oil 800 miles by each mode of transportation?

   Btu's required to ship by pipeline equal

   _____.

   Btu's to ship by waterway equal

   _____.

3. How many Btu's are there in 25 million barrels of oil?

   Btu's in 25 million barrels of oil equal

   _____.

4. Using the answers just calculated, determine what percentage of the energy in the 25 million barrels would be used for transporting the oil 800 miles if only the pipeline were used.

   Percentage of energy used equals

   _____.

REFINING PETROLEUM

After the crude oil is removed from the well, it is transported to a refinery. Refining of oil is a process that separates crude oil into its major components, which are then converted into consumer products. (See Appendix P.) The refining process is an integral part of the petroleum industry.

Today, there are more than 120 United States' refining companies. The choice of a refinery location depends upon such factors as proximity to a large consuming area, proximity to oil fields, and ready access to water, electric power, and transportation. Figure 5-29 shows a typical refinery.

FIGURE 5-29 This refinery breaks crude oil down into different hydrocarbon products to be used by consumers and industries. (Courtesy of Chevron Corporation)

Characteristics of Refining

To analyze the refining process, it is important to characterize crude oil, which is not uniform in substance. Crude oil is mixed with a variety of substances, including water, inorganic matter, and gases. Despite the presence of such impurities, crude oil generally consists of various types of ingredients. Figure 5-30 shows each ingredient with its average percentage of the total.

Crude oil primarily consists of hydrogen and carbon. Some crude oils contain more than 1,000 chemical combinations of hydrocarbons. As a result of the differences in hydrocarbon molecular arrangements, crude oils have a variety of colors. Some are light in color, or pitch black. Other oils are amber, brown, or green. Some may flow like water while others may flow like molasses.

Sulfur Content

The crude oils that contain large amounts of sulfur and other mineral impurities are referred to as *sour crudes*. Those having a low sulfur content are called *sweet crudes*. Note that sulfur content is a negative aspect of crude oil. The sulfur content contributes to corrosion of piping and machinery at the processing plant, and it must be reduced to avoid air pollution problems. When gasoline or diesel fuel containing sulfur is burned in an internal combustion engine or other combustion process, the sulfur pollutes the air. This results in a greater potential for acid rain. As shown in Figure 5-31, crude oils found in the United States have a sulfur content near or below 2 percent. Middle Eastern crude oil sulfur content is close to 5 percent.

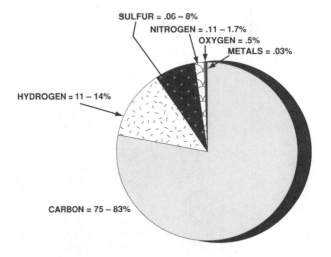

FIGURE 5-30 Crude oil is made up of various ingredients including carbon, hydrogen, and other chemicals.

Boiling Points

The various combinations of hydrocarbons in crude oil differ in regard to their boiling points. Thicker oils have higher boiling points; thinner oils have lower boiling points. The term *boiling point* is used to identify the temperature at which a liquid will start to boil and thus be converted to a vapor. Hydrocarbon boiling points range from −250° F to more than 1300° F. Most hydrocarbons boil at a temperature between 200° and 700° F. Hydrocarbons do not have a single boiling point, but boil over a range of temperatures depending upon the exact characteristics of the given hydrocarbon. For example, a No. 2 diesel fuel begins boiling off some of its hydrocarbons near 350° F. However, most hydrocarbons in this fuel will be converted to a vapor at 650° F. Gasoline begins to boil some of its hydrocarbons around 95° F, but most of its hydrocarbons will be completely vaporized around 390° F. Figure 5-32 presents several common fuels and their ranges of boiling points.

Sulfur Content

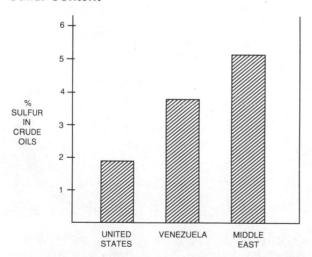

FIGURE 5-31 Crude oils from different parts of the world have different sulfur contents.

Boiling Points for Fuels

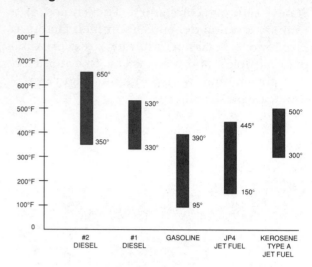

FIGURE 5-32 These are examples of the boiling point temperature ranges of common fuels.

Distillation Processes

Distillation is the most common refinery technology. The various hydrocarbons in a given crude oil are separated into distinct fractions, depending on their boiling temperatures. *Fractions* are portions of the oil mixture that have been separated from the crude oil mixture by distillation. Distillation is also referred to as ***fractional distillation***. This process is done in a fractionating column.

The distillation process can be explained by referring to Figure 5-33. A heating source is placed at the bottom of the distillation column. The heating source is steam, which is pumped through the lower part of this column. The heat from the steam is sufficient to turn the hydrocarbons into a vapor. As the hydrocarbons are vaporized in the column, they start to rise. As the vapors rise, they cool. This cooling results from the vapor moving farther from the heat source. Once cooled from near their boiling points, they are condensed back into a liquid on a series of condensing trays. Light fractions condense into products, such as gasoline. Heavier fractions condense into products like lubricating oil, diesel fuels, and asphalts.

The process of condensing the vapors back to a liquid is done at each of the condensing trays within the fractionating column by a device called a *bubble cap*. Figure 5-34 shows that as the vapors rise in the column, they go through a series of bubble caps on each tray. Those vapors not condensed on a given tray rise to a higher tray (one cooler in temperature) to be condensed. The liquid fractions are then removed for further refining and processing.

Schematic of a Fractionating Column

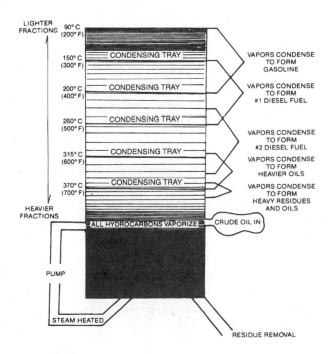

FIGURE 5-33 This is a schematic of a fractionating column. The oil is heated from below. Then as vapors rise through the condensing trays, the vapors cool and condense back to a liquid at different levels in the column.

Conversion Processes

Conversion processes are the refineries' methods of improving various fractions. After crude-oil refining was first developed in the early 1900s, the market for petroleum products changed. Automobiles, which burn gasoline, became more popular. Thus, there was an increase in the demand for gasoline. Problems developed in the refineries since only a limited quantity of gasoline could be extracted from a barrel of crude oil.

Aside from increased demand, the refineries were faced with seasonal demand shifts. For example, more fuel-oil production was required during winter to meet demands for heating in the northern United States. Winter and summer demands and the popularity of the automobile caused refineries to find methods of producing more fuel oil and gasoline by designing various conversion processes.

Cracking

Conversion processes are designed to convert other hydrocarbons to gasoline. These processes are known as *cracking*. The most popular conversions or cracking processes are thermal cracking, catalytic cracking, hydrocracking, polymerization, alkylation, isomerization, and catalytic reforming and treating.

Thermal Cracking

Thermal cracking is the oldest conversion process. It is accomplished by taking heavier fractions of petroleum and subjecting them to higher temperatures and pressures. These conditions cause rearrangement of the hydrocarbon molecules. Many of the resulting molecules have boiling points within the gasoline fraction range.

Bubble Cap

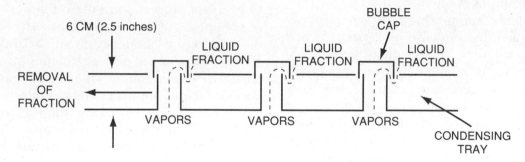

FIGURE 5-34   Bubble caps help the vapors to condense to a liquid in the condensing tray.

Thermal cracking contributed to the improvement of gasoline, heightening its anti-knock characteristics by increasing the octane of the fuel. Octane is a measure of fuel ability to resist burning. (To resist burning means a higher temperature is needed for the fuel to ignite.) As octane was increased, the amount of gasoline derived from each barrel of oil was increased. While a meaningful breakthrough, thermal cracking was only a beginning in the development of better cracking methods by the refining industry.

Catalytic Cracking

In the 1930s, catalytic cracking was introduced in the United States to help improve octane. In this process, a fine-grained catalyst is placed in the presence of heavier hydrocarbons. (A *catalyst* is a substance that alters the rate of chemical reaction and is itself unchanged by the process.) The catalyst causes the hydrocarbons to break apart into lighter fractions in the gasoline range. With the use of a catalyst, the operating pressure does not need to be as high as in thermal cracking. Since higher pressure refining equipment does not need to be used, the overall cost of the cracking process is reduced. Introduction of this process helped

to reduce consumer costs. In addition, the resulting gasoline products were better than those produced by thermal cracking. Figure 5-35 shows catalytic cracking units (towers) used in a refinery.

Although the catalyst utilized in the cracking process can be reused indefinitely, a crust of carbon eventually surrounds the particles of the catalyst, thus fouling the cracking process. The catalyst can, however, be cleaned by reheating it with the heat produced in the process.

Hydrocracking

Hydrocracking is a popular process used to supplement catalytic cracking. The underlying concepts for this process are similar to other cracking methods. However, higher pressures and temperatures are used in the presence of a catalyst and hydrogen. The main advantages of hydrocracking are higher octane ratings and less carbonizing of the catalyst (than in catalytic cracking) because of the higher pressures and temperatures.

Polymerization

Polymerization is a process in which fractions that are lighter than gasoline, such as refinery gases, are forced to unite or *polymerize* and

FIGURE 5-35 These catalytic cracking units in a refinery are used to obtain more gasoline from a barrel of crude oil. (Courtesy of Amoco Corporation)

form liquid fractions that have boiling points near the gasoline range. These fractions are called polymers. The gases are subjected to high pressures and temperatures in the presence of a catalyst. This causes the lighter fractions to unite, producing a heavier fraction.

Alkylation and Isomerization

Alkylation and isomerization also are methods used in the refining industry to alter hydrocarbon molecules. The purpose is to improve the octane ratings of gasoline fractions. In the alkylation process, higher octane fuels are produced by using a catalyst of sulfuric acid or hydrogen fluoride. In the isomerization process, platinum oxide is used as the catalyst at a temperature of 320° F and a pressure of 400 psi.

Reforming and Treating

The catalytic reforming process uses a form of platinum as its catalyst. Price decides the type of platinum used and is based upon economic considerations. The reforming process is used to better the quality of the products being produced.

One final process, called treating, will remove sulfur and impurities, especially from fractions produced by sour crude oil. Various chemicals are added to reduce sulfur. Pollution standards in recent years have caused refineries to implement more treating processes.

PETROLEUM PRODUCTS AND THEIR CHEMISTRY

Several terms are used to help describe the petroleum products used by consumers. Other terms are used to describe the chemical characteristics of products produced from crude oil. Currently, there are over 3,000 petroleum products. The following definitions and descriptions will help in understanding petroleum technology associated with gasoline and diesel fuels.

Octane Number

Octane is a measure of a fuel's ability to resist low-temperature burning in the combustion chamber. Fuels that resist burning longer need a higher temperature to start combustion. Also, the combustion is more controlled and even. Lower-octane fuels cause engines to "knock" or "ping." Knocking or pinging is caused by the fuel igniting too soon inside the combustion chamber. Higher-octane fuels help to reduce knocking. Thus, octane number is a measure of the antiknock value of a gasoline.

The higher the octane number, the higher the antiknock quality. The range of standard octane ratings at a typical gasoline pump is from 80-90 for regular gasoline and up to 110 for higher grade gasoline. Antiknock compounds are formulated into gasoline products during refining to increase octane ratings.

Types of Octane

The three kinds of octane measures generally in use are *research*, *motor*, and *road octane*. *Research octane* is a laboratory measure of gasoline antiknock characteristics under relatively mild engine operating conditions of low speed and temperature. *Motor octane* is a laboratory measure of gasoline antiknock characteristics under severe engine operating conditions, such as high speed and temperature. *Road octane* is a rating that represents realistic road driving conditions; this index is the rating posted on most gasoline pumps. Road octane is generally four or five octane numbers lower than the corresponding research and motor numbers.

Changing Octane Ratings

Gasoline octane numbers have changed as compression ratios in automobile engines have increased. Research octane for regular gasoline in 1935 was near 72. Today, regular octane ratings commonly range between 87 and 90.

Spurred by pollution concerns, which emerged in the 1960s, octane ratings began to drop in the mid-1970s as lead was no longer added to gasoline. The use of the additive tetraethyl lead (TEL) in the refining process improved octane ratings. However, it created concern regarding lead pollution. Therefore, the TEL was removed—causing a drop in octane ratings. This, in turn, caused automobile

manufacturers to reduce their compression ratios from as high as 12 to 1 to an average of 8.5 to 1. Today, however, octane ratings are back up. This increase is a result of other non-polluting types of chemicals being used to increase octane ratings.

Costs and the Reduction of Lead in Gasoline— Energy Issue 5-9

When tetraethyl lead was removed from gasoline to reduce lead pollution, oil refinery companies increased the price of new nonleaded gasoline. Based upon your knowledge so far, why do you think the price raised when an ingredient was removed from the fuel?

Cetane Number

Cetane numbers are used to express the ignition quality of a diesel fuel. The cetane number is a measure of a diesel fuel's ability to burn, low temperature startability, warm up, and smooth combustion in diesel engines. The range of cetane numbers for diesel fuels is 40 to 60 with values of 45 to 50 being most common. These cetane values are satisfactory for medium and higher speed engines. Low speed engines may use fuels with a cetane range of 25 to 40.

ALTERNATIVE FUTURES

Oil Shale

There also has been an interest in a depletable resource known as *oil shale*. Oil shale has often been described as "the rocks that burn." A sample of oil shale is shown in Figure 5-36. It is a fine-grained sedimentary rock (called *marlstone*) often containing an organic solid mater-

FIGURE 5-36 Oil shale has the potential for producing additional resources of oil. (Courtesy of Chevron Corporation)

ial called *kerogen*. When kerogen is heated to about 850° F, a gas and a heavy oil are released, both of which can be upgraded and refined with existing technology.

The use of oil shale dates from the fourteenth century when the Austrians and the Swiss used this resource to produce a form of rock oil. In the United States, Native Americans used this resource over 100 years ago in their campfires. Commercial applications of the resource were introduced in the 1800s. These consisted of burning the crushed rock in boilers and producing fuel gases, and other petrochemicals. This resource, however, declined in use with the introduction of the oil industry. Its decline was a result of not being able to compete economically with the refining of crude oil.

United States' Oil Shale Supply

The oil derived from oil shale in the United States is estimated to equal between 80 billion and 964 billion barrels of oil—both recoverable and technically recoverable. On the basis of these estimates, oil shale represents more than double the current estimates of United States' recoverable petroleum reserves. However, to date there is only one proven technology for converting this resource into a refinery feedstock, and the costs associated with this process do not make it economically feasible. It is likely that large-scale utilization of oil shale will not occur in the foreseeable future.

Oil shale deposits are categorized as high, medium, and low grade, depending upon the impurities and quality of the kerogen available in the resource. High-grade oil shale deposits are estimated to produce an average of 26 gallons of oil per ton of shale. Medium-grade oil shales would yield an average of 10 gallons per ton; low-grade deposits would average between 5 and 9 gallons per ton. Most United States' oil shale deposits are located in the Green River Formation in western Colorado.

Oil Produced from Oil Shale— Math Interface 5-3

Oil shale is considered for possible energy use in the future. The potential for success is partly determined by the costs of mining and processing the fuel source. Solve the following problems:

1. There are two reserves of oil shale. One reserve of high-grade oil shale produces about 26 gallons of oil per ton of shale. Another reserve of medium-grade oil shale produces approximately 11 gallons of oil per ton. Each

reserve contains 150,000 tons of oil shale. How many gallons of shale oil can be taken from each reserve?

Gallons of shale oil from the high-grade reserve equal

_____.

Gallons of shale oil from the medium-grade reserve equal

_____.

2. How many Btu's would be available from the high-grade reserves if one gallon of oil has 130,000 Btu's? From the medium-grade reserves?

Btu's available from high-grade reserves equal

_____.

Btu's available from medium-grade reserves equal

_____.

FIGURE 5-37 Core samples of rock must be taken to help locate oil shale deposits. (Courtesy of Colony Development Operation)

Oil Shale Technology

The technology involved in tapping oil shale resources is similar to that utilized in mining coal. First, core samples are taken, as shown in Figure 5-37. Once oil shale has been found, it must be mined (in the same manner as coal), prepared, processed, and transported. Both underground and surface-mining techniques are used.

To convert this resource into a usable fuel, a *retorting* process is used to remove the oil from the rock. In this process, the oil shale is heated in an atmosphere that does not allow complete oxidation. The retorting process allows the hydrocarbons to be reduced to a liquid or gaseous state—as shale oil or natural gas. Again, it needs to be stressed that, although considerable oil shale technology is currently available, its development is based upon how economically competitive it is with crude oil production. Research is continuing to make oil shale a significant future energy resource.

SOCIAL AND ENVIRONMENTAL COSTS OF PETROLEUM USE

The social and environmental costs of using petroleum are many. Though some of these costs are minimal, others present dangers to plant and animal species. The social and environmental costs presented here should be addressed if petroleum is to be put to use in a more socially and environmentally acceptable manner.

Social and Environmental Costs of Extracting Petroleum

Clearly, all energy sources have social and environmental limits. In some instances the land areas involving resource extraction are too valuable to permit mining, oil drilling, or any other activity. Many individuals in this country have placed an intangible value on land. As such, land may be valued for its scenery, ecology, or its cultural significance. Individuals holding this view would rather see a picturesque land area than one dispoiled by a view of oil wells. In this situation the underlying fossil fuel, like petroleum, may not be valued for its resource if the aesthetic or social value of the land is regarded as more important by a significant number of people.

Besides the social value land may hold, it is important to recognize some dangers of petroleum extraction: the destruction of fish killed by explosions during seismic prospecting, the loss of habitat to animals when wells are drilled, and the ecological damage that results from oil spills and well blowouts. Each of these conditions represents a social and environmental cost of petroleum extraction.

Environmental Costs of Pipelines

The pipeline method of crude oil transportation also has environmental implications. If there are large geologic ruptures in the earth's surface (such as earthquakes), the pipeline could break resulting in an oil spill. The Alaska pipeline is vulnerable to breakage in another way since much of the pipe is placed above ground to prevent the oil from getting too thick and to prevent warming of the permafrost and tundra.

Environmental Costs of Tankers

With the use of super tankers as a form of transportation, oil spills are a danger. Tankers, which can carry more than 2 million barrels of oil, can run into reefs, collide with other ships, or develop leaks. Spills cause great environmental harm. One oil spill was that of the Exxon Valdez. This super tanker was carrying oil from the Valdez terminal in Alaska to the United States. The super tanker struck a reef and spilled about 11 million gallons of oil in and around Valdez, Kenai National Wildlife Refuge, Kodiak Island, and surroundings. At first, the environmental damage was severe. Thousands of animals and birds were killed, shorelines were damaged, and the fishing industry in the area was greatly affected. It was estimated that the environmental damage lasted more than five years after the initial oil spill. Since the oil spill, Exxon has spent millions of dollars cleaning the shorelines in the affected area. Table 5-2 provides a listing of some of the more notable recent oil spills in the United States.

The difference in safety and educational standards among the world's tanker fleets and crews (respectively) causes many oil spills. Based on statistical data collected by the federal government, the United States has averaged about 10,000 oil spills per year. Additionally, the United States has had approximately 10 million gallons of oil spilled each year since 1971. In recent years, safety standards on United States' tankers have improved. New methods have been developed to handle tanker traffic, educate officers of oil tankers and barges, and train workers to load oil and clean up spills properly. Additionally, in response to the Oil Pollution Act of 1990, regulations have been passed to strengthen the authority of the federal government to prevent oil spills in marine and freshwater environments. Under this act, new tanker vessels operating in United States' waters must be built with double hulls, and existing single-hull vessels must be phased out by 2015.

Environmental Costs of Burning Petroleum and Using Petroleum Products

The combustion of fossil fuels, such as petroleum, inevitably results in the release of hydrocarbon products, some of which are known to be carcinogens (agents or substances that produce cancer), mutagens (substances that can cause changes in the genetic structure of later generations of organisms), or teratogens (agents that cause fetal malformation during the first three months of pregnancy). Even in doses as small as milligrams per kilogram of body weight, hydrocarbons are dangerous to humans. Unfortunately, most of these hydrocarbon compounds in question have not been adequately investigated to determine the long-term health effects to humans and other animals.

Table 5-2
Notable United States' Oil Spills

1. **December 15, 1978:**
   Argo Merchant tanker, Massachusetts' southeastern coast. This spill released 7.6 million gallons of crude oil when the ship ran aground.

2. **November 1, 1979:**
   Burmah Agate tanker, Galveston Bay, Texas. This tanker spilled over 10.7 million gallons of oil after colliding with another ship. A fire resulted.

3. **November 22, 1980:**
   Georgia tanker, Pilottown, Louisiana. When this ship ran aground, 2.8 million gallons of oil were spilled.

4. **July 30, 1984:**
   Alvenus tanker, Cameron, Louisiana. When this ship ran aground, 2.8 million gallons of oil were spilled.

5. **August 18, 1984:**
   Triangle Oil Corporation, Jacksonville, Florida. Approximately 2.5 million gallons of oil burned after lightning caused a fire.

6. **October 31, 1984:**
   Puerto Rican Tanker, San Francisco, California. After an explosion in the ship, some 2 million gallons of oil spilled and burned.

7. **November 6, 1985:**
   Exploratory well blowout, Ranger, Texas. Approximately 6.3 million gallons of oil spilled after a blowout of an exploratory well.

8. **January 3, 1988:**
   Ashland Oil Company, Jefferson Borough, Pennsylvania. 3.8 million gallons of oil spilled when an aboveground storage tank ruptured.

9. **March 24, 1989:**
   Exxon Valdez, Prince William Sound, Alaska. When the Exxon Valdez ran aground, 10.92 million gallons of crude oil was spilled into the surrounding waters.

10. **June 8, 1990:**
    Norwegian tanker Mega Borg, Gulf of Mexico. Following an explosion of the tanker, 4.3 million gallons of light crude oil were released into the waters of the Gulf of Mexico.

Other environmental costs of burning petroleum are resource depletion, global warming, acid rain, and the release of waste heat. These environmental costs were discussed in chapter 4 and are not reiterated here.

Social and Environmental Costs of the Automobile

The automobile is not only a necessity for many people, but has become an economic and social symbol of achievement and personal freedom. Among its benefits are greater mobility, the creation of a division of labor, dispersal of commercial and factory centers, urbanization, and giving people a wider choice of what to do and where to go. However, the social and environmental costs of the automobile are equally numerous.

The social and environmental costs of the automobile include the costs of upkeep and development of highways (shared by trucks), streets, and the infrastructure of a transportation system; the damage caused to those injured by automobiles (an automobile injury occurs every 19 seconds in the United States and an automobile related death happens every 11 minutes); and the diversion of funds that might otherwise be available for other goods and services. One could also consider the costs associated with resource depletion and with air pollution from automobile exhausts.

As to the latter, the United States is losing its efforts to curb air pollution problems that result from the automobile. An example of this can be seen by reviewing the trends in New York City. After nearly two decades of attempting to meet federal guidelines for clean-air standards, the city has found that its air quality is getting worse. The problem is so vast that officials in the state of Connecticut are now insisting that they cannot meet their own clean air standards as a result of dirty air crossing Connecticut borders from New York. In fact, Connecticut officials have stated that even if they stopped all manufacturing and traffic in their state they would still not be able to meet federal standards because of automobile air pollution caused in New York City.

Social and Environmental Costs of Oil Shale

Although oil shale has been proposed as a potential energy resource for the future, it does come with several social and environmental costs associated with its extraction and use. From the extraction perspective, one of the major concerns is the disposal of solid wastes. In the extraction of oil shale, massive amounts of solid wastes are generated. Producing a barrel of shale oil generates about one ton of solid shale waste. This waste needs proper disposal in an environmentally friendly manner. A single oil shale refinery producing two million barrels of shale oil per day is estimated to generate enough waste in its operating lifetime to cover 100 square miles of ground to a depth of 400 feet.

Another environmental concern is the possibility of groundwater contamination from shale oil production. Wastes from production have a particle size so small that erosion and siltation of water resources are a concern. In addition, the alkalinity is high enough that it discourages plant growth. The wastes may often contain leachable organic residues hazardous to people, plants, and animals. Should these leachable organic residues come in contact with a large body of water or an underground aquifer, it would be virtually impossible to restore the water resource to its original condition.

Another processing concern of oil shale is air pollution. Currently, there are serious questions about the oil shale industry's ability to meet the federal guidelines of the Clean Air Act. The extraction of oil shale would create massive amounts of airborne contaminants. The United States Office of Technology Assessment estimated that an oil shale production facility could only produce 400,000 barrels per day while meeting the guidelines of the Clean Air Act. Although this is far below the consumption requirements of industry and consumers in the United States, it will help extend current energy reserves.

Finally, oil shale has another disadvantage when compared to other fossil fuel sources. It is considered a dirty fuel source because it contributes heavily to greenhouse gases. The carbon dioxide generated by a facility burning this fuel source or converting it to a feedstock would output more greenhouse gases than those that use coal or petroleum.

POINT/COUNTERPOINT 5-1

TOPIC

Increasing Available Petroleum by Exploiting the Outer Continental Shelf and Other New Discoveries

Theme: There seem to be abundant oil reserves located in several areas—one being the outer continental shelf and the other, Alaska. This discussion deals with identifying the social and environmental effects of continuing to exploit these areas as sources of crude oil. In a classroom discussion or debate, attempt to answer the following questions:

1. What would be the purposes of getting more oil from these areas?
2. What are the environmental effects of increasing offshore wells or wells in Alaska's wildlife areas?
3. If society is not willing to exploit these energy resources, from where will the oil come if demand continues to increase?
4. What are the positive and negative social effects of having this oil available to society?
5. What are the positive and negative social effects of not having this oil available to society?
6. Could controls be placed on the large oil-production companies to limit oil taken from these reserves? Who should pay for the enforcement of these limits?
7. If more petroleum is available from these and other new discoveries, will the United States ever become oil independent?
8. If consumers continue to demand oil, is it a wise decision to go to foreign countries to purchase this resource—especially if there is oil available, but unexploited, within the boundaries of the United States?
9. What role should the government have in determining if these oil reserves should be exploited?
10. What role should consumers have in determining if these oil reserves should be exploited?
11. Economically and politically, what advantages and disadvantages are there if these new discoveries are exploited?

SUMMARY & REVIEW

Summary/Review Statements

1. M. King Hubbert predicts that oil production will continue to increase until about the year 2000, at which time one-half of the total world oil will (by his calculations) have been consumed.

2. Oil consumption increased by approximately 2 percent per year between 1985 and 1990. The demand for oil is projected to increase by 1.35 percent per year until 2010.

3. The demand for oil is expected to be most rapid in the developing nations in Asia and the Middle East.

4. The economies of developing countries now need 40 percent more energy than do industrial ones to produce the same value of goods and services.

5. The Organization of Petroleum Exporting Countries (OPEC) has three-fourths of the world's oil reserves. The countries that form OPEC include Algeria, Ecuador, Gabon, Indonesia, Iran, Iraq, Kuwait, Libya, Nigeria, Qatar, Saudi Arabia, the United Arab Emirates, and Venezuela.

6. As a result of a lack of surplus of oil production outside of OPEC, increases in price will be needed to balance the demand and supply of petroleum.

7. The consumption of oil is expected to be very close to production levels; and by 2010, world oil consumption could outpace production.

8. The United States is the world leader in petroleum consumption. Citizens of the United States use more oil per person than any other group of people in the world. On the basis of the data collected, the populace of the United States consumed approximately 3 gallons of petroleum per person per day in 1990.

9. In response to increasing oil prices and with the goal of becoming petroleum independent, the United States government established the Strategic Petroleum Reserve. This reserve is to be used in the event of supply disruptions from exporting oil countries.

10. The United States has not fulfilled its goal of becoming oil independent. By 1991, 42 percent of the petroleum used in the United States came from imports.

11. The transportation sector is projected to remain the leading end user of petroleum products through the turn of the century. By 2010, this sector is estimated to consume 66 percent of all the petroleum used in the United States.

12. Proven reserves are the estimated quantities that geological and engineering data show with reasonable certainty to be recoverable in future years. These supplies have been located and are known to be recoverable with existing facilities, present technology, and at current costs and price levels.

13. Undiscovered resources are those found outside known gas and oil fields in which the presence of resources has been confirmed by exploratory drilling.

14. Technically recoverable resources are those reserves that should be able to produce crude oil but may not be economically profitable.

15. United States' domestic oil reserves are often termed *existing reserves*. Many United States' petroleum companies and governmental agencies predict that, at current rates of consumption, existing oil reserves will quickly decline over the next 15-year period.

16. Approximately half of the estimated undiscovered oil resources are on the outer continental shelf.

17. The total estimated, undiscovered recoverable crude-oil reserves of the United States range between 33.3 billion and 69.9 billion barrels of crude oil.
18. Three types of oil formations are known to hold crude-oil reserves—anticlines, faults, and stratigraphic traps.
19. Petroleum formations can be located by four devices: the seismograph, the magnetometer, the gravimeter, or core sampler.
20. Several enhanced recovery methods are used either to maintain the pressure or to remove more oil from an otherwise exhausted well. The most popular methods include steam stimulation and in situ combustion (or fire flooding).
21. Refining of oil is a process that separates crude oil into its major components. These major products are then converted into consumer products.
22. Distillation, also referred to as fractional distillation, is the most common refinery technology. The various hydrocarbons in a given crude oil are separated into distinct fractions by their boiling temperatures. Fractions are portions of the oil mixture that have been separated from the crude-oil mixture by the distillation process.
23. There is a potential for more oil to be developed from the use of oil shale.

Discussion Questions

1. What are some factors that cause oil reserve estimates to vary?
2. What effects would be felt, both positively and negatively, if oil shale were significantly developed in the next ten years?
3. Oil importers are highly vulnerable to rapid price increases and energy shortages. Such fluctuations have occurred three times on a global scale during the past 20 years. What causes these increases, and what steps could be implemented (if any) to correct this situation?
4. Project how a society might function if oil reserves continually declined 10% each year, over a period of 20 years. List changes that might be made to satisfy continuing consumer demand for energy.
5. What is the purpose of cracking in the refining industry? Why was cracking technology developed?
6. How could worldwide population control offset the growing demand for petroleum and the environmental effects from its use? What are the social implications for developing countries?

chapter 6

Natural Gas Resources

PURPOSES

Goals of Chapter Six

The goal of this chapter is to investigate the use of natural gas as an energy source. This investigation includes a brief historic review of natural gas and an indepth look at its availability, characteristics, environmental and economic implications, future outlook, and associated technologies. At the completion of this chapter, you will be able to

1. Identify the availability of natural gas energy resources in terms of both supply and demand.
2. Determine the environmental, economic, and legislative implications of using natural gas.
3. Examine the status of future reserves of natural gas.
4. Define the energy characteristics of natural gas.
5. Identify how natural gas is stored and transported.
6. Describe the basic natural gas refining technology.
7. Identify the potential for using methane as a fuel.
8. Analyze the environmental and social costs of natural gas use.

Terms to Know

By studying this chapter, the following terms can be identified, defined, and used:

Associated gas
CCF
Coal gas
Cogeneration
Combined-cycle electrical generation
Conventional resources
Devonian shale deposit
Dry gases
Ethylene
Flared gas
Fractionation process
Heavy hydrocarbons
Interstate gas
Intrastate gas
Light hydrocarbons
LPG
Nonassociated gas
Normal paraffins
Pig
Polyethylene
Price controls
Saturated hydrocarbons
Steam-injected gas turbines
TCF per year
Therm
Unconventional resources
Wet gases

INTRODUCTION TO NATURAL GAS

The use of natural gas dates to the twelfth century. Its use, however, was restricted to those areas where natural gas fields were found. The Chinese of that time dug wells several hundred feet deep and used bamboo pipes to direct the gas to specific locations. Although other civilizations also used this resource, it was not until the nineteenth century that gas became important on a large scale. This usage can be traced to the first experiments with lighting from coal gas.

Early Use of Coal Gas

During the nineteenth century, experiments with gas had shown that it was possible to light single buildings with coal gas. The idea of using this resource to light entire areas, as well as streets and cities, was proposed. *Coal gas* is a form of natural gas found around or near a coal reserve. One of the first individuals to make a significant contribution to this effort was Friedrich Albrecht Winzer of Germany. Having seen a demonstration of gas lighting by Philippe Lebon in Seignelay, Paris, Winzer was convinced that gas could be applied on a large scale. In 1803, Winzer moved to England and changed his name to Frederick Albert Winsor. Winsor was able to establish the world's first gas company, the National Light and Heat Company, in 1806.

Winsor installed gas lighting on Pall Mall in London in 1807, but his technical skills with the use of this energy source and the lighting devices did not match his business skills. The use of coal gas required much skilled work to overcome problems encountered in providing acceptable service. Coal gas has a strong odor, is toxic, and is explosive when not used correctly. Provisions had to be made for produc-ing, distributing, and using it safely and economically. In addition, fumes and smoke from the use of insufficiently purified gas and poorly designed burners created a distrust of gas lighting for many years.

By the 1840s, the use of efficient burners allowed coal gas to be used on a wider scale. These burners introduced air into the flow of gas just prior to combustion. The net effect was to allow the gas to be burned more completely, safely, and efficiently. With these technical changes to the gas burner, the use of gas spread steadily in the urban communities of England and later throughout Europe.

Early Use of Coal Gas in the United States

The use of coal gas in the United States followed a similar pattern of development as that in Europe. Baltimore, Maryland, was the first to use this gas as a lighting source for streets, beginning in 1816. As in Europe, experiments with this energy source also led to its use in the lighting of buildings. By 1925, gas light companies were established in Baltimore, Boston, and New York City.

Early Use of Natural Gas in the United States

The United States had an advantage over other countries since it possessed supplies of natural gas. The earliest report of natural gas was by a local salt production company in Pennsylvania in 1820. While it was recognized that this resource could provide an energy source for lighting, it was not used for that purpose. The second discovery of natural gas, however, in Fredonia, New York, was used for house lighting.

Due to isolated wells and the lack of long-distance transmission capabilities, natural gas served primarily as a fuel for nearby industries rather than for lighting. This resulted in the manufacture of coal gas or *town gas* (as it was primarily known) for lighting. As such, natural gas was considered a waste product or, at best, a cheap fuel to be used only in the oil fields or their immediate vicinity.

During the late 1860s and 1870s there were isolated experiments in which natural gas was used to fuel iron and steel industries near Pittsburgh. The fuel was also used on a small scale for the making of firebricks and as a source of lampblack (a fine black soot that can be used as a pigment) for printers' ink.

The late 1880s saw increased discoveries of natural gas in Pennsylvania, West Virginia, and Ohio. These supplies led many illumination companies to emphasize the promotion of gas for cooking, water heating, and space heating.

Having discovered the benefits of natural gas deposits, the United States began to capitalize on this new-found energy source. Natural gas soon displaced the use of town gas for lighting—even in the nation's capital. Technical improvements to the gas burner soon followed. Chief among them was the incandescent mantle developed by Carl von Welsbach in 1886. This device enabled the gas burner to produce a brilliant white light, and gas lighting became more widely used than the newly introduced electric lamp.

Numerous experiments were conducted during this early period of development with the underground storage of natural gas. In Ontario, Canada, a partially depleted gas reservoir was successfully reinjected with natural gas in 1915. The following winter the gas was able to be withdrawn to meet the seasonal requirements of gas heating. This innovative storage mechanism was adopted in the United States. The first use of this storage technology in the United States occurred in New York in 1916,

followed by Kentucky in 1919, and Pennsylvania in 1920. By the end of 1930, there were nine storage pools in six states with a total storage capacity of 18 billion cubic feet.

Improvements in Natural Gas Leading to Widespread Use

The major improvement that led to the widespread use of natural gas in the United States occurred during the late 1920s. This improvement was not connected to burner design or improved safety conditions, but rather the transportation of the fuel source. Until the late 1920s, natural gas was transported by means of small-diameter pipes joined with screw couplings. This resulted in high costs for construction and a limited distance over which gas could be economically transported: an average of only 250–300 miles. As a result, the vast quantities of natural gas that were being discovered remained largely unused.

It was during the Great Depression (1929–1939) that the practice of welding was applied to pipeline joints. This practice allowed the construction of large-diameter and seamless pipes. Simultaneously, the digging of the trench for the pipeline and handling the pipe became mechanized. By 1935, it was possible to transport gas economically for 1000 miles. While some longer lines were laid, it remained for the Big Inch and the Little Inch pipeline projects to demonstrate the practicability of long-distance gas transmission. These two lines—24 and 20 inches in diameter, respectively—were laid during World War II. These pipelines transported petroleum from the southwestern oil fields to the northeast as a defense measure to relieve the burden on tanker transport of petroleum.

Following the close of World War II, the Big and Little Inch were eventually used for natural gas transmission, and the great gas boom began. By 1947, there were 77,000 miles of main line gas transmission (by 1963 this had risen to over 200,000 miles). With this expansion and the post-war economic recovery came further improvements in construction technology that completely mechanized the laying of the pipe at greatly reduced costs. This resulted in all but two states being served by natural gas by 1955.

NATURAL GAS IDENTIFICATION AND CHARACTERISTICS

Origin of Natural Gas

Crude oil and natural gas are found in similar geological environments—predominately in sedimentary rock formations. Like crude oil, natural gas is formed by the decomposition of organic materials that occurred millions of years ago. This sharing of common origins has led many authorities to classify natural gas as a gaseous petroleum. Some authorities even refer to natural gas under the ambiguous term of crude oil.

Characteristics of Natural Gas

Natural gas has been identified as a clean-burning fuel. In comparison with other fossil fuels, natural gas is extremely clean burning. Generally, if the combustion process has adequate air, the exhaust includes three main ingredients. These include thermal energy measured in Btu's, carbon dioxide, and water. Because of the clean-burning characteristics of natural gas, it can be used in homes (furnaces, clothes dryers, stoves, and so on) and in industrial buildings in lift trucks with minimal concern for health damage to individuals.

Types of Natural Gases

When analyzing the technology associated with natural gas, knowledge of its characteristics becomes important. This is also true when discussing its applications and future socioeconomic implications.

Ordinarily, raw natural gas is combustible, but noncombustible components (such as carbon dioxide, nitrogen, and helium) are often present. Natural gas is composed of different types of hydrocarbons, similarly to crude oil. These hydrocarbons each have different heating values, impurities, burning characteristics, and chemical weights. The molecules that form natural gas may be simple or complex in their structural arrangement or linkage of atoms. The hydrocarbons found in natural gas are called *normal paraffins* or *saturated hydrocarbons* because each carbon atom is linked to a greater number of hydrogen atoms. For example, a saturated hydrocarbon with three carbon atoms would also contain eight hydrogen atoms.

Hydrocarbons with only one to six carbon atoms are referred to as *light hydrocarbons* in the refining industry. Natural gases that contain more than six carbon atoms are called *heavy hydrocarbons.* While each hydrocarbon has a specific formula and chemical name, in industrial practice they are also frequently referred to by the number of carbon atoms that they contain. For example, methane (CH_4) can be referred to as C_1; ethane (C_2H_6), as C_2; propane (C_3H_8), as C_3; and butane (C_4H_{10}), as C_4. Each of these gases represents a part of the chemical composition of natural gas. Figure 6-1 illustrates these gases as either light or heavy hydrocarbons. Note that there are other additional natural gases including C_5 (pentane) and C_6 (hexane) not shown in the figure.

Natural Gas Hydrocarbons

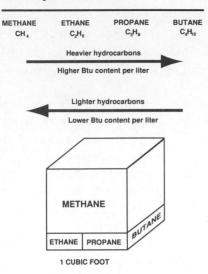

FIGURE 6-1 Natural gas is composed of four major hydrocarbons. The majority of a cubic foot of natural gas is methane, with smaller quantities of ethane, propane, and butane.

Just as with crude oil, there is a direct relationship between the molecular weight of the hydrocarbon fraction and its Btu content. In the case of natural gas, the lighter hydrocarbons of methane and ethane have the lowest heating values in Btu's per cubic foot. The heavy hydrocarbons of propane and butane have more Btu's per equal volume. Natural gas that contains significant amounts of C_3, C_4, C_5 (i.e., pentane), and C_6 (i.e., hexane) is a member of the *wet gases*. Natural gas that contains mostly methane and some ethane (C_1 and some C_2) is a member of the *dry gases*. The terms wet and dry denote whether or not the gas contains hydrocarbons heavier than butane, which may be in liquid form.

In most natural gas reserves, methane is generally found in the greatest quantity, about 75 to 95 percent. As shown in Figure 6-1, amounts of ethane, propane, and butane make up the remaining percentages of a cubic foot of natural gas. Figure 6-2 illustrates the characteristics of several natural gas reserves (i.e., samples) and the approximate Btu content per cubic foot for each. Also, refer to Appendix L to compare natural gas energy content to other forms of energy.

Sample #1 of Figure 6-2 is made up of mostly methane, and the heating content is low (about 1015 Btu's per cubic foot). As more ethane, propane, and butane (heavier hydrocarbons) are added, the heating content (Btu's) increases. Sample #3 illustrates such an instance since there is less methane, and more ethane, propane, and butane. The heating value has increased to 1230 Btu's per cubic foot. On the other hand, impurities have a

Selected Samples of Natural Gas

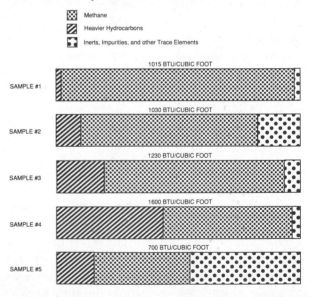

FIGURE 6-2 Different samples of natural gas are made of varying combinations of methane and heavier hydrocarbons such as ethane, propane, and butane. Generally, the more ethane, propane, and butane in the sample, the higher the heating content.

tendency to reduce the heating content of natural gas. This is shown in sample #5, where there are only 700 Btu's per cubic foot.

Associated and Nonassociated Natural Gas

Natural gas can also be categorized as associated or nonassociated. These terms are used to express the way in which the gases occur underground. Associated gas is that which is found mixed with crude oil. Nonassociated gas is found in a reservoir that contains a minimum quantity of crude oil or no crude oil.

These differences are only important to the drilling and initial refining processes. For example, *associated gas* usually must be sent through a dryer immediately after it is removed from the well so that the wet impurities can be removed for ease of refining. *Nonassociated gas* does not have to undergo this drying process, and consequently the total cost of refining nonassociated gas is reduced. Since the collection of natural gas involves the same drilling technologies as crude oil, they will not be repeated in this chapter.

NATURAL GAS AVAILABILITY

World Natural Gas Reserves

The total proven gas reserves worldwide equal over 4,000 trillion cubic feet, as illustrated in Figure 6-3. From this figure, you can see that the former Soviet Union holds the greatest share with some 1,600 trillion cubic feet or 38.02 percent. The second largest shareholder of natural gas reserves is the countries of the Middle East (shown as OPEC on Figure 6-3). Together the former Soviet Union and Middle East account for two-thirds of the total reserves. This distribution of resources suggests that trade between these regions and the major con-

sumers of natural gas (particularly Europe and Japan) will increase in importance over the next 15 years. The expanded use of natural gas will help to diversify energy imports and reduce some of the environmental problems associated with other fossil fuels in the consuming countries. This important aspect of natural gas will be presented later in this chapter.

The future distribution of natural gas will be influenced in a large degree by a number of factors. These include its price relative to prices for other energy sources, the availability of capital to develop new sources and build distribution systems, the development of competitive markets, and growth in economic activity. Note that proven natural gas reserves worldwide increase yearly.

World Natural Gas Consumption

World consumption of natural gas is projected to increase dramatically by 2010, as illustrated by Figure 6-4. The projections suggest that natural gas could be the fastest growing energy source in the world. The overall share of natural gas consumption would go from just over one-fifth of the world energy consumed to just under one-fourth of all energy consumed by 2010. As can be seen by Figure 6-4, natural gas consumption is anticipated to increase from 65.8 trillion cubic feet in 1987 to about 111.1 trillion cubic feet in 2010. Such an increase represents about 2.5 percent annual growth.

One reason for this anticipated growth rate is that estimates of world natural gas resources continue to increase. Additionally, technical advancements have added further to the expectations for natural gas use. These advancements (to be presented later in this chapter) have increased the efficiency of gas-fired technology; therefore, electric utilities are now more likely to use natural gas as a fuel source.

World Natural Gas Reserves (Trillion Cubic Feet)

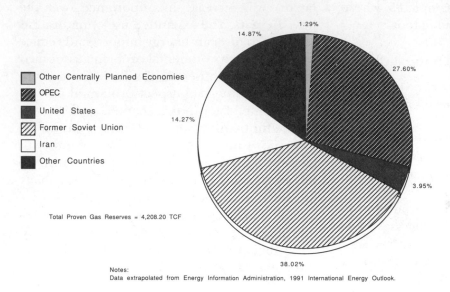

Other Centrally Planned Economies
OPEC
United States
Former Soviet Union
Iran
Other Countries

Total Proven Gas Reserves = 4,208.20 TCF

1.29%
14.87%
27.60%
14.27%
3.95%
38.02%

Notes:
Data extrapolated from Energy Information Administration, 1991 International Energy Outlook.

FIGURE 6-3 The total proven natural gas reserves in the world equal 4,208.20 trillion cubic feet. The former Soviet Union holds the greatest share of these reserves followed by the countries of the Middle East (labeled as OPEC on chart).

World Natural Gas Consumption (Trillion Cubic Feet) with Projections to 2010

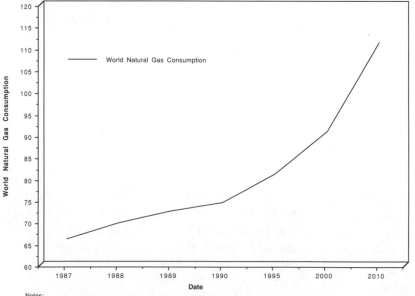

World Natural Gas Consumption

Notes:
Data extrapolated from Energy Information Administration, 1990 International Energy Outlook,
and Energy Information Administration, 1992 International Energy Outlook.

FIGURE 6-4 The world consumption of natural gas is projected to increase at an annual rate of 2.5 percent through 2010.

Worldwide, gas-fired power plants have become more popular with electric utilities over the last 15 years. This has occurred because smaller gas plants can be built faster and more economically than can the larger coal plants. These smaller plants also show no loss in performance in terms of energy or environmental efficiency.

Another incentive for the increased use of natural gas is concern for the environment. As already noted, natural gas is clean burning as compared to other fossil fuels. The increase in natural gas is expected to come at the expense of coal-fired and nuclear power plants—particularly in densely populated areas where environmental considerations are significant.

The international trade in natural gas is also projected to increase substantially over the next 20 years. Some countries, like Japan, are totally dependent upon imports. Europe also imports a considerable amount of natural gas from the former Soviet Union and Algeria. Note that the former Soviet Union is the world leader in natural gas traded internationally. Figure 6-5 illustrates the consumption of natural gas by various countries and regions. As can be seen by this figure, the former Soviet Union is also the world's leader in natural gas consumption. Overall, natural gas accounts for approximately 42 percent of the energy consumption in the former Soviet Union.

As with petroleum, natural gas is anticipated to grow most rapidly in developing countries between 1995 and 2010, which can be seen in Figure 6-5. This rate of growth is over 3 percent per year as compared to about 1 percent per year for the United States. Natural gas reserves are abundant in numerous developing countries, particularly in the Middle East. As elsewhere, natural gas is expected to be used in developing countries for the production of electricity. However, natural gas use in developing countries outside the Middle East will be dependent upon foreign investment. Unfortunately, this investment may be difficult for developing countries to acquire because of

World Natural Gas Consumption 1987 with Projections to 2010

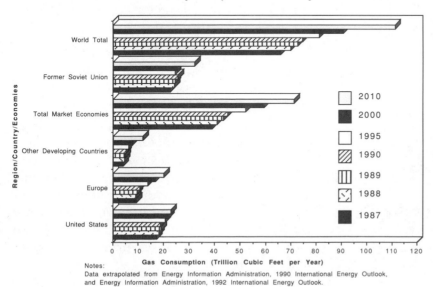

Notes:
Data extrapolated from Energy Information Administration, 1990 International Energy Outlook, and Energy Information Administration, 1992 International Energy Outlook.

FIGURE 6-5 This figure illustrates the current consumption of natural gas by various countries, regions, and economies with projections to 2010. Developing countries and the former Soviet Union are expected to have the fastest growth rates of natural gas use over the next 15 years.

their existing debt. This is particularly true for much of Latin America.

Natural Gas and Developing Countries—Energy Issue 6-1

Petroleum geologists have found large reserves of natural gas in over 45 developing countries. Most of the reserves are located in the oil producing countries of Algeria, Indonesia, Mexico, Nigeria, and Venezuela. These countries, however, have treated natural gas as a waste by-product of petroleum production. Therefore, these countries have burnt it off without utilizing any of its useful energy. This practice represents an incredible waste of a natural resource that could go far in helping to meet the energy needs of developing countries. For example, Nigeria flared (i.e., burned off) 21 billion cubic meters of gas in 1990. This was enough natural gas to meet all of Nigeria's commercial energy needs, plus those of neighboring Benin, Cameroon, Ghana, Niger, and Togo. Another country burning off large resources of natural gas was India. In 1990, the country flared more than 5 million cubic meters of gas. If India had recovered this natural gas and used it as an energy source, the country would have saved nearly $700 million on oil import bills. Attempt to answer the following questions after reviewing the use of natural gas in other developing countries:

1. *What mechanisms, technical and social, might be needed to encourage the use of natural gas in developing countries?*

2. *What types of technology transfer might industrial countries promote to developing countries?*

3. *What is (or could be) the role of the World Bank in helping developing countries to become energy efficient and oil independent?*

United States' Natural Gas Consumption

The history of domestic natural gas consumption differs from that of domestic crude oil consumption. The market for natural gas expanded as low prices spurred demand throughout the 1950s and 1960s. Figure 6-6 shows natural gas consumption in *TCF per year* or trillion cubic feet per year. Natural gas reached an all-time high consumption of 22 trillion cubic feet in 1972. After this date the uncertainties about supply and rising energy prices helped to reduce overall consumption. The use of this energy source declined further from 1979 to 1986 and was reduced to its lowest level in 20 years before the trend reversed in 1987. As can be seen in Figure 6-6, from 1986 to 1990 the consumption of natural gas increased greatly. This trend is expected to continue through the year 2000. Overall, the annual rate of projected growth in natural gas consumption is 0.9 percent. However, this level of consumption is not expected to reach the 1972 historic high of 22.1 trillion cubic feet before the year 2000. Based on the data collected, approximately 23 percent of the United States' energy demand is currently met by natural gas resources.

United States' Natural Gas Production

Most domestic natural gas production has been in the lower 48 states, which account for about 98 percent of all United States' gas production. The largest producers of natural gas in the United States are Oklahoma, Louisiana, and Texas. Of these, Texas is the leading producer, followed by Louisiana and Oklahoma. These three states account for about 67 percent of the total natural gas produced in the United States. The majority of these wells are located onshore and out-produce offshore wells by almost 4 to 1.

United States Natural Gas Consumption 1949 with Projections to 2010

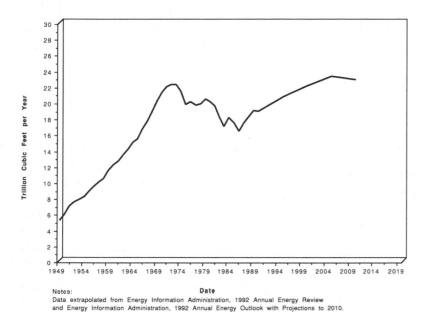

FIGURE 6-6 The consumption and projected use of natural gas in the United States is illustrated in this figure. The annual rate of projected growth in natural gas consumption is 0.9 percent through 2010.

Notes:
Data extrapolated from Energy Information Administration, 1992 Annual Energy Review and Energy Information Administration, 1992 Annual Energy Outlook with Projections to 2010.

Today, there are more than 270 thousand gas wells in operation in the United States. These wells account for three-fourths of all the natural gas removed for consumer use. The remainder of the wells supplying natural gas are oil wells. United States' gas wells peaked their production in 1971 with the average gas well producing some 435 thousand cubic feet per day (United States' natural gas production levels, with future projections, are illustrated in Figure 6-7). By 1991, the average productivity of gas wells in the United States had declined to a record low of 161.7 thousand cubic feet per day. Like oil, this decline is a result of aging fields (i.e., older fields produce less natural gas).

The United States has placed increased emphasis on the use of imported natural gas. Previously the country's gas trade was limited to the border countries of Mexico and Canada. With the beginning of shipping natural gas in liquefied form (i.e., liquefied natural gas [LNG]) in 1969, the United States was able to obtain imports from Algeria. Projections indicate that natural gas imports will continue to increase as the productivity of United States' wells declines. The annual growth of imports through 2010 is anticipated to be at 4.4 percent.

The greatest share of United States' imports is expected to continue to be Canadian. These sources are forecasted to rise through 1997 with the construction of more pipelines to the United States. These imports are projected to reach 2.2 trillion cubic feet per year sometime after 1997. LNG imports are also expected to increase substantially through 2010. These imports are predicted to grow from 79 billion cubic feet in 1990 to levels of 1.3 trillion cubic feet in 2010.

UNITED STATES' NATURAL GAS DEMAND BY SECTORS

Natural Gas Demand in the Residential and Commercial Sectors

The need for natural gas in the residential sector is projected to remain stable through 2010. This demand is shown graphically in Figure 6-8. In the residential sector, more homes will use natural gas in the future. However, the gas consumed per home in the residential sector should decline by the turn of the century because of improvements in the efficiency of equipment and buildings, and because of increased conservation by consumers. The demand for gas in the residential sector is also offset by increases in energy derived from other sources. These include both electricity and renewable fuels.

The need for natural gas in the commercial sector is expected to increase slightly through 2010. (Review Figure 6-8.) This sector—which uses natural gas for large office buildings, hospitals, fast-food restaurants, and retail establish-

ments—is already a heavy user of natural gas for space heating. The future increases are a result of expansions to the application of natural gas in heating and cooling uses. The commercial sector's average annual growth rate for natural gas is forecasted to be one percent until 2010.

Natural Gas Demand in the Energy Sector

The choice of natural gas for use by electric utilities will be a major factor contributing to increased consumption of this energy source in the United States. The annual growth rate for natural gas in the energy sector through 2010 is projected to be approximately 3.5 percent. The growth of natural gas use by electric utilities results from an increase in innovative technologies currently in development. These technologies include the use of combined-cycle electrical generation and the use of steam-injected gas turbines.

Combined-cycle electrical generation is a technology applied to existing and new electric plants. With this technology, gas turbines convert the energy content of natural gas to elec-

United States Natural Gas Production 1947 with Projections to 2010

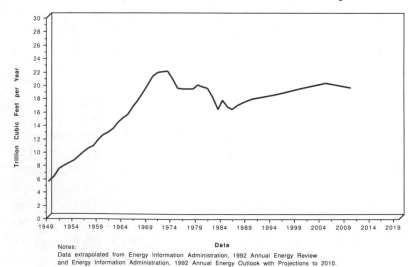

FIGURE 6-7 United States' production of natural gas, with future projections, is illustrated in this figure. Note how production of natural gas was, until recently, in a general state of decline from its 1971 high.

Notes:
Data extrapolated from Energy Information Administration, 1992 Annual Energy Review and Energy Information Administration, 1992 Annual Energy Outlook with Projections to 2010.

Natural Gas Consumption by Sector with Projections to 2010

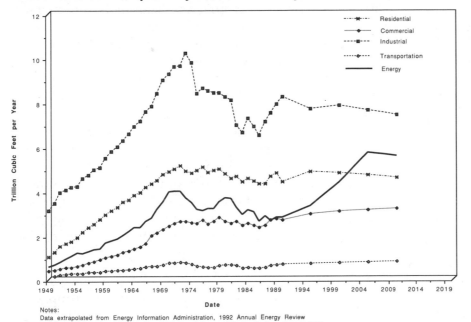

Notes:
Data extrapolated from Energy Information Administration, 1992 Annual Energy Review
and Energy Information Administration, 1992 Annual Energy Outlook with Projections to 2010.

FIGURE 6-8 This figure illustrates the use of natural gas in various sectors of the United States and provides future projections. Although most of the sectors are expected to experience small growth through 2010, the energy sector is projected to increase by an annual growth rate of 3.5 percent.

tricity. The turbines, similar to jet engines, use gas combustion to rotate the turbine blades. These turbines are directly connected to generators that produce electricity by placing rotating conductors in a strong magnetic field. When operated in a combined-cycle system, the gas turbine is combined with a steam generator. This steam generator acts as a heat recovery system—using the residual heat energy to drive a steam turbine. Combined-cycle systems have obtained efficiencies as high as 50 percent under ideal operating conditions. Additionally, there is research continuing on the use of steam-injected gas turbines. In a *steam-injected gas turbine*, some of the steam produced from burning natural gas or other fossil fuels is injected directly into the combustion chamber. These electrical generating units have reported efficiencies greater than 50 percent.

Natural Gas Demand in the Transportation Sector

The use of new technologies in the transportation sector may cause natural gas usage in this sector to increase dramatically in the future. Currently, however, very little increase is expected through the year 2010. In some areas of the United States, industry, local governments, and gas distributors are experimenting with cars, trucks, and buses fueled by compressed natural gas—(CNG), are also referred to as liquefied natural gases. These experiments are a result of stricter air quality guidelines imposed by the Clean Air Act Amendments and other environmental protection laws. With the use of CNG as an energy source in the transportation sector, reductions in toxic exhaust emissions may be possible. However, until further results are obtained on the economic and environmental benefits as well as safety concerns of CNG in the transportation sector, little increase of natural gas is projected. Additional material related to LPG in transportation is presented later in this chapter under the heading of Liquid Petroleum Gas.

Natural Gas Demand in the Industrial Sector

Natural gas use should increase gradually in the industrial sector—from 7.2 trillion cubic feet in 1990 to between 7.5 and 8.0 trillion cubic feet by the turn of the century. (Review Figure 6-8.) This projected rise in the use of natural gas most likely will be used in power systems using *cogeneration*. A cogeneration facility is one that produces electricity and another form of useful thermal energy (such as heat or steam) for use by an industrial process. Papermaking plants are an example of an industry that uses cogeneration to produce both electricity and steam in the making of paper. While the use of cogeneration in the industrial sector is similar to that of the electric utilities, its growth is anticipated to be far lower.

NATURAL GAS SHORTAGES

Natural Gas Shortages in the United States

Natural gas supplies have not always kept abreast of demand. Bitterly cold weather during severe winters seems to trigger shortages of natural gas within the commercial, residential, and industrial sectors of society. This has been especially true in the eastern two-thirds of the United States. Midwestern factory employees working in overcoats, plants shutting down for lack of fuel, schools closing, and implementing emergency price deregulation policies are all indications that supplies of natural gas often fall short of the demands. Such shortages have had several major effects on U.S. society. These include:

1. Air pollution standards were relaxed because other less-clean fuels were needed in place of natural gas.
2. Many industries were forced to switch to other fuels, often at higher prices. This had a tendency to increase energy prices—costs which were passed along to the consumer.
3. Large buildings, such as schools and apartments, were forced to use other fuels not as clean burning as natural gas.
4. Individual homeowners were denied the fuel of their choice in some geographical areas, increasing their monthly heating and cooling bills by 50 to 76 percent.

Causes of Natural Gas Shortages

To gain an understanding of total supply and demand of natural gas and how it changes throughout the years, refer to Figure 6-9. As illustrated in this figure, natural gas production exceeded consumption until 1964. Prior to that year, the production had consisted of existing or proven reserves and a small quantity of overland imports from Canada and Mexico. Shortly after 1964, the supply of natural gas approximately matched the consumption until about 1973. In 1973, the production of natural gas fell significantly in relation to constantly increasing consumption. The shortage of natural gas after 1973 was due to several factors. The first and most important reason was the *price control* of natural gas prices at the wellhead. This means that the federal government allows only a certain amount of price increases each year, usually less than inflation. The federal government began price controls of natural gas because of large increases in inflation. From an economic standpoint, price controlling of any product or commodity at a level below that which would prevail in a free, competitive market can be expected to have certain effects. With regard to natural gas the supply, or availability, will decline due to decreased profits earned by the gas companies. Essentially, the companies cannot increase the cost of natural gas to stay even with inflation. If the supply is low and the demand is greater, the price of the existing or available natural gas will increase substantially.

Natural Gas Shortages—Energy Issue 6-2

Identify some of the reasons the United States may have a natural gas shortage in the future related to consumption, living patterns, weather, and so on.

The shortage of natural gas remained until 1978, as can be seen in Figure 6-9. Up to that time the increase in the price of natural gas at the wellhead had been far below that of inflation and other comparable fuels. Then, in late 1978, price controls on natural gas began to be lifted. The lifting of these prices helped to encourage more exploration of natural gas by drilling companies. This increased emphasis on exploration helped to bring the production of natural gas equal to demand by 1982.

During the years of the natural gas shortage, price controls were placed on only a certain type of gas pipelines. There are two types of natural gas pipelines. *Intrastate gas* is gas that is sold and consumed in the state in which it was produced. *Interstate gas* is gas that is sold and consumed in a state other than that in which it was produced. Intrastate gas was not subject to federal price regulations and normally was sold for a profit. However, price controls were applied to interstate pipelines. Since most natural gas is interstate, price regulations had a huge negative effect. The net result was that

United States Natural Gas Consumption, Production, and Imports, 1949 with Projections to 2010

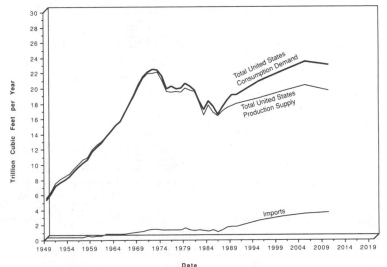

Notes:
Data extrapolated from Energy Information Administration, 1992 Annual Energy Review and Energy Information Administration, 1992 Annual Energy Outlook with Projections to 2010.

FIGURE 6-9 The supply and demand for natural gas have not been balanced in the United States. This resulted in gas shortages during the 1970s and, since then, an increasing dependence on imports.

natural gas companies had little incentive to drill for more natural gas during those years. This was the major cause of the natural gas shortage in the 1970s.

Since 1978 additional price controls were removed, and by 1990, all natural gas price controls were removed. Prices for natural gas were also adjusted, which eventually caused the natural gas companies to explore for more reserves. The United States, however, has become dependent upon imports to meet its demand for natural gas. This dependence began in 1984 and is expected to continue into the future.

Price Controls and Natural Gas— Energy Issue 6-3

Price controls often have certain effects on the price and availability of a product. List three advantages and three disadvantages of price controlling a specific energy product, such as natural gas.

SUPPLIES OF NATURAL GAS

Proven, Undiscovered, and Technically Recoverable Resources

Just as with crude oil supplies, the United States classifies natural gas reserves under the categories of proven, undiscovered, and technically recoverable resources. Since these categories were presented in chapter 5, they will not be reiterated here.

Conventional and Unconventional Resources of Natural Gas

Aside from the categories of proven, undiscovered, and technically recoverable resources, the federal government uses two additional categories for natural gas. These are conventional and unconventional resources. *Conventional resources* are those that are producible by natural pressure, pumping, or by injection of water or gas into the well site. Conventional resources would not include resources produced by enhanced oil recovery methods. *Unconventional resources* include usable gas produced from tar deposits, heavy oil deposits, oil shales, coal and peat formations; gas in low-permeability (i.e., tight) reservoirs, gas in pressurized shales and brines, or natural gas in various compounds associated with water. To better illustrate the differences between conventional and unconventional resources, several unconventional resource extraction methods are presented in more detail.

One unconventional natural gas resource is that of synthetic natural gas. This can be made from coal and peat formations with a coal gasification process. Another unconventional natural gas resource is extracted from coal deposits. These deposits, when originally formed, may have trapped natural gas from the decay of plant and animal matter. Extraction of this gas is accomplished by drilling a gas well. Natural gas has also been found in tight sand and sandstone formations. When the low-porosity sandstone is fractured by pressure, the trapped gases flow out and can be captured. Natural gas can also be removed in the same way (i.e., by fracturing the formation) from tight clay formations. This resource is referred to as a *Devonian shale deposit*. Finally, unconventional natural gas resources can be extracted from reserves of geopressured natural gas. Geopressured natural gas reserves are found below 12,000 feet and are usually associated with water at that level. However, the water and natural gas must be separated; this increases the costs of extraction. Various government and private industries are actively involved in testing the feasibility and commercialization of these resources of natural gas.

Existing and Estimated Natural Gas Reserves

There are several types of conventional natural gas reserves, two of the essential being existing and estimated future reserves. As with oil, *existing reserves* are considered proven reserves, or those that have been found and identified through exploration and development drilling. *Future reserves* are those expected to exist based upon geological surveys, but for which no exploratory wells have been drilled. The latter are also referred to as *potential reserves*. These types of reserves are typically found in the southern United States.

NEW DISCOVERIES

New Natural Gas Resources

Scientists have found two very different zones of hydrocarbon fields. One is a shallow area containing both oil and natural gas, and the other zone—much deeper (at or below the 15,000-foot level)—tends to contain more natural gas. Large quantities of crude oil and natural gas are located below the 15,000-foot level. Scientists hypothesize that extremely high temperatures that exist at this depth help to crack the crude oil into hydrocarbon gases.

The discovery of this deep zone is not the only recent development that has increased natural gas reserves. In addition, new exploration procedures have improved in the accuracy of finding reserves. Today, computers can sort through thousands of seismographic vibrations, echo by echo, to help find more natural gas reserves. The result is that exploration companies find exact locations of natural gas

reserves that could not be detected several years ago.

> **Making Natural Gas Into Consumer Products— Energy Issue 6-4**
> *What environmental and social consequences might occur if the natural gas industry were to exploit greater amounts of the new reserves for consumer products?*

Problems in Estimating Natural Gas Resources

Numerous methods of estimating undiscovered natural gas resources have been developed. These methods vary greatly as to their complexity and sophistication. The one used in any given instance depends upon the kind and quantity of data available for that given area. Many mineral rights are privately owned in the United States (except for minerals underlying federal onshore and offshore lands). It is, therefore, difficult to apply a specific method of calculating undiscovered reserves with any degree of certainty because privately owned lands are not analyzed by governmental agencies for the mineral resources that could exist there. Consequently, national oil and gas resource assessments are subjectively generated by experts on both federal and privately owned land. These subjective estimates are based upon an analysis of available evidence regarding a particular area (e.g., seismographic, magnetometric, and gravimetric studies) and on comparisons with geologically similar areas in which oil and gas resources are known to exist. Again, this subjective approach to undiscovered oil and gas resources provide the federal government with a range of estimates for the possible recovery of undiscovered resources.

Estimates of Undiscovered and Technically Recoverable Conventional Natural Gas Resources

Federal estimates of undiscovered and technically recoverable conventional gas resources are made available by the U.S. Department of the Interior's United States Geological Survey. This survey data provides estimates for the onshore and the state offshore waters. Estimates for federal offshore waters is provided by the United States Mineral Management Service. Based upon these estimates, the United States has a 5 percent probability of obtaining an additional 520.87 trillion cubic feet of natural gas by 2010. This figure represents the high-end estimate for undiscovered and technically recoverable conventional natural gas resources. The low-end estimate (a 95 percent probability) is equal to 296.96 trillion cubic feet of natural gas. Table 6-1 presents this data with probability estimates for other potential natural gas finds. Remember that the data presented in Table 6-1 represents only an estimation of the resources available. These estimates may increase or decrease because of improved drilling technology, exploration, or extraction of the resource.

Estimates of Unconventional Natural Gas Resources

Besides the estimates of undiscovered and technically recoverable conventional natural gas resources, the federal government has also prepared data on the recovery of unconventional gas resources. This data, illustrated in Table 6-2, shows that the United States expects to recover some 849.34 trillion cubic feet of natural gas via unconventional extraction by 2010. When the proven, undiscovered,

Table 6-1
Estimated Undiscovered and Technically Recoverable Conventional Natural Gas Resources

Probability of at Least the Amount of Resource Shown (in percentage)	Trillion Cubic Feet
5	520.87
10	489.52
15	469.44
20	454.06
25	441.28
30	430.10
35	420.00
40	410.65
45	401.80
50	393.29
55	384.96
60	376.66
65	368.27
70	359.62
75	350.52
80	340.65
85	329.49
90	315.97
95	296.96

Notes:
Data extrapolated from the Energy Information Administration, *1993 Assumptions for the Annual Energy Outlook.*

TABLE 6-1 This table illustrates the estimated undiscovered and technically recoverable conventional natural gas resources.

technically recoverable, and unconventional natural gas resources are added together, it is possible to find the total natural gas reserves expected to be available in the United States. These reserves equal approximately 1536 trillion cubic feet of natural gas (high-end estimates have been used for the undiscovered and technically recoverable conventional natural gas resources). However, a note of caution is needed. Much of the total (i.e., undiscovered and technically recoverable resources) represents estimates of the gas resources that *may* exist or *may* be recoverable in the future. These are not proven resources and, clearly, not all of these will be recovered.

Table 6-2

Estimated Unproven Unconventional Natural Gas Resources

Reserve Category	Trillion Cubic Feet
Tight Gas	329.79
Devonian Shale	22.94
Coalbed Methane	71.94
Other Unconventional	424.67
Total	**849.34**

Notes:
Data extrapolated from the Energy Information Administration, *1993 Assumptions for the Annual Energy Outlook.*

TABLE 6-2 Estimated unproven unconventional natural gas resources are illustrated in this table. Although these reserves may help to meet the increasing demand for natural gas, they require the use of technologies that may not make them economically feasible to recover until the price of natural gas increases dramatically.

How Long Will Natural Gas Last?— Math Interface 6-1

The following math questions will help the reader to better understand the problem of living on a world with finite resources.

1. Based upon the assumption that the United States could recover all of its estimated natural gas resources (1536 trillion cubic feet), how long would these resources last if the United States relied solely on them (i.e., no imports) and continued to extract 23 trillion cubic feet per year until the resources were exhausted? Note that this assumes no growth and is based on current usage rates shown in Figure 6-9.

   Resources would last _____ years.

2. Again, based upon the assumption that the United States could recover all of its estimated natural gas resources (1536 trillion cubic feet), how long would these resources last if the United States relied solely on them (i.e., no imports) and continued to extract 23 trillion cubic feet per year with an annual growth rate of 0.9 percent until the resources were exhausted?

   Resources would last _____ years.

3. Based upon the amount of natural gas used in the last year before resources were exhausted (in problem 2), assume that a new natural gas field were discovered to contain 250 trillion cubic feet. How long would this discovery last at 5 percent annual growth?

   Resource would last _____ years.

MEASURING NATURAL GAS

Natural Gas—Units of Measurement

There are several units commonly used to measure natural gas: the *therm*, cubic foot, or 100 cubic feet (referenced as *CCF*). Natural gas is also sold by the cubic foot or 100 cubic feet or by the therm. Figure 6-10 illustrates a portion of a monthly energy bill with the CCF's highlighted. The relationship between Btu's, therms, and CCF's are as follows:

   1 therm or CCF = 100,000 Btu
   1 therm or CCF= 100 cubic feet of
      natural gas
   1 cubic foot of natural gas = 1030 Btu's

Applying Natural Gas Measurements

The measurements illustrated in the foregoing section can be used to make calculations to show the hourly Btu demand of a household.

Natural Gas Utility Bill Example

	Reading Dates	Days	Meter Readings	Energy Use		ACTUAL	BUDGET
GAS RESIDENTIAL			METER # - S000008150				
	01-27	30	9335	172 CCF			
	12-28		9163				
ADJ FOR: HEAT CONTENT 1.0074				173 CCF			
GAS BILLING						$92.31	
PURCHASE GAS ADJUST			172 CCF	AT	$.05855-	10.13CR	

SUBTOTAL						82.18	
CITY FEE AT 1.50%						1.23	
CURRENT GAS RESIDENTIAL						83.41	

FIGURE 6-10 This utility bill shows that natural gas is measured by the CCF (100 cubic feet).

For example, a household may be charged for 56 CCF (therms) of gas over a typical monthly billing period. This means that for the 30-day period the customer used 5,600,000 Btu's (56 × 100,000) or 186,667 Btu's per day (5,600,000/30), which is equivalent to 7,778 Btu's per hour (186,667/24 hours).

Heating Homes with Natural Gas—Math Interface 6-2

Solve these sequential math problems concerning natural gas used in a residential dwelling:

1. A residential dwelling demands 20,000 Btu's per hour in January to maintain an inside temperature of 68° F. How many Btu's are needed during this month for heating the dwelling?

   Btu's needed in the month equal

   _____.

2. How many cubic feet of natural gas are needed in January in this dwelling (problem 1) if each cubic foot of natural gas contains about 1030 Btu's? For this problem assume the furnace is 95 percent efficient. (This means that the furnace can only extract 95 percent of the available Btu's in a cubic foot of natural gas.)

   Cubic feet of natural gas needed in January equal _____.

3. How many CCF (therms) are used in January in the dwelling (problem 1-2)?

   CCF (therms) used in January equal

   _____.

4. If natural gas sells for $.65/CCF (therm), how much would it cost to heat the dwelling in January?

   Cost to heat the dwelling in January equals _____.

ASSOCIATED NATURAL GAS TECHNOLOGY

Liquid Petroleum Gas (LPG)

Liquefied petroleum gas (LPG) has experienced an increased popularity in the past 10 to 15 years. Since LPG is such a clean burning fuel, it is considered one possible fuel source for motor vehicles. This is especially true in urban areas, such as Los Angeles, where air pollution is a serious issue. LPG currently is used extensively as a rural fuel for home heating.

LPG is made from the propane fraction identified earlier in this chapter and is one of the heavier natural gas chemicals. It is generally pressurized into a liquid form for ease of storage and transportation.

Propane vaporizes (turns to a gas) at or near −41° F. For comparison, butane vaporizes at about 32° F. Natural gas fractions with lower vapor points weigh less per gallon than the usual liquid fuels such as diesel fuel. For example, propane at its vapor point only weighs 2.05 lbs per gallon while #1 diesel fuel vaporizes at 465–650° F and weighs 6.97 lbs per gallon. However, the heavier hydrocarbons, such as diesel fuel and gasoline, produce more Btu's per cubic foot. This is because these fuels weigh more per volume.

Changing Natural Gas into a Liquid

As mentioned earlier, propane vaporizes at −41° F. As long as the temperature of the propane remains above this point, the fuel will be in a gaseous state. When the temperature falls below its vapor point, the gaseous propane turns into a liquid. However, if propane is placed under pressure, its vapor point will rise. This vapor point can be raised high enough to cause the gas to liquefy. Thus, in a contained vessel under pressure, propane can exist in liquid form. Because of this, LPG can be transported, stored, and sold as a liquid quantity measured in gallons or pounds. Figure 6-11 shows a specially designed ship used to transport liquid petroleum gas.

In this liquefied condition (LPG), the petroleum gas will take up only 1/600 of its original

FIGURE 6-11 Specially designed ships, such as the one shown, transport liquid petroleum gas. (Courtesy of the American Petroleum Institute)

space. This allows the storage of more fuel within a specific volume; therefore, LPG is suitable for use in the transportation sector. However, the containers that hold natural gas must be made strong so as to eliminate the potential for an explosion during an accident. Since methane and ethane have even lower vapor points than propane, it would not be economically practical to pressurize these fuels for transport applications. To maintain ethane and methane in a liquid form, the pressure would have to be so greatly increased that both economic and safety considerations would make it prohibitive.

Ethane and methane are commonly used in stationary applications where the fuel can be piped in or where there is room to store large quantities of gaseous fuels. Space heating and stationary boilers are the most common applications for the use of ethane and methane natural gases.

Distribution and Storage of Natural Gas

The distribution and storage of natural gas are
mostly done through pipeline and underground storage, respectively. As shown earlier,
natural gas that comes from a well is considered either associated (wet) gas or nonassociated (dry) gas. Figure 6-12 illustrates both types
of gas wells and how the natural gas is distributed from the wells to the customer.

When the nonassociated gas is tapped from
the well site, it is sent through a cleaner and
immediately fed into the pipeline. When the
associated gas is removed from the well site, it
includes both oil and gas that must be separated. The separated gas is then fed directly into
the pipeline distribution system.

At gas-processing plants, shown in Figure
6-13, both associated and nonassociated gases
are cleaned and separated. The gases are separated into two basic categories. These categories are those that remain in a gaseous state
and gases that can be pressurized and put into
liquid storage (such as propane and butane).

Once the gases are separated, a series of
compression stations (see Figure 6-12) move
the gas from the processing plant into various

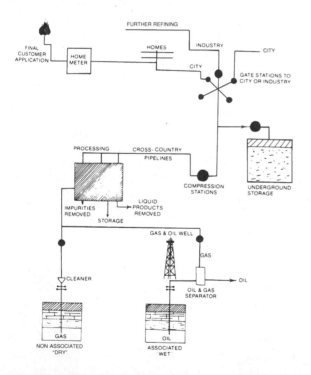

FIGURE 6-12 Natural gas, both wet and dry, is sent
into processing and compression stations. It is then
stored underground and finally distributed to homes and
industrial plants for use.

FIGURE 6-13 This natural gas processing plant separates natural gas into different products—such as
methane, ethane, propane, and butane. (Courtesy of
Chevron Corporation)

underground storage systems throughout the country. These systems allow the natural gas to remain in a gaseous form with no high-pressure containerization needed for its storage. These storage facilities were developed to allow the pipeline to be operated at essentially a constant transmission rate throughout the year. Pipelines are designed for a delivery rate roughly equal to the average demand rates. Any excess gas delivered during low-demand periods is stored for use during peak periods. Thus, the use of storage areas allows the pipeline to run at near capacity most of the time and helps to minimize the cost of natural gas transportation.

The underground reservoirs used to store excess natural gas are often depleted oil and/or gas fields, coal mines, salt domes, and water caverns. These reservoirs must have the capacity to hold large volumes of natural gas and must also be gastight. Figure 6-14 shows the locations of most underground gas storage reservoirs throughout the United States. Each is either a salt dome, water cavern, coal mine, or depleted oil and/or gas field.

From the storage facility, the natural gas is sent through thousands of miles of underground pipe to the local gas companies and industries that use natural gas for their products. A typical natural gas pipeline, as shown in Figure 6-15, generally consists of one or more lines of pipe 10 to 45 inches in diameter. The pipe is normally buried two to four feet below the surface of the ground. Pipes are cleaned by inserting a 15,000-pound urethane ball or *pig* into the pipelines. Gas moving through the pipeline pushes the pig, which scours out liquids and sediment. Although natural gas is compressed at the compression station, additional compressors are located 50 to 100 miles apart to help move the gas in the pipelines.

Location of Underground Gas Storage

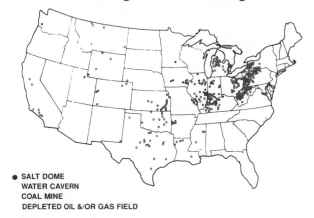

• SALT DOME
  WATER CAVERN
  COAL MINE
  DEPLETED OIL &/OR GAS FIELD

FIGURE 6-14 Natural gas is stored in underground reservoirs made from salt domes, water caverns, coal mines, and depleted oil and/or gas fields.

FIGURE 6-15 Natural gas is sent through miles of cross-country pipeline to get from the processing plant to the storage facility. (Courtesy of Northwestern Public Service)

The major interstate natural gas pipeline systems movements are shown in Figure 6-16. The width of the arrows roughly estimates the quantity of natural gas in cubic feet per year that is transported.

Transporting Natural Gas by Pipeline—Energy Issue 6-6

Most natural gas is moved or transported by pipeline. Identify the disadvantages and advantages of shipping natural gas by this method. Answer in terms of political, environmental, and economic aspects.

Petrochemical Industries

The petrochemical industries are those that use natural gas to manufacture various types of plastics, alcohol, and so on, for use in making end products. Although natural gas is used mostly in the residential and commercial sectors of society, a significant amount is fed directly into petrochemical plants. Some experts within the energy field believe that the petrochemical industries utilize natural gas to its best possible advantage. This is because these industries produce a final product (as opposed to burning the fuel) that can later be converted to a fuel source. However, more recently the petrochemical products have been identified as a major source of environmental pollution. Products such as plastic, for example, simply do not decompose well; therefore, they represent a solid-waste problem. An example of this is the use of plastic food containers from fast-food restaurants. These containers have an average user life of about 10 minutes, but they will last in a landfill for some 500 years before decomposition occurs. Currently, there is a strong environ-

Interstate Natural Gas Movements

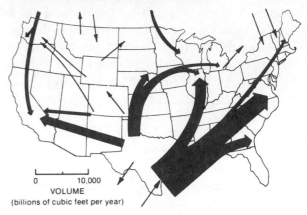

FIGURE 6-16 The arrows and their widths show the direction and approximate volume of natural gas movements in billions of cubic feet per year.

mental movement toward responsible use of plastics and recycling used materials to help reduce pollution. In addition, many cities have instituted plastic-recycling programs in which the plastic (along with many other products) is recycled for further use.

Using Natural Gas—Energy Issue 6-7

Identify the positive and negative consequences of using natural gas for both combustion processes and for petrochemical products.

Petrochemical industries are designed to convert certain hydrocarbons found in natural gas into consumable products. Figure 6-17 is a schematic diagram of a petrochemical plant. This layout represents only one type of petrochemical industry, and many variations on the basic process and products exist.

As the natural gas is pumped into the pipeline by the compressor station, a portion (approximately 8 percent) of its volume is

Petrochemical Industry Layout

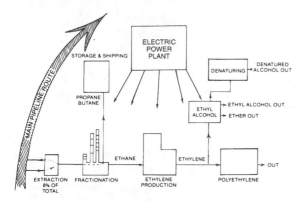

FIGURE 6-17 The petrochemical industries extract ethane; and through several refining processes, produce polyethylene for many of the plastic products used today.

removed at the extraction station in the plant. This portion of natural gas is then sent into the fractionation process. The *fractionation process* separates natural gas into its components, including methane, ethane, propane, and butane.

The fractionation process is similar to the crude oil refining process discussed in chapter 5. Propane and butane, both of which are very low in quantity in comparison with methane, are stored and shipped to various consumers. These fuels are supplied to the buyer in cylinders that range in size from less than 1 pound to about 110 pounds. These fuels can be used for space or water heating or industrial applications. Some larger producers of LPG even ship these gases on tanker cars, which range from 3 to 50 tons.

Ethane, which is also extracted from natural gas, is sent directly into the ethylene section of the processing facility. Once at this section, it is chemically converted into the product *ethylene*. This chemical process is accomplished through various cracking units at temperatures of near

1,500° F. This unit also purifies the ethylene for use in making other products that will be manufactured within the plant.

The purified ethylene is sent into the polyethylene section where it is converted into the product *polyethylene*, with resins of different forms and densities. These plastic resins, shown in Figure 6-18, are commercially sold to companies that manufacture different plastics for their final products—such as toys, plastic bags, plastic industrial products, fertilizers, and the like.

FIGURE 6-18 There are numerous products that can be made from the petrochemical industry's polyethylene. Shown are several plastic resins that will be used to make plastic toys, bags, and the like.

Rather than being used to create polyethylene, some of the ethylene is also sent directly into the ethyl alcohol unit that produces different products. These include ether and ethyl alcohol. The ethyl alcohol is then sent to a denaturing (meaning to change the natural quality of a chemical) unit, which prepares it for certain cosmetics, drugs, and other industrial purposes. The denaturing unit allows the ethyl alcohol to be modified to meet the particular needs of the product to be manufactured.

Various types of equipment are used in a petrochemical plant. Furnaces must be capable of developing temperatures of 350° F to 3,000° F. Pumping equipment must be capable of producing pressures that range from a vacuum to near 43,500 psi. To meet these requirements, centrifugal, gear, sliding vane, and reciprocation pumps are used. Reciprocating, centrifugal, and rotary compressors produce pressures to convey the chemicals both in liquid and gaseous states. Besides the various processing units at the petrochemical plant, an electrical generating power plant also supplies energy in the form of steam and electricity, so that the equipment may operate continuously. Figure 6-19 illustrates a typical petrochemical plant.

Life Without Plastic—Energy Issue 6-8

What social consequences would occur if the United States were to eliminate petrochemical industries in a period of three years? Identify the industries and service sectors that would be most affected by this elimination.

FIGURE 6-19 Petrochemical plants, such as this, are used to make various products from natural gas. (Courtesy of Amoco Corporation)

ALTERNATIVE FUTURES

Using Methane Gas from Landfills

Numerous research projects are now in progress to extract methane gas from landfills. This gas is formed as household food waste and other organic matter decomposes beneath the ground. It should be remembered from chapter 3 that methane is one of the greenhouse gases contributing to global warming. Normally, methane gas collects in landfills and must be burned off with flares so that it does not build to explosive levels. These gases could also be released from landfills by vent pipes or naturally as the gas rises to the surface. In either situation, this release of methane into the atmosphere is harmful to the environment.

In the extraction of methane gas, the gas is collected by gas wells with pipes in the ground and pipes that line the surface of the landfill. These surface pipes are used in conjunction with a blower to move the gas from the wells to a compressor. Once the compressed gas has been cooled, it is filtered to remove condensation, grit, and debris. The methane gas can then be burned as a fuel to drive generators that produce electricity or as an industrial fuel for an industry at the landfill site. Currently, this use of methane gas to generate electricity is done profitably in California, Illinois, Minnesota, Wisconsin, and in numerous states on the East Coast. Though there are only 70 such methane collection systems in the country, this resource is potentially large when one considers that there are over 9,000 landfills in the United States, and worldwide over 4.5 million tons of waste is thrown away per day.

One example of the use of landfill methane gas to produce electricity can be found at the

Los Angeles landfill in California. This is the largest landfill in the United States and receives over 13,000 tons of waste products a day, including sludge from wastewater treatment plants. Using the decomposition of both garbage wastes and sludge, this landfill is the largest producer of methane gas in the United States. The power plant at the landfill produces one percent of Los Angeles' electricity. This is enough electricity to power some 100,000 local residential dwellings. Note that the United States is currently the world leader in promoting the use of this technology.

SOCIAL AND ENVIRONMENTAL COSTS OF USING NATURAL GAS

Before illustrating some social and environmental costs of natural gas, it is important to reiterate that natural gas is the cleanest fossil fuel. For example, combustion of natural gas creates between 30 and 55 percent of the CO_2 (carbon dioxide) of coal and petroleum. Additionally, natural gas produces virtually no sulfates and less NO_x (nitrogen oxides) as compared to other fossil fuels. From an ecological point-of-view, this makes gas an ideal choice for the production of electricity in the nation. Based upon the technology available, natural gas can be substituted for oil and coal to produce electricity. However, it is not clear that domestic supplies would be adequate to satisfy electrical demand beyond the year 2040.

Natural gas could also be substituted for gasoline in the transportation sector. This fuel source has been tested and marketed in numerous areas of the country with success. Natural gas in the transportation sector has demonstrated a reduction in tailpipe emissions and reduced maintenance costs in fleet vehicles. The only limitations to its widespread use appear to be the costs of converting gasoline engines to use this fuel source. This limitation could be overcome with the mass manufacture of engines specifically designed to utilize this fuel. Again, the net result of natural gas use in the transportation sector would be cleaner air and reductions in the depletion of crude oil.

Social and Environmental Costs of Extracting Natural Gas—Similar to Coal and Petroleum

The extraction and use of natural gas has social and environmental costs similar to those of coal and petroleum. These costs include the aesthetic destruction of land by wells and pipelines, loss of habitat, destruction of fish during seismic prospecting, ecological damage from well blowouts, and the release of hydrocarbon products from the burning of these fuels.

Social and Environmental Costs of Flaring Natural Gas

As earlier noted, geologists have found large reserves of natural gas in over 45 developing countries. These countries, however, have treated natural gas as a waste by-product of petroleum production. As a waste product the natural gas has been burned off—*flared gas*. (Flaring is also a safety device in the refining process to keep the internal pressures equal throughout the plant.) This flaring of natural gas results in a loss of a natural resource, increased dependency on petroleum, and an increase in greenhouse gases. Note that the practice of flaring is not limited to developing countries. Industrial countries also practice flaring at the well site or in the processing of crude oil to remove unwanted natural gases. Estimates worldwide indicate that the process of flaring has resulted in the loss of quadrillions of cubic feet of natural gas.

In flaring natural gas at the well site or in the processing of crude oil, other gases are burnt off as well. One of these gases, helium, has numerous industrial applications with no other substitute available.

The practice of flaring is even more questionable when one considers that natural gas can be substituted for nearly any energy source and with the benefit of higher operating efficiencies in many instances. Clearly, the practice of flaring should be eliminated as much as possible to better use the finite energy resources of the world.

Social and Environmental Costs of LNG Tankers

The use of LNG tankers poses a significant threat for a catastrophic accident. These concerns are a result of the fact that the gas is very flammable and heavier than air. In the event of a major leak from a LNG tanker at a harbor area, the gas could spread out over a wide area, suffocating both humans and animals. In a worst-case scenario, the gas could be ignited after having been released and spread over the harbor/city area. The tanker, if fully loaded, could have enough energy content to be equivalent to a medium-sized hydrogen bomb. The resulting explosion could kill hundreds of thousands of people. Although the chance of such an accident happening is quite low, experts note that it is not negligible and represents a danger where LNG tankers are unloaded in densely populated areas, like Boston, Massachusetts. Similar concerns can also be applied to the transportation of natural gas by railroad cars.

POINT/COUNTERPOINT 6-1

TOPIC

Keeping the Supply and Demand of Natural Gas in Balance

Theme: During the past 10 years the supply of natural gas has been in balance with its demand. This was not true during the 1970s when there was a shortage. Keeping the supply and demand of natural gas in balance will help to keep the cost competitive and the supply stable. In a classroom discussion attempt to answer the following questions:

1. Why is it important to keep the supply and demand of an energy source in balance?
2. What technical applications that now use natural gas could be made to use other forms of energy? What economic or environmental effects would be evident?
3. What are the negative environmental effects of increasing the use of natural gas?
4. What political consequences would result from reducing the use of natural gas energy resources?
5. If the federal government decided to stabilize inflation by controlling the price of energy resources, how could drilling and exploration industries still be provided an incentive to develop more natural gas?
6. How could consumers be convinced to help reduce the demand for natural gas in their daily lives? What specific convennience products made from natural gas could be used less?
7. What role should the federal government play in helping to stabilize the supply and demand of natural gas?
8. Economically, what are the advantages and disadvantages of increasing the demand for natural gas, and thus causing more exploration for new gas deposits?
9. If the United States were to reduce its dependence on natural gas, what other energy resources could be used to make up for the reduction in the use of this energy source? What replacement would do the least amount of damage to the environment?
10. Should the petrochemical industries be forced by the federal government to develop plastics that decompose in landfills? Who would pay the research and development costs for these new products?

Summary/Review Statements

1. The major improvement that led to the widespread use of natural gas in the United States during the late 1920s was the creation of pipelines to transport the gas.
2. Crude oil and natural gas are found in similar geological environments—sedimentary rock formations. Also like crude oil, natural gas is formed by the same decomposition of organic materials that occurred millions of years ago.
3. In comparison with other fossil fuels, natural gas is extremely clean burning.
4. The hydrocarbons found in natural gas are called normal paraffins or saturated hydrocarbons because each carbon atom is linked to a greater number of hydrogen atoms.
5. Hydrocarbons with only one to six carbon atoms are referred to as light hydrocarbons in the refining industry. Natural gases that contain more than six carbon atoms are called heavy hydrocarbons.
6. The lighter hydrocarbons, methane and ethane, have the lowest heating values in Btu's per cubic foot. The heavy hydrocarbons of propane and butane have more Btu's per equal volume.
7. A natural gas that contains significant amounts of C_3, C_4, C_5, and C_6 belongs to the class of wet gases. Natural gas that contains mostly methane and some ethane (C_1 and some C_2) belongs to the class of dry gases.
8. Associated gas is that which is found mixed with crude oil. Nonassociated gas is found in a reservoir that contains a minimum quantity of crude oil or no crude oil.
9. Usually, four types of hydrocarbons are extracted from a sample of natural gas: methane, ethane, propane, and butane.
10. The total proven gas reserves worldwide is equal to over 4,000 trillion cubic feet.
11. The former Soviet Union holds the greatest share of natural gas reserves—some 1,600 trillion cubic feet. The second largest shareholder of natural gas reserves is the countries of the Middle East.
12. The overall share of natural gas consumption is expected to go from just over one-fifth of the world energy consumed to just under one-fourth of all energy consumed by 2010.
13. Based upon the data collected, approximately 23 percent of the United States' energy demand is currently met by natural gas resources.
14. Most domestic natural gas production has been in the lower 48 states. These states account for about 98 percent of all United States' gas production.
15. The decline in productivity of United States' gas wells has resulted in an increased emphasis on the use of imported natural gas. The annual growth of imports through 2010 is anticipated to be at 4.4 percent.
16. Natural gas supplies have not always kept abreast of demand in the United States.
17. Natural gas supply/demand ratios can be adversely affected by controlling the price at artificially low levels.
18. Natural gas is shipped by both intrastate or interstate pipelines.
19. Natural gas reserves are classified by proven, undiscovered, technically recoverable, existing, and future reserves.
20. The federal government uses two additional categories (see number 19) for natural gas: conventional and unconventional resources.
21. Conventional resources are those that are producible by natural pressure, pumping, or by

injection of water or gas into the well site. Conventional resources do not include those produced by enhanced oil recovery methods.

22. Unconventional sources include tar deposits, heavy oil deposits, oil shales, gas in low-permeability (i.e., tight) reservoirs, gas in pressurized shales and brines, or natural gas in various compounds associated with water.

23. Methane constitutes the greatest quantity or volume in a natural gas sample, but has the least Btu's per cubic foot.

24. Butane constitutes the least quantity or volume in a natural gas sample, but has the most Btu's per cubic foot.

25. A therm of natural gas is equivalent to 100,000 Btu's. It is also equal to 100 cubic feet of natural gas (CCF).

26. Butane is used primarily to produce LPG (Liquid Petroleum Gas) products.

27. Ethane is used primarily in petrochemical plants to produce various plastic products.

28. Research projects are now in progress to extract methane gas from landfills.

Discussion Questions

1. What are several factors that have limited the supplies of natural gas in the past?

2. What factors helped to increase the supply of natural gas in the past 10 years?

3. Create a scenario for the future of natural gas, its reserves, applications, trends, and new technologies through the year 2040.

4. With a diagram, trace the flow of natural gas from the well to the consumer. On the diagram, identify the interstate and intrastate portions. Also identify the locations in which energy is used to get natural gas from the well to the consumer.

5. What effects would be felt socially and environmentally if natural gas were removed as an energy resource in the next 15 years?

chapter 7
Nuclear Energy Resources

PURPOSES

Goals of Chapter Seven

Another nonrenewable energy source is that derived from the fission (or splitting) of certain heavy unstable atoms. This energy form was defined as nuclear energy in chapter 1 of this text. Although fusion is another type of nuclear energy, this chapter focuses primarily on the technology associated with the fission type of nuclear energy. A brief historic analysis of nuclear energy is also presented. Basic chemistry of nuclear fission, types of reactors, and nuclear waste are covered in detail. At the completion of this chapter, you will be able to

1. Trace the historic development of nuclear energy.
2. Trace the historic use of nuclear energy in terms of supply and demand.
3. Use the terminology pertaining to basic nuclear chemistry.
4. Analyze the nuclear fuel cycle, the associated technology, and its social and environmental implications.
5. Compare and contrast the different types of nuclear reactors, including the light-water reactor, high-temperature gas reactor, and breeder reactors.
6. Analyze the process and implications of nuclear waste disposal.

Terms to Know

By studying this chapter, the following terms can be identified, defined, and used:

Alpha decay
Amortization period
Background radiation
Beta decay
Beta particle
Breeding fuel
Control rods
Curie
Decommissioned
Deuterium
Disassembly
Emergency core-cooling systems
Encasement
Enrichment
Entombment
Fast reactor
Fission
Fissle

Fuel rods/Fuel bundles
Fusion
Gamma decay
Gamma ray
Gaseous diffusion
Half-life
High-level waste
Isotopes
Licensing basis
Low-level waste
Meltdown
Millirem
Moderated
Monitored Retrievable Storage Facility
Mothballing
Natural radiation
Nuclear fuel cycle
Nucleons
Person-rem
Plutonium
Rad
Radioactive decay
Radioactivity
Reasonably assured resources
Rem
Slow neutrons
Speculative resources
Tailings
Theory of relativity
Thermal pollution
Thermal reactor
Transmuted
Yellow cake

INTRODUCTION TO NUCLEAR POWER

Distinct Trends in the Development of Nuclear Power

Before presenting information on the historic development of nuclear energy, note that this technology was a result of weapon development conducted by the military during World War II. Its development illustrates several distinct trends in the history of technology including: the shrinking interval between a scientific discovery and its application in society, the blurring of traditional roles of scientists and technologists, and the important role of the federal government in fostering scientific and technological advancement. These trends will be illustrated in the historic analysis of nuclear energy.

Discovery of Barium

One of the first steps in the military development of atomic energy was the discovery of barium in late 1938 by two German scientists, Otto Hahn and Fritz Strassman. Hahn and Strassman found barium after their bombardment of uranium with neutrons. The discovery suggested that the neutrons had not only dislodged a few particles from the uranium nucleus but had split it almost in half. Since the total mass of the resulting particles was significantly less than that of the original uranium nucleus, it could be concluded that the process released a large amount of thermal energy.

This discovery led to numerous experiments involving the bombardment of uranium with neutrons in many other countries. From these experiments it was demonstrated that each fissioning uranium nucleus emitted an average of two or more neutrons, which theoretically were capable of splitting other uranium nuclei. This theory of being able to create chain reactions of fissionable material led scientists to theorize that they could use uranium to produce energy. However, the process eluded scientists for many years. One problem scientists had to overcome was how to maintain the chain reactions.

A scientist who helped to create continuous chain reactions was Enrico Fermi, an Italian physicist who immigrated to the United States in 1939. He discovered that the neutrons emitted by the fissioning of the uranium nuclei had to be slowed down or *moderated* to allow the chain reaction to continue. A failure to moderate the collisions caused the bombarding atoms to be absorbed by the uranium nuclei, thus stopping the chain reaction. This discovery led Fermi and a colleague, Leo Szilard at Columbia University, to use a highly pure form of graphite as a moderator. This, however, was only a first step on the long road to use of nuclear energy.

Problems in Splitting Uranium Nuclei

The fundamental problem for scientists working with the splitting of uranium nuclei was that they were uncertain that they could establish a significant continual chain reaction needed to produce usable energy. First, physicists did not possess sufficient quantities of uranium and a moderator. Second, like many elements found in nature, uranium exists as a mixture of several isotopes, and scientists did not know which isotope would best lend itself to the fissioning process. (An *isotope* is an element that exists in one of two or more forms distinguished by small differences in atomic mass and physical and chemical properties but having the same atomic number.)

It was not until March of 1940 that John R. Dunning and his associates at Columbia University demonstrated that the uranium-235 isotope (^{235}U), not natural uranium-238 (^{238}U), was easily fissionable with slow neutrons. However, this discovery was tempered by the fact that ^{235}U constitutes less than 1 percent of naturally occurring uranium. This would create the need for separating the isotope ^{235}U from the more plentiful ^{238}U. Fermi concluded that such a process would be difficult at best because separation must be done by a physical process and not a chemical one. The separation process must also be extremely sensitive to the small differences in the atomic weights of the two elements.

Scientists in the United States thought it would be possible to assemble large stocks of ^{238}U and a moderator to sustain a chain reaction with ^{235}U present in the ^{238}U. Such a process would not call for the separation of the two uranium isotopes. U.S. physicists concluded in 1940 that it would be possible to build a pile of graphite (to act as a moderator) interspersed with lumps of uranium metal to produce energy or a weapon of mass destruction. However, using this device for a bomb was not practical since it would weigh hundreds of tons and would not be transportable with the aircraft of the time.

In Great Britain, Rudolph Peierls concluded that ^{235}U would fission with fast, as well as slow, neutrons. This meant that it might be possible to produce a chain reaction in a small mass of sufficiently pure ^{235}U without incorporating the added weight of ^{238}U or a moderator. Such a device would allow nuclear energy to be used as an energy source or weapon. British scientists also concluded with laboratory experiments that they could separate the uranium isotopes by diffusing a gaseous compound of uranium through a series of porous membranes or barriers. In this process the lighter ^{235}U atoms passed through the barrier more easily than the ^{238}U. Thus, purer amounts of ^{235}U could be obtained by pumping the gas through large numbers of barriers. (This process will be presented in more detail later in this chapter.)

Collaboration of Scientists and Technologists

In the early 1940s, the United States in conjunction with its allies became convinced of the need to beat the Germans in the race to develop an atomic bomb. President Roosevelt of the United States had learned as early as 1939 from a group of scientist-exiles of German Chancellor/Führer Hitler's plan to develop a new bomb of unimaginable destructiveness. This bomb was based on the newly discovered process of uranium fission. It, therefore, became important for allied scientists and technologists to work more closely together and abandon all speculation about the peaceful uses of atomic energy.

The collaboration of scientists and technologists was realized with the start of the Manhattan Engineer District of the United States Army Corps of Engineers. What became known as the Manhattan Project was initiated shortly after the Japanese attack on Pearl Harbor. It was headed first by Vannevar Bush and then by General Leslie Groves at Oak Ridge, Tennessee. The project had one main charge as established by President Roosevelt— to develop an atomic bomb before Germany succeeded in building their own weapon. The Manhattan Project proved that the federal government could have a direct effect on advancing technology by creating the proper conditions for research and development. While work was conducted at Oak Ridge, research was done at various universities under the direction of Arthur Compton of the University of Chicago and at Los Alamos, New Mexico, directed by J. Robert Oppenheimer.

Scientists and technologists working on the Manhattan Project took a different approach than did the British. Instead of trying to separate the different isotopes of uranium, United States scientists worked on identifying other fissionable elements that would work as a weapon. In February 1941, Glenn T. Seaborg and associates at the University of California, Berkeley, identified a new human-made element—*plutonium* (^{239}Pu). By May of 1941, Seaborg reported that plutonium was even more fissionable than ^{235}U. Plutonium could be created by the non-fission absorption of neutrons in ^{238}U. The scientists and technologists working with Seaborg concluded that ^{235}U in a pile of ^{238}U could be used to sustain a chain reaction. With such a process, the ^{238}U would be *transmuted* (i.e., the change of one element into another) into ^{239}Pu. This plutonium could then be isolated by a chemical means and used in the production of an atomic weapon.

The First Nuclear Chain Reaction

The key to any nuclear chain reaction is the atomic *pile*, or reactor, as it became known by the end of the war. The reactor was the achievement of Enrico Fermi and scientists and technologists working at the University of Chicago. Beginning in early 1942, Fermi had established a laboratory at an abandoned squash court under the stands at the University of Chicago's Stagg Field. At this site scientists and technologists had built a crude nuclear reactor of 57 alternating layers of uranium (i.e., ^{235}U in a pile of ^{238}U) and graphite bricks. These bricks were placed one atop the other and were propped up with wooden four-by-six boards. It was called Chicago Pile 1.

By the morning of December 2, 1942, the construction on the reactor was complete, and Fermi ordered the electrically controlled moderators of graphite to be withdrawn from the pile. Once removed, the moderators would allow the splitting of atoms to occur. Fermi and the scientists involved in the experiment determined that a chain reaction had occurred, and more importantly, that this chain reaction was self-sustaining.

While Chicago Pile 1 generated 200 watts of power, it had no practical importance in the production of bombs or nuclear power. However, the experiment provided the basic information for the design of six other reactors built during World War II. Three of these plutonium-producing reactors were at Hanford, Washington, and were central to both the war and postwar development of nuclear energy.

The Hanford reactors produced the plutonium for the first weapon device tested in Alamogordo, New Mexico, on July 16, 1945, and for the first weapon dropped on Nagasaki, Japan. These reactors also supplied experience in the continuous operation of nuclear technology. This experience, though, did not contribute to the peacetime use of nuclear power. Although the Hanford reactors were designed to operate at 250,000 kilowatts, the operating temperature was too low to make power generation efficient, and the emphasis remained on plutonium production until the war ended.

Postwar Developments in Nuclear Power

While the possibility of using heat energy from the nuclear reactor to drive electrical generators was technically possible, the translation of this concept into useful energy production remained difficult. Scientists and technologists would need to develop reactors with operating temperatures higher than those already developed. In addition to the technical problems of power plant design, there were also administrative obstacles to overcome.

At the end of World War II, nuclear research had been organized to produce fissionable materials and nuclear weapons. Scientists had no legal or practical means of fostering reactor designs for peaceful purposes. There seemed to be a consensus within the United States that the government would have to control atomic energy activity. However, control would not be established until Congress spent the first half of 1946 debating the question of military or civilian control of this new power source. This question over who would control nuclear power was answered when President Harry Truman signed into law the Atomic Energy Act on August 1, 1946.

The Atomic Energy Act

The Atomic Energy Act created a five-member Atomic Energy Commission that held absolute authority over all sources of information, fissionable materials, and equipment for utilizing these materials (including weapons and reactors). This commission was, therefore, responsible for the development of both military and commercial uses of atomic energy. Under the Atomic Energy Act, the United States government could keep secret much of the technical knowledge gained during the military development of atomic energy. Historically this legislation was far reaching—never before in peacetime history had Congress established an administrative agency with such sweeping authority and responsibility.

The overall purpose of the Atomic Energy Act was to ensure both the military secrecy and peaceful use of nuclear energy. The Act specified that the use of atomic energy would be directed to improving the public welfare, increasing the standard of living, strengthening free competition among private enterprises, and helping to foster world peace. Although the promise of atomic energy for peaceful purposes was established, it was not until 1951 that such an endeavor was realized.

On December 20, 1951, a group of 16 scientists, engineers, and technologists produced the world's first usable electricity from nuclear fission. Note that this event happened only 13 years after the early discoveries of barium. The collaboration of scientists and technologists had been an influencing factor in successful development of this energy source. The achievement of generating electricity occurred at the Experimental Breeder Reactor I at the Argonne National Laboratory in Idaho. This first use of atomic energy in generating electricity allowed the scientists to light four light bulbs. Other technical advancements soon followed, and by July 17, 1955, electricity was produced by atomic energy to light the town of Arco, Idaho.

AVAILABILITY OF RADIOACTIVE ELEMENTS

World Uranium Reserves

Various sources can be found showing the amount of uranium available worldwide. These sources include data from the Nuclear Energy Agency of the Organization for Economic Cooperation and Development and the International Atomic Energy Agency. Estimates for centrally planned economies are provided with hard statistics for the rest of the world. Uranium resources available are classified as reasonably assured or speculative resources.

Reasonably Assured and Speculative Uranium Resources

Reasonably assured resources are those that are considered recoverable with existing facilities, present technology, and at current costs and price levels. *Speculative resources* are those

believed to exist in the judgment of geological engineers, but there is no data about discoverability, availability, or economics associated with their extraction. Currently, there are estimated 2 million metric tons of reasonably assured uranium deposits and more than 2.2 million metric tons of speculative reserves outside of centrally planned economies. The largest known resources are located in the United States, Canada, South Africa, and Australia. Although other resources are known to exist in centrally planned economies, these resources are estimated to be far less than the 4.2 million metric tons of the rest of the world. Centrally planned economies are estimated to hold approximately 0.3 million metric tons of reasonably assured uranium deposits. Of these deposits, the largest reserves can be found in the former Soviet Union, and the Peoples Republic of China.

World Demand for Uranium

The uranium market is a reflection of interconnected economic parameters. These include reactor-related uranium demand and supply, and uranium prices. Each of these parameters is presented in the following paragraphs.

Reactor-Related Uranium Demand

Reactor-related uranium demand is based upon nuclear electricity generating capacity. It is estimated that the uranium demand for the *world outside centrally planned economies* (WOCPE) increased from about 4,000 tons of uranium in 1965 to 41,500 tons in 1989. This was equivalent to an annual growth rate of over 10 percent. The rate of growth for uranium demand is expected to increase through 2005, though at a smaller rate. Figure 7-1 illustrates the projected growth for uranium through 2005. The estimated annual uranium demand is forecasted to be approximately 53,000 tons by 2005. This reflects an annual rate of growth of about 1.5 percent.

Supply of Uranium

The geographic concentration of uranium resources, located outside of centrally planned economies, can be found in three main areas: Canada, the United States, and Australia. These areas have a combined share of over 57 percent of the uranium produced worldwide. Also, only five countries (Canada, United States, Australia, Namibia, and France) produce 77 percent or more of the world uranium outside of centrally planned economies. As can be seen in Figure 7-2, Canada is the largest producer of uranium (11,000 tons), followed by the United States (4,600 tons), and Namibia (3,600 tons).

The supply projections for 1995-2005 are based upon the continued production capability of existing and committed mines and mills that use low-cost resources (i.e., recoverable at a United States cost of $80 per kilogram [kg] or below). It is also assumed that uranium demand will continue to increase through 2005. And, it is expected that these increases in demand will be accompanied by a significant decrease in production. Supply and demand projections are illustrated in Figure 7-3. By 2005, the demand for uranium is forecasted to outpace production by some 20,000 tons. However, the decreases in production of uranium do not represent a supply deficit. Nor are these decreases in production forecasted to limit the growth of electricity generated by nuclear power. Decreases are

World Uranium Demand with Projections to 2005

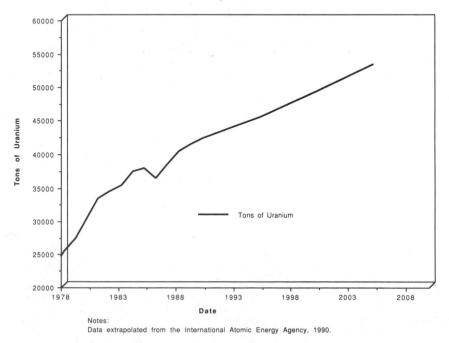

Notes:
Data extrapolated from the International Atomic Energy Agency, 1990.

FIGURE 7-1 The estimated uranium demand is projected to be approximately 53,000 tons by 2005. This reflects an annual rate of growth of about 1.5 percent through 2005.

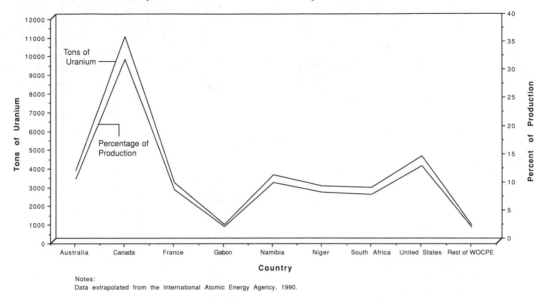

Uranium Production (World Outside of Centrally Planned Economies)

Notes:
Data extrapolated from the International Atomic Energy Agency, 1990.

FIGURE 7-2 Only five countries produce 77 percent or more of the world uranium outside of centrally planned economies. Canada is the largest of these producers.

expected to be offset by the large uranium inventories of both producers and consumers of nuclear-based fuels. Currently, there are some 70,000 tons of uranium exceeding the amount needed for buffer stocks through the forecasted period. In addition, uranium stocks are expected to increase further as their use for defense purposes begins to enter the civilian market. This is particularly true as the former Soviet Union and the United States make further reductions in their nuclear weapons programs. These reductions and the buffer stocks already existing will affect the selling price of uranium.

Selling Price of Uranium

Total uranium stocks should be able to fill the production gaps projected through 2005 without large increases in price. As with certain other energy resources (such as petroleum), there are two prices for uranium: the price for short-term deliveries and contract prices for longer-term periods. The volume of uranium sold under these price conditions varies greatly. However, most of the material sold is under long-term contracts with a lower price than that sold for short-term deliveries. The short-term delivery prices have ranged from a low of $16/kg in 1972 to a high of $112/kg in 1978 (in United States dollars). Short-term delivery price has held at about $30/kg during the mid to late 1990s.

The future of uranium supplies will depend upon both the supply available and the costs associated with mining and processing the fuel. With the large inventories available, there will be an abundance of uranium in the marketplace. This abundance is expected to help to keep uranium prices low through the turn of

World Uranium Demand and Supply with Projections to 2005

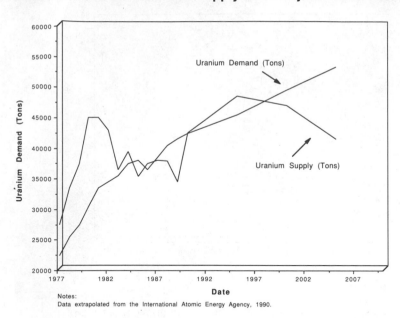

FIGURE 7-3 World uranium supply and demand (with projections to 2005) is illustrated in this chart. Although demand is expected to outpace production near the end of the decade, the decreases in production are not expected to represent a supply deficit because of large inventories held worldwide.

Notes:
Data extrapolated from the International Atomic Energy Agency, 1990.

the century, unless nuclear energy consumption increases substantially in the future.

World Nuclear Energy Consumption

Worldwide there are more than 400 nuclear power plants operating in 25 countries. These power plants supply almost 17 percent of the world's electricity. Several countries—most notably France, Belgium, the Republic of Korea, and Hungary—have relied more heavily on the use of nuclear power in meeting their electricity demands than have other countries. Additionally, several countries have recently announced major construction efforts to increase the role of nuclear power in the generation of electricity.

Examples of this renewed interest in nuclear power include the efforts of the Dutch government and recent construction plans in Japan, South Korea, and France. The Dutch government has plans to increase its share of

nuclear power from around 1 percent to 25 percent in the next century. This goal is being undertaken to reduce carbon dioxide emissions and to reduce the dependence on imported natural gas. Japan is also committed to an increased dependence on nuclear power. Today, nuclear power provides Japan with about 27 percent of its electrical production; Japan hopes to expand this to 45 percent by 2010. South Korea has a goal of making nuclear power its main electricity generator by 2000. France hopes to export nuclear-generated electricity to other European nations. Worldwide, over 80 new nuclear-powered energy plants are now under construction. Figure 7-4 provides a graphic representation of countries with the highest production of nuclear energy. France and Belgium derive the largest share of their electricity from nuclear energy worldwide.

World Growth Rates for Nuclear Power

Throughout the 1990s and into the next century, the consumption of nuclear power is not foreseen to increase as fast as it did in the 1970s and 1980s. Figure 7-5 illustrates the growth of nuclear energy from 1965 and its projected increase to 2010. The world annual growth rate for electricity produced by nuclear power is expected to be between 1 and 3 percent through the turn of the century.

The lower rates of growth expected in the late 1990s and through 2010 are a result of reduced public acceptance of nuclear energy, especially since the 1986 Chernobyl disaster; relative costs, when compared to lower-priced fossil fuels; and unique environmental problems, particularly those dealing with the disposal of nuclear waste. Another meaningful factor in the reduced public acceptance of nuclear power is the political changes taking place in Eastern Europe. Because of political and economic reforms occurring in these countries, the nuclear power industry is facing uncertainty resulting from heightened safety standards and public opposition. Specifically, political changes in the region have led to a reassessment of Soviet-designed and -supplied nuclear reactors. This reassessment is a direct result of the Chernobyl disaster, which will be presented in more detail later in this chapter.

In the former Soviet Union, the political and economic reforms have allowed a greater freedom of expression by the populace, which has resulted in a greater concern about the environmental problems inherent in the use of nuclear power. While these concerns were always present in the past, they were suppressed. The greater freedom has made it possible for anti-nuclear groups to assert their political clout to suppress the construction of new nuclear power plants.

Assessment of Nuclear Power Plants in Eastern Europe

A recent study by the United Nations International Atomic Energy Agency of nuclear power plants in Eastern Europe found more than 1,000 specific problems that could lead to a disaster. The major concern is the threat of a Chernobyl-like accident at one of Eastern Europe's outdated, Soviet-designed reactors. The reactor, part of Bulgaria's Kozloduy nuclear complex, has been described by experts as the most dangerous civilian nuclear power plant on the planet. The CSFR's Bohunice power plant located in Trnava, Zapadoslovensky, has also had several near-accidents that have caused political concern in neighboring Austria.

Problems in Closing Poorly Designed or Operated Nuclear Reactors

Despite the concerns and dangers associated with nuclear power plants, there has been a great reluctance to shut down these plants. The difficulty in shutting down poorly operated or unsafe nuclear power plants in Eastern Europe and the former Soviet Union is a consequence of several factors. First, there is a need for electrical power. Shutting down nuclear power plants would call for the building of fossil-fuel generated electrical plants. Coal, being abundant in the former Soviet Union, would possibly be the fuel of choice. However, this fuel choice has numerous environmental problems. Second, nuclear power currently provides a critical source of electricity to the country's energy intensive and inefficient industrial sector. In fact, the Czech and Slovak Federal Republic relies on nuclear power for nearly a third of its electrical power. This is also true for the newly formed republics

Countries with the Highest Production of Nuclear Energy

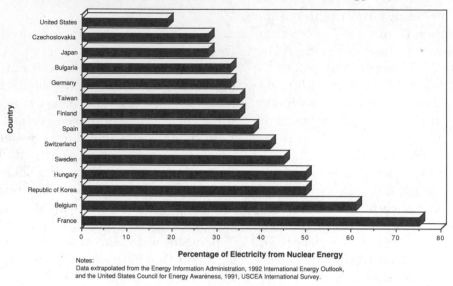

Notes:
Data extrapolated from the Energy Information Administration, 1992 International Energy Outlook, and the United States Council for Energy Awareness, 1991, USCEA International Survey.

FIGURE 7-4   This figure illustrates those countries with the highest production of nuclear energy.

World Nuclear Energy Production 1965 with Projections to 2010

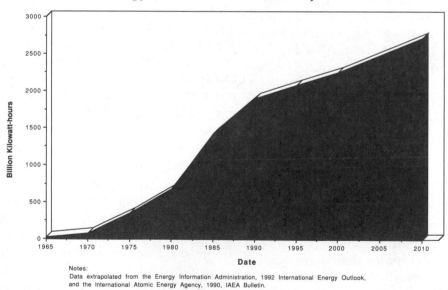

Notes:
Data extrapolated from the Energy Information Administration, 1992 International Energy Outlook, and the International Atomic Energy Agency, 1990, IAEA Bulletin.

FIGURE 7-5   The world annual growth rate for electricity produced by nuclear power is forecasted to be between 1 and 3 percent through the turn of the century. Note how this growth rate is not as fast as it was during the 1970s and 1980s.

of Lithuania and Ukraine. The governments in each of these areas fear that a shutdown of nuclear power plants would worsen the already common electricity shortages. Another consequence of shutting down nuclear power plants would be to retard economic growth, which is linked to the conversion of inexpensive energy sources.

One possible solution to the problems of shutting down poorly operated or unsafe nuclear power plants is to postpone building of new plants until the older ones can be retrofitted with Western-designed safety features. However, this approach would not be without hardship and expense. This approach would require both economic and technical aid from Western countries and would put a further strain on meeting the energy requirements of these Eastern European countries. Clearly, more technical and economic assistance is needed to help resolve the problems of unsafe nuclear power plants.

NUCLEAR POWER IN THE UNITED STATES

United States' Nuclear Energy Consumption

The importance of nuclear energy to the United States economy cannot be underestimated. At the time of the 1973 oil embargo by the countries of the Middle East, petroleum accounted for about 17 percent of the United States' electrical production. Nuclear energy accounted for 5 percent of electrical production during this same period. In 1990, petroleum accounted for only about 4 percent of the United States electrical supply, nuclear energy about 21 percent. With the use of nuclear energy, the United States has reduced some of its dependence on imported oil. Had the U.S. not

developed nuclear power for the generation of electricity, other fossil fuels would have been used to meet its economic needs. This is particularly important when one considers that since 1973 the United States' population has increased from nearly 211 million to about 250 million. During this same period, the economy has grown about 50 percent.

Today, there are over 100 nuclear power plants in operation in the United States. These plants supply over 570 billion kilowatts of electricity to the country. This accounts for nearly 22 percent of the electricity produced in the United States. Figure 7-6 provides an historic record of the increase in nuclear power in this country and projections for the future. In 1957, there was only one operable nuclear power plant, and this was providing less than 0.05 percent of the electricity consumed in the country. Nuclear power is forecasted to continue to be a part of our energy mix in the future. As illustrated in Figure 7-6, nuclear power is expected to supply about 630 billion kilowatts of electricity by 2010. In addition, the nuclear power industry would like to build several new power plants to meet the increasing demand for electricity.

United States' Future Power Plants

Since the early 1990s, the United States nuclear power industry has made the option of nuclear power attractive to utility companies. This industry has consolidated several of its existing organizations and formed new ones to better address nuclear power issues. Two examples of this reorganization effort can be found in the formation of the Nuclear Management and Resources Council and the Nuclear Power Oversight Committee. To further attract utilities, industry representatives wrote the *Strategic Plan for Building New Nuclear Power Plants*, the

United States Nuclear Power - Operable Units, Nuclear Electricity Generation (1957 with Projections to 2010)

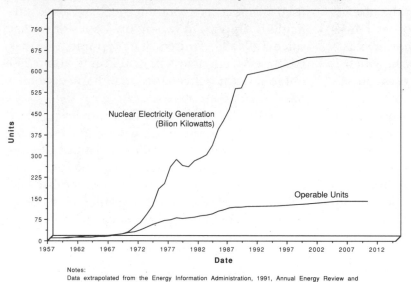

Notes:
Data extrapolated from the Energy Information Administration, 1991, Annual Energy Review and the Energy Information Administration, 1992, Annual Energy Outlook with Projections to 2010.

FIGURE 7-6 Nuclear power in the United States accounts for about 22 percent of the electricity produced. Nuclear power is forecasted to supply about 630 billion kilowatts of electricity by 2010.

overall goal of which is to have orders for new nuclear power plants by the late 1990s. Although no new nuclear power plants have been ordered as of the mid-1990s, the industry has made progress in several other activities.

With nuclear industry backing, the Department of Energy has developed a plan of action to enhance the prospects of new nuclear power plants. This action plan includes

1. Providing $20 million for engineering support for new reactor designs.
2. Developing a procedure for early site certification for nuclear power plants. (This important aspect provides the utility company with an early site permit and provides a degree of certainty for obtaining a new order for a nuclear power plant.)
3. Providing a process for renewing a current nuclear power plant license for 20 years beyond its current operation period.

Although the action plan developed by the Department of Energy may help utilities to propose the construction of new power plants, there are considerable obstacles to overcome.

Obstacles to Construction of New Nuclear Power Plants in the United States

Before a new nuclear power plant is built in the United States, numerous environmental and economic issues must be resolved. One of the first issues that must be addressed concerns a permanent storage site for spent high-level nuclear waste. The federal government promised to have a national disposal site established by 1983, but this promise has not materialized. Officials from the Department of Energy now estimate the earliest date such a facility could be operational would be 2010. Without such a storage facility, nuclear power plants already in operation may be forced to close as their on-site storage areas are filled to capacity. Meanwhile, to store spent high-level nuclear waste on a temporary basis, the concept of a **Monitored Retrievable Storage Facility** (MRSF) has been proposed.

The MRSF is called *monitored* because the United States Nuclear Regulatory Commission with local government agencies will be required to oversee the storage area constantly. By *retrievable,* the facility is meant to store the spent high-level waste only temporarily. At some later date, it will be removed and sent to a permanent repository. Although the MRSF makes a significant point in demonstrating that the government is serious about disposing of nuclear waste safely, note that many states are against having an MRSF within their borders. The concern expressed by residents in these states is that such a facility will become a permanent storage area, and they simply do not wish to live in a state or area with spent high-level nuclear waste. An example of this was the political and taxpayer concerns expressed in 1994 in the proposal to establish an MRSF in Prairie Island, Minnesota. The proposed site was in response to the Prairie Island nuclear power plant running out of on-site storage space for spent nuclear fuel.

A second issue in the construction of new nuclear power plants in the United States involves the licensing process, which is both comprehensive and rigorous. Throughout the history of nuclear power in the United States, the utility industry has had to adjust to an ever-changing regulatory environment. This constant changing of regulations for nuclear power plant construction and operation was a result of goals to make plants safer and to minimize their effect on the environment. The regulations were also an attempt to keep electricity charges to consumers at an acceptable level. In addition to federal regulations, nuclear power plants have had to adapt to increased state involvement in nuclear policy making. State energy departments, in liaison with Public Utility Commissions, often subjected utility investment plans to greater examination. In addition, most notably in California and New York, the Public Utility Commissions began to place a greater emphasis on renewable energy sources. This focus helped to slow the development of nuclear power by reducing the rate that the utilities could charge for the work in progress when applying for construction permits or building the power plant.

To meet the ever-changing nature of both federal and state regulations, attempts have been fostered at the federal level to expedite the licensing process. The proposals supported by the nuclear power industry would allow a one-step licensing process, which would provide for a combined construction permit and operating license. Other proposals currently recommended include the preapproval of standardized designs and advanced site certification. However, each of these initiatives would require legislative approval before they could be implemented.

A third issue of nuclear power involves its relative economic advantage. Nuclear power must be proven to be competitive with other renewable energy sources proposed to generate electricity. Officials in the nuclear power industry have expressed confidence that they can become competitive with other forms of electrical generation. This goal of being more competitive is to be fulfilled by reducing the construction time for nuclear power plants. In the 1980s, the average construction time for nuclear power plants was 14 years. Industry spokespersons believe that by reducing the construction time to 6 years, nuclear power can demonstrate an economic advantage as compared with other energy sources.

The last major issue of nuclear power concerns the lack of cooperative agreements among states for the disposal of low-level nuclear waste. The nuclear power industry is a large producer of low-level radioactive waste. Annually, over 1.5 million cubic feet of low-level nuclear wastes are produced in the United States. The nuclear power industry accounts for approximately 50 percent of these wastes in the form of protective clothing, machine parts and tools, filters, and deposits that collect in valves and pumps or collect on surfaces inside the plant. The lack of disposal sites for low-level radioactive wastes and cooperative agreements between states to accept this material is a major obstacle to any utility considering a new nuclear power plant.

Nuclear Plant Construction—Energy Issue 7-1

Nuclear power plants under construction have been canceled. It is also very difficult to get a permit to construct a nuclear power plant. Identify other reasons why more nuclear power plants are not built to produce increased amounts of electricity demanded by society.

BASIC CHEMISTRY AND DESIGN

The understanding of nuclear energy hinges upon certain scientific principles and knowledge of atomic structure and energy levels. The following information will help in understanding the basic chemistry and design of fission power plants.

Nuclear Energy and Costs—Energy Issue 7-2

Often, there is conflicting information about the costs of using electricity generated from a nuclear power plant. Some experts report that it is less costly to produce electricity with the use of nuclear power as compared to that produced with other fossil fuels. What factors make nuclear power statistics, such as this, conflicting? What factors will change how the cost of nuclear energy is determined?

Atomic Weight

The atomic weight, also called the mass number, is the sum of the protons and the neutrons within the atom. For example, cobalt, containing 27 protons (atomic number 27) and 33 neutrons, has an atomic weight of 60. (See Figure 7-7.) Note that the atomic weight is found by adding the protons and neutrons.

Cobalt Identification

(a)	ATOMIC WEIGHT $_{PROTONS}CO_{NEUTRONS}$	$_{27}^{60}CO_{33}$
(b)	$_{PROTONS}CO^{ATOMIC\ WEIGHT}$	$_{27}CO^{60}$
(c)	$CO^{ATOMIC\ WEIGHT}$	CO-60

FIGURE 7-7 Elements can be identified in several ways. The protons and the neutrons when added equal the mass number (atomic weight) of an element.

Isotopes

Some elements consist of atoms that can vary in atomic weight while the atomic number remains constant. For example, the number of neutrons in atoms of the same element can vary. As earlier noted, such forms of the same element are called isotopes. For instance, hydrogen has three isotopes. As graphically shown in Figure 7-8, hydrogen, deuterium (heavy hydrogen), and tritium are isotopes of hydrogen. The only difference among these atoms is the number of neutrons, which determines the respective atomic weight of the isotopes. All three isotopes are electrically the same since they contain the same number of protons and electrons.

Chemically, elements often exist naturally as a mixture of isotopes. One such element is uranium, which occurs as ^{235}U, a common nuclear fuel, with several other isotopes, including ^{234}U and ^{238}U, present in the uranium element.

Radioactivity

Many natural and manufactured isotopes are radioactive. **Radioactivity** is defined as the spontaneous disintegration or decay of the nuclei of atoms. As the disintegration occurs the atom comes apart, which is called **radioactive decay**. The term *radioactivity* was coined by Marie Curie. Curie was a physicist who won Nobel Prizes in both physics and chemistry for her pioneering work with radioactive elements. In this process of disintegration, the emission of alpha, beta, and sometimes gamma particles occurs. The emission of these particles is often referred to as alpha decay, beta decay, and gamma decay, respectively.

Alpha Decay

In **alpha decay**, a nucleus emits two protons and two neutrons. These are bundled together as a helium-4 nucleus (^{4}He). The ^{4}He is called an

Isotopes of Hydrogen and Uranium

	ATOMIC WEIGHT	SYMBOL	PROTONS	NEUTRONS	ELECTRONS
HYDROGEN	$^1_1H_0$		•	NONE	•
DEUTERIUM	$^2_1H_1$		•	•	•
TRITIUM	$^3_1H_2$		•	• •	•
U-234	$^{234}_{92}U_{142}$		• = 92	• = 142	• 92
U-235	$^{235}_{92}U_{143}$		• = 92	• = 143	• 92
U-238	$^{238}_{92}U_{146}$		• = 92	• = 146	• 92

FIGURE 7-8 Hydrogen and uranium both have several isotopes. These isotopes are made by the addition of extra neutrons.

alpha particle (α), a name dating back to the turn of the century when it was not yet known that the particles emitted were helium nuclei. Alpha decay is common among large unstable nuclei that need to rid themselves of excess protons. While undergoing alpha decay, ^{238}U emits an alpha particle that dislocates two protons, dropping the atomic number of the remaining nucleus by 2. Since the alpha particle is made up of a total of four nucleons, the mass number drops by 4. (Together, neutrons and protons are called **nucleons**.) Thus, in an alpha decay, ^{238}U emits an alpha particle leaving a nucleus of thorium-234 (^{234}Th). Figure 7-9 graphically illustrates this process. In this illustration note how the sum of the protons and neutrons equals the mass number.

Alpha particles are heavy and fast moving; however, they travel only about 1 inch in air and can be stopped by a sheet of paper or clothing. Those alpha particles that get past

Alpha Radioactivity

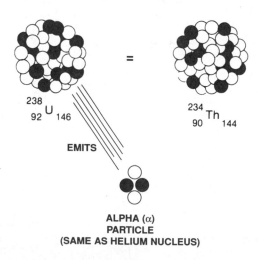

ALPHA (α) PARTICLE (SAME AS HELIUM NUCLEUS)

FIGURE 7-9 As ^{238}U emits an alpha particle, it changes its weight and structure to ^{234}U. This process or emission and restructuring continues during the decay of such a radioactive material.

clothing will be stopped by the first few layers of skin. Since the top layers of skin are dead (except for the lips and eyes), no significant damage should occur. The danger of alpha particles occurs when they enter the body. This can happen through a skin puncture, ingestion, or breathing. Once inside the body the alpha particles can exist for a long time and adversely affect live cells.

Beta Decay

In *beta decay*, unstable nuclei with too many neutrons purge themselves of neutrons or gain protons. In this process, a neutron changes itself into a proton and an electron also known as a *beta particle* (β). This beta particle moves out, leaving the nucleus with one more proton and one fewer neutron than it originally had. As a result of having another proton, the atomic number has increased by 1. The total number of nucleons remains the same, leaving the mass number unchanged. A beta particle is a radioactive particle consisting essentially of electrons. Although there are other beta particle components, they will not be discussed in this text.

Beta particles are fast-moving electrons that can penetrate up to 1 inch of wood and travel several feet in the air at sea level before stopping. Unlike alpha particles, beta particles can penetrate the skin. Shielding can be accomplished by light plywood or sheets of aluminum. Like alpha particles, beta emissions can cause cell damage.

Gamma Decay

A nucleus can be struck by another particle that bounces off it or moves right through it without causing a chemical or physical change. In this instance, the nucleus retains its identity, but it may acquire some excess energy in the process. A nucleus with excess energy is said to be *excited*. The nucleus is like a quiet tuning fork that has been struck, but now is vibrating causing a sound. Unlike the struck tuning fork that causes a sound by emitting its excess energy, the nucleus may store its excess energy for a time. When the energy is released, it is in a sudden burst as an energy bundle called a **gamma ray**. Once the nucleus has emitted its excess energy by this process of **gamma decay**, it returns to its original unexcited state.

Gamma rays are essentially the same as high-intensity X rays. Their high energy gives them high penetrating powers. Gamma-ray fields exist in nuclear reactors; therefore, humans must be protected against them by a few feet of concrete or a few yards of water. More protection can be achieved with the use of heavy metals, such as lead.

Gamma radiation is electromagnetic and, as previously noted, is similar to X rays. It is the most penetrating type of radiation. In gamma decay, both alpha and beta particles may also be given off. Gamma particles (γ) have the greatest energy (and deepest penetration) and can be the most harmful particle to humans. Biological effects from the exposure to high-level gamma radiation can include cancer, genetic altering of cells, or death. The biological effects of radiation will be presented in further detail later in this chapter.

Half-life

The most convenient method of describing or measuring the rate at which a material loses its radioactivity is in terms of its half-life. As earlier noted, several isotopes that exist in nature are unstable and are subject to decay. When an isotope decays to another form through the emission of alpha, beta, or gamma rays, the new form may also be unstable and decay further into a new element. Eventually, a stable form will be reached, but many stages of decay can occur before this is achieved.

The term **half-life** refers to the time needed for half of the unstable atoms originally present in a radioactive substance to decay into their new form (as illustrated in Figure 7-10). Therefore, after a period corresponding to one half-life, half the original atoms remain; after two half-lives, a quarter remain; after three half-lives, an eighth remain; and so on. After 10 half-lives, only 0.1 percent of the original material remains.

While the decay of certain elements is extremely slow, a very large amount of decay has already occurred when one considers the age of the earth. For example, the half-life of ^{238}U is about 4,500 million years. Since scientists have estimated that the earth may have been in existence for about this same time, some of the ^{238}U has already undergone a period of one half-life. Note that the half-life of an element can range from several seconds to millions of years. Table 7-1 gives the half-life of various elements.

Nuclear Energy and Half-Life— Math Interface 7-1

1. Krypton 85 has a half-life of 10 years. How many years would it take for the Krypton to reduce its radiation to below 1.56 percent of its initial radiation value?

Years to reduce below 1.56 percent equal _____.

Half-Life Periods

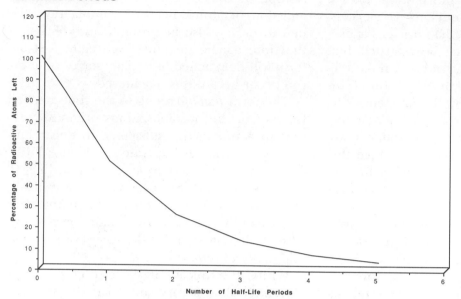

FIGURE 7-10 This chart shows how radioactive elements decay during half-life periods.

---

2. If 500 grams of a radioactive material were present at noon on Wednesday and 125 grams were present at noon the following Monday, what is the half-life of the material?

Half-life of the material would equal

_____.

Measuring Radiation

There are three units used to quantify radiation emitted by a radioactive element: the curie, rad, and rem. These units are used by chemists, physicists, and medical personnel to describe the characteristics of an element or to determine energy deposited in living tissue.

A piece of radioactive material contains many nuclei undergoing decay. One method of characterizing radioactivity in an element is to state how many decays occur in a given time.

This number, called the activity of the given piece of material, is measured in curies. The *curie* is the unit of measure for all radioactive materials. It is equal to an atomic nucleus that undergoes 37 billion radioactive disintegrations per second, which is approximately the activity of one gram of pure Radium 226.

Radioactivity is measured in curies or fractional curies. Consumer products typically have activity levels measured in millionths of a curie. Smoke detectors, for example, often use small amounts of radioactive materials and give off minute amounts of radiation. On the other end of the spectrum, a medium-sized nuclear bomb produces approximately 100 million curies of Iodine 131. The curie is simply a measure of the property of a source of radiation. This measurement provides no information about the objects exposed to this radiation.

Scientists use the unit of measure known as a rad to provide information on the effect that radiation has on an object. The *rad*, radiation

Half-Life of Various Elements

Barium-140	12.8 days
Carbon-14	5,730 years
Cerium-141	32.5 days
Cerium-144	590 days
Cesium-137	30 years
Iodine-129	17 million years
Iodine-131	8 days
Plutonium-239	24,400 years
Plutonium-242	379,000 years
Radium-226	1,600 years
Rhodium-103	57 minutes
Ruthenium-106	1 year
Strontium-89	54 days
Strontium-90	29 years
Uranium-235	704 million years
Uranium-238	4.5 billion years

TABLE 7-1   The half-life of various elements is illustrated.

*a*bsorbed *d*ose, describes the amount of energy the radiation has deposited in or on an object. Since different types of radiation vary in their ability to penetrate tissue and alter cell structure, the rad is weighted for the type of exposure an organism receives. For example, a rad of exposure from beta particles would produce a greater biological effect than one from a rad of X rays alpha particles. This exposure is measured in the rem. The *rem* is independent of the type of radiation that an organism is exposed to. Thus, a rem of exposure would produce a constant biological effect, regardless of the radiation type. Often more meaningful when discussing low-level exposure of radiation is the *millirem* (mrem)—which is equal to 1/1000 of a rem.

Scientists and health-care providers often use the term *person-rem* when measuring the nucleation exposure of a population. The *person-rem* is defined as the cumulative doses to each of the exposed individuals. The assumption is that there is a linear relationship between dose and biological effect. For example, 50 persons exposed to 2 rem each would be 50 person-rem. This effect is assumed to be identical to one person receiving 50 rem, which would also represent 50 person-rem. Table 7-2 summarizes the rem, rad, and curie.

Background Radiation

Each year the average United States citizen is exposed to about 360 mrem of radiation. This type of radiation is typically called *natural radiation* or *background radiation*. About 80 percent of it comes from the sun, outer space, the atmosphere, the earth, and other natural sources. Altitude affects the amount of background radiation exposure. People living in the mountains, at 10,000 feet, absorb more

Units of Radioactivity and Radiation

Units	What is being measured
curie (Ci)	Activity; rate of radioactive decay in a sample of radioactive material (1 Ci = 37 billion decays per second).
rad	Energy that is absorbed in material exposed to radiation.
rem	Radiation dose in terms of its net biological effect. (A rem of exposure would be expected to produce a constant biological effect, regardless of the type of radiation.) For low-level exposure, the millirem is often used (1 mrem = 1/1000 of a rem).

TABLE 7-2 The various units of radioactivity and radiation are illustrated.

than twice the radiation of people living at sea level. Airplane travel at 40,000 feet from coast to coast will also add 5 mrem of radiation exposure to an individual.

The remainder of background exposure comes from human-made sources. Of these, the largest contributor is medical X rays. One full set of dental X rays or one chest X ray can add about 55 mrem to one's yearly exposure to radiation. Other sources of human-made radiation are color televisions, microwave ovens, military bomb tests, and electrical power plants. Note that coal-fired power plants also emit radioactivity because of the presence of radioactive materials in coal. Figure 7-11 illustrates the percentage contribution of various radiation sources to United States citizens. For more information, refer to Appendix I, Radiation Exposure in the United States.

Mass and Energy

A great amount of thermal energy can be extracted from a very small quantity of uranium fuel. This is a function of the relationship between mass and energy. The thermal energy that can be taken from atoms is indicated in a formula developed by Albert Einstein in his *theory of relativity*. It states that: $E = mc^2$.

Einstein's equation ($E = mc^2$) theorizes that energy (E) equals mass (m) multiplied by the square of the speed of light (c) in a vacuum. The equation suggests that mass (m) and energy (E) are merely two different forms of the same physical reality. They are interchangeable, which means that all mass can be transformed into energy and energy can be transformed into mass. In a nuclear power plant, mass is converted to thermal energy. Theoretically, neither matter nor energy can be created or destroyed. Since light travels at almost 186,000 miles per second, it becomes apparent from the formula that a small amount of mass would create an incredible amount of energy.

When a conversion from mass to energy happens in a nuclear reaction, it is expressed in millions of electron volts (MeV), the unit of energy used in nuclear physics. An electron volt is a unit of energy equal to that gained by an electron in passing through a difference in potential of one volt (one electron volt is equal

Percentage Contribution of Radiation to the Total Average Effective Dose (United States)

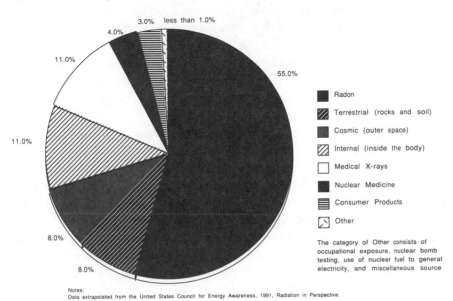

FIGURE 7-11 The percentage contribution of various radiation sources to United States citizens is shown in this figure. Note that 55 percent of exposures are a result of radon.

Radon
Terrestrial (rocks and soil)
Cosmic (outer space)
Internal (inside the body)
Medical X-rays
Nuclear Medicine
Consumer Products
Other

The category of Other consists of occupational exposure, nuclear bomb testing, use of nuclear fuel to general electricity, and miscellaneous source

Notes:
Data extrapolated from the United States Council for Energy Awareness, 1991, Radiation in Perspective.

to 4.451×10^{-26} kWh or 10^{-6} MeV). For example, if one kilogram of ^{235}U were completely fissioned, 200 MeV would be produced.

Fusion and Fission

Energy can be released from atoms by either a fusion or fission process. *Fusion* is defined as the joining of nuclei to form heavier atoms. Again, it was found that when two atoms were joined, the resulting mass was slightly less than the mass of the former two atoms. It was then realized, as defined by Einstein's theory of relativity, that the difference in mass was converted into thermal energy.

It was also found that if an atom's nucleus were to be split apart, a process called *fission*, the parts weighed less than the whole. The difference in weight, again, was converted to thermal energy. From this research it was found that there are two types of nuclear energy:

Fusion: binding together to convert mass to energy
Fission: splitting apart to convert mass to energy

Fusion

Today, all nuclear powered systems work on fission, not fusion, principles. Nuclear fusion is still under development and could become a commercial source for producing electrical energy in the future—around 40 to 75 years from now. However, in order for such a system to be commercially available, several obstacles must be overcome. These include joining two light nuclei to form a heavier one, heating the fusion fuel particles to extremely high temperatures (near 100 million degrees Kelvin), and keeping the reaction contained so as to maintain the fusion process. Overcoming all the obstacles simultaneously has proven to be extremely difficult for scientists. Particularly, trying to obtain these high temperatures in a controlled situation and maintaining the temperatures in a confined area has been difficult.

Fission

The nucleus of an atom is held together by extremely concentrated nuclear forces. This fact is the basis for all nuclear fission processes. Nuclear fission was not possible until scientists and technologists found that energy could be released by splitting the nucleus and causing a chain reaction to occur. Uranium is used as a fuel because it is one of the few known elements in which nuclear fission induces a self-sustaining chain reaction, which in turn will cause more output energy than input energy.

Figure 7-12 illustrates how the process of nuclear fission occurs. First the ^{235}U nucleus (shown on the left) must be struck by a neutron bullet. The nucleus then splits into two nuclei (e.g. krypton and barium, as shown) of medium atomic weight, releasing in the process vast amounts of thermal energy. This release of energy is equivalent to about 190 MeV. When the nucleus splits, it also emits two or three more neutrons, which in turn can penetrate neighboring ^{235}U elements. This can, in effect, cause further fission, more thermal energy, and freer neutrons. Thus, a self-sustaining reaction begins. Because the fission process happens in less than a millionth of a second, enormous thermal energy output can be achieved quickly.

Note that the speed of these free neutrons is considered slow; therefore, they are called *slow neutrons*. These neutrons have a lifetime in the chain reaction between 0.6 and 80 seconds.

Fission Reaction

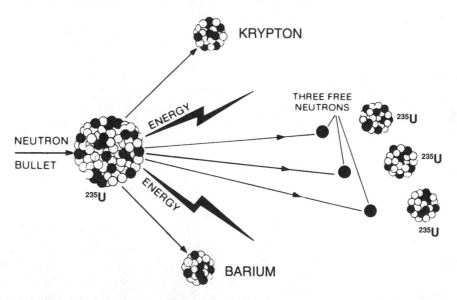

FIGURE 7-12 When a neutron bullet strikes a ^{235}U atom, it splits in two and releases thermal energy. Other neutrons are also released to help continue the fission process.

Since these neutrons arise from the decay of fission products, if this reaction is left uncontrolled, the resultant energy released is similar to that of an atomic bomb.

The two medium-weight atoms that have been created in our example are krypton (Kr) and barium (Ba). Traces of other radioactive materials are also created in the fission process. The total atomic weight of barium and krypton plus the other materials is equal to less than the original weight of the ^{235}U. Again, the difference in weight has been converted to thermal energy during fission. However, fission occurs in only one out of every 139 uranium atoms because only one in every 139 uranium atoms is ^{235}U. The rest consists of other isotopes of uranium, primarily ^{238}U. If all the atoms in 1 pound of ^{235}U were split or fissioned, the energy released would be equivalent to that from burning 1,500 tons of coal (see Figure 7-13).

Materials that are fissionable are said to *fissile*. There are only a few fissionable elements. These include ^{235}U, ^{239}Pu, ^{233}Pu, and ^{233}U. Plutonium 239 and 233 are artificial elements. Uranium and plutonium are the elements most widely used in nuclear reactors.

As noted earlier, uranium occurs naturally in the earth as a combination of ^{238}U and ^{235}U. Of the uranium mined, 99.28 percent is ^{238}U while only 0.72 percent is the fissile ^{235}U used as a fuel for nuclear reactions. If reactors today and in the future continue to use ^{235}U as a fuel, it is possible that someday in the future, there could be a shortage of this uranium. Depending upon usage rates, ^{235}U is anticipated to be available (as a finite fuel) between 100 and 200 years.

Under certain conditions ^{238}U can be fissioned; however, it requires a great deal more energy to induce fission. ^{238}U also needs faster

Energy Equivalency

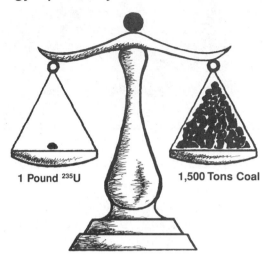

1 Pound ^{235}U   1,500 Tons Coal

FIGURE 7-13   On average, if one pound of ^{235}U were completely fissioned, it would produce the equivalent amount of thermal energy derived from 1,500 tons of coal.

neutron bombardment. The technology needed to speed up neutrons can be accomplished; however, speeding up neutrons is uneconomical to produce energy. Thus, ^{238}U is generally not used directly for fission.

Btu's and Uranium—Math Interface 7-2

To understand better how much energy is in ^{235}U, solve the following problems:

1. A ton of uranium removed from the earth has an abundance of 0.72 percent of ^{235}U. How much energy is released if this is fissioned in a nuclear reactor? Refer to Appendix L, Energy Content of Fuels, to obtain this answer.

   Btu's in the ton of uranium would equal
   _____.

Nuclear Breeding

Breeding Fuel

There is a way, however, that ^{238}U can be used in a nuclear power plant. It is called *breeding fuel*. When a slow neutron strikes the ^{238}U atom, the atom does not split but absorbs the neutrons. This causes a very unstable ^{239}U element. Because of its instability, the ^{239}U atom quickly changes to a new or artificial plutonium atom called ^{239}Pu. The ^{239}Pu atom then behaves much like the ^{235}U atom. ^{239}Pu is fissionable, and can be used in the nuclear reaction. Figure 7-14 illustrates graphically how this happens. As a neutron strikes the first ^{238}U with 92 protons and 146 neutrons, it instantly turns into an unstable ^{239}U with 92 protons and 147 neutrons. Then, it immediately turns into ^{239}Pu with 94 protons and 145 neutrons.

Figure 7-15 shows how three more neutrons are emitted when a free neutron strikes a ^{235}U atom. One neutron continues the chain reaction with ^{235}U while the other two neutrons convert ^{238}U, which will not fissile, into ^{239}Pu, which is fissionable. In this series of reactions, more nuclear fuel is produced. Thus, the term *breeder* is used. This principle is used in the development and operation of breeder reactors.

Notice there were three ^{235}U atoms initially, but six ^{239}Pu atoms were produced (or bred). Again, this means more fuel is developed than consumed by the fissioning process. These ^{239}Pu atoms can then be used in a continued nuclear fission process. Research suggests that, in the future, ^{239}Pu could be removed and used in other reactors.

A second fuel, thorium (^{232}Th), behaves much the same as ^{238}U. ^{232}Th will absorb any slow neutrons and be converted into an isotope of uranium—^{233}U. This isotope of uranium (i.e., ^{233}U) is also fissionable and can be used as a nuclear fuel. Note, however, that thorium is not as plentiful naturally as ^{235}U.

Breeder Reactors

Today, the United States has an ongoing research program to learn more about breeder

Development of Plutonium

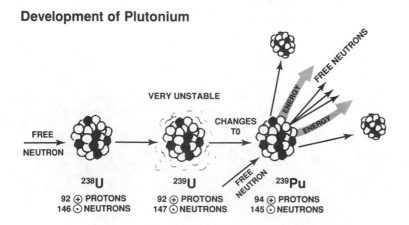

FIGURE 7-14 When ^{238}U is struck by a neutron bullet, it converts to ^{239}U. The ^{239}U will immediately change to a synthetic element, ^{239}Pu, which behaves much the same as ^{235}U if hit by a free neutron.

Breeding of Nuclear Fuel

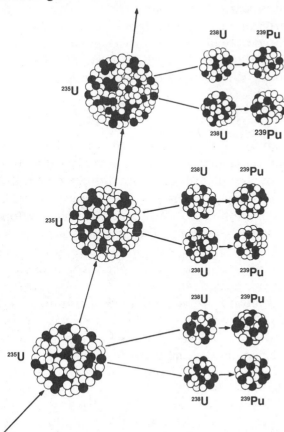

FIGURE 7-15 When nuclear fuel is bred, more ^{239}Pu is produced than there was ^{235}U to begin with.

reactors. However, depending upon public opinion and interest by the federal government, money for research has often been lacking. The United States has not yet developed commercially usable breeder reactors. It did however, have a demonstration plant in Tennessee. In addition, several nations with large nuclear programs have also invested research money to develop breeder-reactor technology. France, the United Kingdom, and the former Soviet Union have already complet-ed demonstration plants and are looking forward to developing commercial breeder reactors in the future.

Enrichment

The content of ^{235}U in natural uranium samples cannot produce a continuous self-sustaining reaction. This is because there are not enough ^{235}U atoms in the fuel. Therefore, uranium fuel is enriched with more ^{235}U so that a self-sustaining reaction can occur. This process, called enrichment, is detailed later in this chapter.

Moderators and Controls

A second method of increasing the likelihood of a self-sustaining reaction is by controlling the speed of the bombarding neutrons. Neutrons in a fission reaction travel at approximately 9,980 miles per second. When only 1 in 139 atoms is ^{235}U, the chances for fission to take place are low. If the neutron could be slowed to travel at 0.6 to 1.2 miles per second, the chance of hitting a ^{235}U atom would be increased significantly. With this fact in mind, researchers have used moderators to slow down the neutrons' speed without affecting the neutrons themselves.

Control of the fission process is also essential. Certain materials are used in nuclear reaction processes to absorb the excess neutrons, thus stopping any chain reaction at the appropriate time. These materials are usually in the form of rods, called *control rods*. The control rods are inserted into a nuclear core to slow down the rate of a nuclear reaction. They are withdrawn from the reactor when an increase in the output is needed. The materials most often used for moderators and control rods are listed in Figure 7-16. Common moderator

Materials Used as Moderator or Control Element in Nuclear Reaction

Moderator	Control
1. Light Water	1. Any Moderator
2. Heavy Water	2. Cadmium, Silver, and Indium Alloys
3. Graphite	3. Boron Mixtures
4. Beryllium	

FIGURE 7-16 These are examples of materials used as moderators and as control rods in a nuclear reaction.

materials include water, graphite, and beryllium. Control rods can be made from any moderator or from cadmium, silver, indium alloys, and various boron mixtures.

NUCLEAR FUEL CYCLE

Before it is finally used within a nuclear power plant, nuclear fuel undergoes several major processes collectively called the *nuclear fuel cycle*. Although this cycle varies with the type of fuel used and the quality of the uranium ore mined, the basic processes are similar. The nuclear fuel cycle stages, processes, and prod-

ucts are shown in Figure 7-17. In the following section, mining, milling, UF_6 production, enrichment, and fuel fabrication will be presented in detail. Fuel reprocessing and disposal of spent nuclear fuel will be illustrated after presenting various types of reactor designs.

Mining Nuclear Fuel

Uranium in the earth's crust constitutes about four parts per million. This means that uranium is as plentiful as lead and more plentiful than silver or mercury. The amount of uranium commercially available for fuel development is based upon the price of the ore. As with other fuels, the higher the price, the greater the commercial incentive to explore and mine the fuel.

Uranium ore is a mixture of uranium and other materials. Approximately 0.25 percent (1/4 of 1 percent) of the ore is pure uranium. The remaining part is made of rock, soil, and so on. Therefore, to provide enough fuel for existing nuclear reactors, large volumes of uranium ore must be mined. A ton of uranium-bearing ore contains approximately 5.28 pounds of uranium oxide (U_3O_8). From this, about 0.027 pounds of ^{235}U can be obtained.

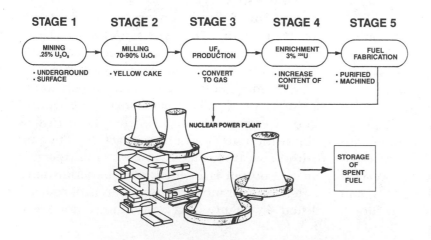

STAGE 1   STAGE 2   STAGE 3   STAGE 4   STAGE 5

MINING .25% U_3O_8 — MILLING 70-90% U_3O_8 — UF_6 PRODUCTION — ENRICHMENT 3% ^{235}U — FUEL FABRICATION

· UNDERGROUND   · YELLOW CAKE   · CONVERT   · INCREASE   · PURIFIED
· SURFACE   TO GAS   CONTENT OF   · MACHINED
                                              ^{235}U

NUCLEAR POWER PLANT

STORAGE OF SPENT FUEL

FIGURE 7-17 Each stage of the nuclear fuel cycle is shown in this figure.

There are two types of mining processes used to extract uranium ore from the earth: underground and surface mining. Shallow deposits of uranium ore are mined by surface, or open pit, mining; deeper deposits are mined through underground techniques.

Formerly, it was uneconomical for companies to explore and mine new reserves of uranium ore. The price of U_3O_8 was so low that there was little incentive even to mine the existing reserves. More recently though (in the past 20 years), uranium prices have increased because of the increased demand from nuclear power plants. Thus, more exploration and mining operations were developed. Unfortunately, it takes about eight years to explore, locate, and begin to mine uranium. There are, to recap, three factors that control the amount of uranium being removed from the ground. These include

1. the price of uranium ore;
2. improvements in exploration and mining technology; and,
3. the demand for uranium, based upon production of nuclear-fueled power plants.

Milling Uranium

Once mined, the second step in the fuel cycle is milling. The purpose of milling is to refine the uranium ore so that it can be shipped more economically to the next stage of the cycle. The ore is pulverized and processed to remove the uranium, which leaves behind vast piles of crushed rock known as *tailings*. The milling stage concentrates the uranium into approximately 70 to 90 percent U_3O_8. The concentrated uranium ore is called *yellow cake*. Figure 7-18 shows samples of yellow cake, which contains only about 0.7 percent (7/10 of 1 percent) ^{235}U. An operation called leaching dissolves out the uranium, which is then separat-

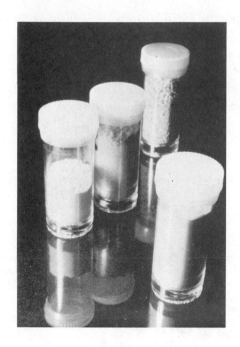

FIGURE 7-18 These are examples of the uranium concentrate called yellow cake (Courtesy of the American Petroleum Institute))

ed and solidified to make the final concentrated yellow cake.

UF$_6$ Production of Uranium

The next step in the nuclear fuel cycle is to convert the yellow cake into a gaseous state necessary for the lengthy and difficult *enrichment* process—separating the uranium isotopes that are chemically similar and that have nearly the same mass. The difference in mass, 235 versus 238, is the only distinguishable characteristic that an enrichment process can use to separate the two isotopes.

In the United States, uranium enrichment is done by *gaseous diffusion*. This process relies on the fact that lighter molecules in a gas move faster. To get uranium into a gaseous form, the yellow cake is combined chemically with fluorine to form uranium hexafluoride (UF_6). This gas is then passed through approximately 500 membranes. Since ^{235}U molecules move slightly faster than the more numerous ^{238}U and strike the membranes more often, more ^{235}U molecules will be deposited on the other side. Thus, this process can produce an enriched form of ^{235}U, as shown in Figure 7-19.

Uranium Enrichment

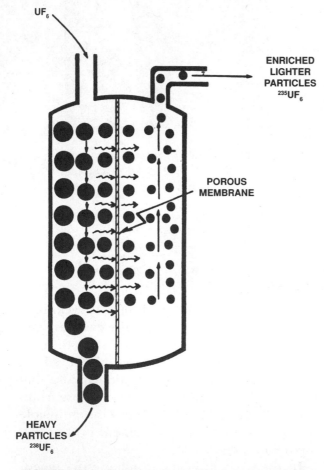

FIGURE 7-19 To concentrate ^{235}U in nuclear fuel, it must be sent through many stages of a gaseous diffusion process. One such porous membrane, shown above, stops the heavier ^{238}U from moving to the other side, thus enriching the fuel.

Gaseous diffusion became the United States' preferred process for uranium enrichment because of the need to produce bomb-grade uranium during World War II. Other countries throughout the world have developed other enrichment processes, such as the use of gas centrifuge chambers that spin at high speed to force heavier material outward, causing it to separate. Another process in development for enriching nuclear fuel is the use of laser enrichment. This process uses lasers to selectively remove electrons from atoms of ^{235}U but not ^{238}U. Once removed, the charged ^{235}U is then separated by electric forces. Note that the process of enriching nuclear fuels is politically sensitive since these fuels can be used to make nuclear weapons. While most nuclear power plants call for fuel that is enriched to about 3 percent, nuclear weapons may require enrichment of several thousand times this amount, depending upon the size and explosive power of the device. For nuclear power plants, a single 1,000 Mw reactor would require only approximately 6,000 tons of uranium for lifetime requirements.

Fuel Fabrication

Once there is enough ^{235}U present in the UF_6 gas, it is then processed chemically to make uranium dioxide (UO_2). This UO_2 is then formed into pellets about 0.5 inches long and 0.375 inches in diameter. Figure 7-20 shows an example of the simulated uranium pellets used in a nuclear power plant.

Each of the pellets (shown in Figure 7-20) contain the energy equivalent to approximately 150 gallons of gasoline or one ton of coal. Once manufactured, the pellets are loaded into zirconium-alloy tubes to form *fuel rods*. Fuel rods are then assembled into groups known as *fuel bundles* for delivery to nuclear power plants.

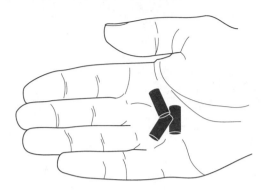

FIGURE 7-20 These are simulations of nuclear fuel pellets used in fuel rods.

TYPES OF REACTORS

Nuclear Reactor Variables

Reactors vary widely in their design. Several designs have been effectively tested and commercially used since the early 1940s. Just as there are different types of engines that power applications—such as jet, gasoline, diesel, and rotary engines—there are also different types of nuclear reactors used to produce power. The type of nuclear reactor used depends on several variables:

1. The purpose for which the reactor is to be used, for example, commercial electrical power, radiation research, or military applications.
2. The desired speed of neutron bombardment. Some use fast neutrons, which indicates there is no moderator; others use moderators to slow down the neutrons.
3. The type of moderator and control rods used. Graphite, light water, heavy water, beryllium, and other materials can be used as moderators. Graphite can also be used as a control rod material. Light water is filtered H_2O; heavy water is an isotope with an additional neutron in the hydrogen.

4. The coolant used within the reactor to carry the heat from the nuclear reaction to a heat exchanger. Light water, carbon dioxide, helium, and liquid sodium are the most popular coolants.
5. The type of fuel, such as ^{235}U, ^{233}U, ^{232}Th, or ^{239}Pu. Of these types of fuel assemblies, ^{235}U is the most common.
6. The arrangement of fuel within the nuclear reactor. It can either be isolated from or mixed with coolant.
7. Structural materials used for retention of fission products and radiation protection.

Given these variables, it is not surprising there are six types of nuclear reactors either being tested or in use today. They are categorized as shown in Figure 7-21.

Within the United States, nuclear power plants are of the light-water reactor type. High-temperature gas reactors and fast breeder reactors are still being researched for possible use.

Light Water

Light-Water Reactor (LWR)

The first type of nuclear reactor discussed is the light-water reactor (LWR) which forms the category to which the pressurized-water reactor (PWR) and the boiling-water reactor (BWR) belong. All LWRs are called thermal reactors because they use slow neutrons (or thermal neutrons) to maintain a controlled chain reaction.

The light-water reactor gets its name from the use of plain (or light) water, which is highly filtered and mineral free. Light water and heavy water are different because there is an extra neutron in the hydrogen nucleus of heavy water. For purposes of discussion, light water will be considered ordinary tap water.

The light-water reactor uses ^{235}U as a fuel, enriched to approximately 3 percent. Al-

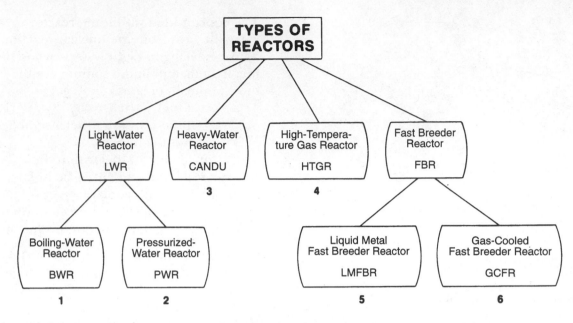

FIGURE 7-21   This illustration shows the categorization of all nuclear reactor types.

though this is its major fuel, the ^{238}U atoms also contribute to the fission process by converting to ^{239}Pu—about one-half of which is consumed within the reactor. Light-water reactors are generally refueled every 12 to 18 months. At which time, about 25 percent of the fuel is replaced.

Light-Water Reactor Efficiency

The overall energy efficiency of nuclear power plants is determined by the ratio of electricity output to total thermal energy produced in the nuclear reaction process. Light-water reactors have efficiencies of about 32 percent, as compared to 38 to 40 percent for more updated fossil-fuel plants.

Fuel Rods in the Light-Water Reactor

As noted, the fuel in light-water reactors is made of enriched uranium oxide (U_3O_8) pellets. The pellets are fitted into metal tubes to form rods running the length of the reactor core. Figure 7-22 shows how the fuel rods are stacked with nuclear fuel pellets. The fuel rods are combined into a fuel assembly. The fuel assemblies are then packaged into the reactor core. Holes are placed in the reactor core to allow room for the control rods to be inserted. The coolant then flows through these fuel assemblies in the reactor vessel to pick up the heat generated from the nuclear reaction.

Boiling Water

Boiling-Water Reactor (BWR) Operation

The boiling-water reactor, illustrated in Figure 7-23, is one of the simplest designs. The control rods are made of a moderator to slow down the neutrons. The control rods are hydraulically controlled and move in or out of the core to change the output of the nuclear reactor. In the BWR the moderator and coolant used is light water.

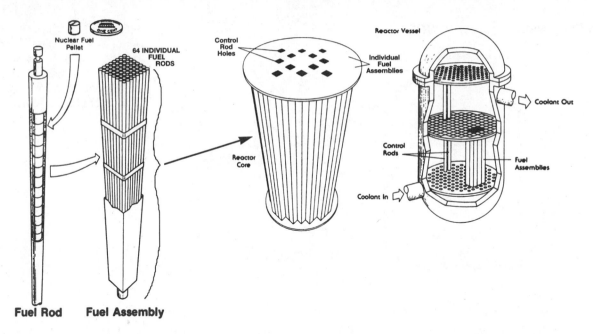

FIGURE 7-22 This illustration shows the placement of the fuel pellets into the fuel assembly, and the fuel assembly into the reactor vessel.

When the nuclear reaction begins, the water is heated to nearly 545° F, under about 1,000 psi. In the direct-cycle coolant system, water boils on its way through the core. This results in steam that is piped directly to the turbine. The turbine exhaust is then condensed and pumped directly back into the pressure vessel to be reevaporated.

The steam turbine technology in a nuclear reactor is very similar to that of any other turbine currently used to produce electricity. If the reactor is to produce commercial electricity, then the thermal and mechanical energy in the steam is extracted by the turbine, which turns an electrical generator to produce electricity. If the purpose of the reactor is for a marine application, then the turbine turns a propeller.

After thermal energy has been extracted by the turbine, the coolant is further reduced in temperature by a condenser to convert the remaining steam into a liquid. The water used

in this condenser is usually taken from a near-by river or lake, cleaned, demineralized, treated, and sent into the reactor for use. The condensing action is accomplished by using cooling towers as shown in Figure 7-24.

In the condensing process, the thermal energy that is removed is normally pumped back into the river or lake, which has been a concern for environmentalists. Both plant and animal life can be affected by slight increases in water temperature. This *thermal pollution*, as it is called, could be put to efficient use in greenhouses and special heating applications in waste heat recovery projects.

The cooled water from the condenser or cooling tower is then pumped back into the reactor to have the cycle repeated again. Several isolation valves are incorporated in this reactor to shut off the steam in the pressurized lines if a rupture should occur. One major concern of the BWR is the accidental loss of

Boiling-Water Reactor (BWR)

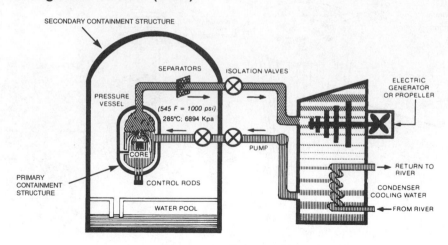

SECONDARY CONTAINMENT STRUCTURE

SEPARATORS   ISOLATION VALVES

PRESSURE
VESSEL

(545 F = 1000 psi)
285°C; 6894 Kpa

ELECTRIC
GENERATOR
OR PROPELLER

CORE

PUMP

PRIMARY
CONTAINMENT
STRUCTURE

CONTROL RODS

WATER POOL

RETURN TO
RIVER

CONDENSER
COOLING WATER

FROM RIVER

FIGURE 7-23 In the boiling-water reactor, water is heated by the fission of radioactive material. The thermal energy in the water is then sent to the turbine to produce electricity.

coolant from a ruptured high-pressure line. Aside from the loss of the coolant, which could cause the core to melt down, the prospect of radioactivity escaping from the coolant is also a danger.

The nuclear reactor is designed to have both a primary and secondary containment structure, secondary or back-up lines, secondary valves, extensive water flooding systems, and pressure regulators to reduce the possibility of a meltdown. Almost one-third of the reactors operating in the United States are BWRs.

Size of the Boiling-Water Reactor

The reactors are quite large, averaging 72 feet in height and 19 feet in diameter. Vessel walls are more than 6 inches thick, and the containment structure weighs about 700 tons. An example of a boiling water reactor power station is shown in Figure 7-25.

Pressurized Water

Pressurized-Water Reactor (PWR) Operation

The coolant in this reactor is ordinary filtered water under very high pressure (about 2,250

psi). The difference between the boiling-water reactor and the pressurized-water reactor is that the latter has both primary and secondary cooling loops, which transfer the thermal energy from the reactor core to the output tur-

FIGURE 7-24 Cooling towers, such as the one illustrated, are used to cool and condense the steam back into a liquid. (Courtesy of Middle South Services, Inc.)

FIGURE 7-25 This boiling-water reactor is located in Monticello, Minnesota, and owned by Northern States Power Company. (Courtesy of Northern States Power)

bine. Figure 7-26 shows a diagram of the internal design of the pressurized-water reactor.

In this system, primary water enters the reactor vessel through nozzles near the middle or top of the pressure vessel. The water then flows down an annular passage between the pressure vessel and a sleeve called the core barrel, to the bottom of the reactor core. This water is then forced to reverse its direction and flow up through the fuel bundles, where it is heated by fissioning. The fissioning heats the water to 600° F, but the water does not boil since it is under high pressure. This heated water exits the core from nozzles near the middle or top of the pressure vessel and flows to the steam generators. Transfer of the generated heat to the secondary coolant water happens through the thin walls of the tubes in the steam generator. The water is then cooled in the primary loop and is pumped back to the core to repeat the cycle. The secondary loop water, heated by the transfer by the steam generator, is turned to steam which is used to spin the turbine. The torque that is created produces electricity in a generator or turns a propeller in a large nuclear-powered ship.

Size of the Pressurized-Water Reactor
The pressurized-water reactor is more compact than the boiling-water reactor. Inside the reactor vessel, the reactor core is between 6 and 10 feet in diameter and between 8 and 15 feet in height. The overall structure of the vessel may be up to 30 feet high and 20 feet in diameter.

Pressurized-Water Reactor (PWR)

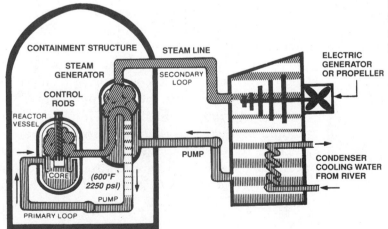

FIGURE 7-26 In the pressurized-water reactor, the nuclear reaction heats water to steam. The thermal energy in the steam is transferred to a secondary loop by the steam generator. This steam is then used to turn a turbine, generator, or propeller.

The control rods in the pressurized-water reactor are operated from the upper part of the reactor vessel. Figure 7-27 shows an open reactor core during refueling.

Again, as in the boiling-water reactor, all of the systems used to operate and allow the nuclear plant to function have both a primary and secondary, or backup system. These redundant systems are designed as a safeguard against failure and possible meltdown. Almost two-thirds of the commercial nuclear power reactors in the United States are PWRs.

Heavy-Water Reactor

Another type of nuclear reactor used in some countries is the heavy-water reactor (HWR). Heavy water can be used as a moderator

FIGURE 7-27 The reactor core is used to hold the fuel assemblies. (Peter Menzel/Stock, Boston)

instead of light water because it absorbs fewer neutrons. Unlike common water (H_2O), heavy water (^2H_2O or D_2O) has a hydrogen isotope whose nucleus contains a proton and a neutron. This isotope is known as **deuterium**. The hydrogen isotope has, therefore, a mass twice that of its neutron. Heavy water is, consequently, inefficient at absorbing neutron energy. By using heavy water in a nuclear reactor there is an abundance of neutrons. This allows nuclear fission to occur using unenriched uranium. Several countries—including Canada, India, and Romania—have developed HWRs. Overall, HWRs offer several advantages over PWRs and BWRs, including the use of lower operating pressures and temperatures, which somewhat reduces the high construction costs associated with nuclear plants. However, the reduction in pressures and temperatures reduces the overall efficiency of HWRs to about 29 percent. Another economic advantage of HWRs is the refueling process. In LWRs the reactor must be partially disassembled before refueling can occur. In some HWR designs, most notably the Canadian CANDU design, the refueling process can occur without disassembly of the reactor.

High Temperature

High-Temperature Gas Reactor (HTGR)

One of the main disadvantages of the light-water reactor is the limit of high temperatures generated within the nuclear reactor. If the temperatures within the reactor core could be increased, then more thermal energy could be extracted from the nuclear reactor. The result would be increased efficiency from the power plant. To this end, several companies have developed the high-temperature gas reactor (HTGR) and tested it within the United States and France. Today, no HTGRs are in operation in the United States. Once there was an HTGR in the United States, the Peach Bottom 1 plant in Peach Bottom, Pennsylvania. This nuclear power plant opened in June of 1967 and generated 40 MWe in conjunction with the Philadelphia Electric Company. This plant was closed in 1974. Another HTGR was at the Fort St. Vrain facility in Platteville, Colorado. This plant produced 330 MWe but was officially closed in August of 1989.

HTGR Operation

While the United States does not have an HTGR currently in operation, the plants previously in operation provided important research information for the future. HTGRs are different from LWRs in two significant ways. First, as its name implies, the HTGR is cooled, not by water, but by a gas. Helium is pumped through the core to extract the thermal energy generated by the fission process. The helium then travels in a loop to a steam generator, where the heat is transferred to a secondary loop of water. This secondary loop and heat transfer system are similar to a PWR.

The HTGR core is also quite different from that of an LWR. In the HTGR the moderator is a graphite block in which the fuel rods are placed in a configuration that will permit a critical chain reaction. About 2,000 fuel rods are needed to fill a single fuel element (fuel assembly); as many as 800 fuel elements may be used in one reactor. (Figure 7-28 shows a cutaway drawing of the internal parts of the high-temperature gas reactor.) The graphite block also contains holes so that the helium can pass through to be heated. In this system the he-

High-Temperature Gas Reactor

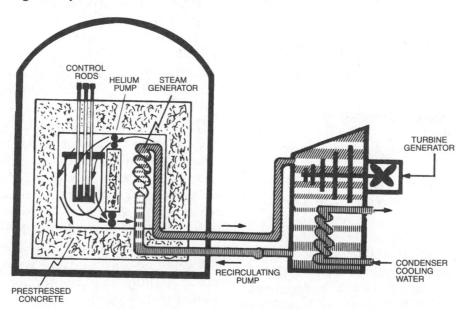

FIGURE 7-28 The high-temperature gas reactor uses helium as the fluid to transfer the thermal energy to the water in the steam generator.

lium is pressurized at 700 psi and enters the core at about 645° F. The heated helium leaves the core about 1380° F, which makes the HTGR a high-thermal efficiency reactor.

The HTGR fuel mixture is also different from that of LWRs. The HTGR's fuel assemblies contain both fissionable ^{235}U and fertile ^{232}Th. As the ^{235}U fissions, some neutrons are absorbed by the ^{232}Th. The ^{232}Th is then converted to fissionable ^{233}U which can be recovered and used as a future fuel.

Size of the HTGR

The reactor core is housed in a carbon-steel vessel that is 35 feet high and 14 feet in diameter. The reactor vessel weighs approximately 150 tons because of the large amount of concrete used to shield the outside environment from radiation.

Advantages of the HTGR

Generally, the advantages of this type of reactor design compared with other LWRs are

1. They can produce nearly 20 percent more electrical energy per equal reactor size.
2. They will require about 30 percent less cooling water.
3. They release correspondingly less heat to the environment.
4. They use 40 percent less uranium ore over their lifetime.

With the HTGR, therefore, efficiency is increased, radioactivity is reduced, and the cost of the nuclear plant will be lower. However, much research still needs to be completed on the design, operation, maintenance, and capacity. Of the two plants that were in operation, neither had met the expectations of its designers. However, newer HTGR designs are benefiting from the experience gained from these previous plants. Some nuclear engineers expect that advanced HTGRs will become safer and more reliable than the LWRs currently in service.

High-Temperature Gas Reactors— Energy Issue 7-4

The HTGR may have advantages over boiling-water and pressurized-water reactors. After researching the HTGR further, identify why power companies have not constructed more of these power plants. What public, political, and environmental concerns need to be addressed to allow this technology to be developed?

Breeder

Fast Breeder Reactor (FBR)

The last category of nuclear reactor to be considered is the fast breeder reactor. Research and development have occurred on two types—the liquid metal fast breeder reactor (LMFBR) and the gas-cooled fast breeder reactor (GCFBR).

The term *fast reactor* indicates there is no moderator to slow down the neutrons. This is in contrast to a *thermal reactor*, which uses slow neutrons. When using fast neutrons, the core of the reactor can be smaller.

The term *breeder* indicates that new fuel will be generated or bred during the nuclear reaction. Most breeder reactors breed fuel at a ratio of approximately 1:1.2 to 1:1.5. The ratio means that for every ten uranium atoms fissioned, there will be about twelve to fifteen new plutonium atoms created. If a breeder reactor operated for 10 to 15 years, enough ^{238}U would be converted to ^{239}Pu to replace the original plutonium in the reactor and provide another reactor of equivalent size with its needed fuel. The result of breeding fuel is that less uranium would have to be mined; therefore, breeder reactors could be used to reduce dependence on foreign oil supplies to generate electricity. Also, the use of breeder reactors could reduce the overall cost of producing nuclear energy because of their high operating efficiencies.

On the negative side, if the excess plutonium bred over the next 50 years ever found its way to terrorists, world security could be threatened. Sabotage will always remain a serious concern when breeding nuclear fuel in any country.

Breeder Reactor Efficiency
Nonbreeding reactors use only 1 to 2 percent of the potential energy in the uranium. Breeder reactors, however, enable approximately 60 percent of the energy from the uranium resource to be converted to usable thermal energy. This produces an overall plant efficiency of about 40 percent as compared to 32 percent for LWRs and 38 percent for HTGRs.

Liquid Metal Fast Breeder

Liquid Metal Fast Breeder Reactor
The first breeder reactor designed was the Experimental Breeder Reactor I, or EBR-I, as it was called, at the Argonne National Laboratory in Idaho. While this was also the first reactor to produce electricity from nuclear power, the scientists working on this experiment were confident that it could create more fuel than it consumed. EBR-I was operated until 1963, when EBR-II became operable and helped to develop fuels and components for more advanced breeder reactors. The United States does not have a commercial LMFBR currently in operation.

One major difference between the liquid metal fast breeder reactor and other reactors is that besides its breeding of fuel, the reactor also uses a liquid sodium coolant. Liquid sodium has tremendous heat-carrying capabilities when compared to water and helium. Sodium melts at 210° F and boils near 1,600° F.

Using liquid sodium meant that new material technology had to be developed to avoid safety problems. Because it is highly corrosive, liquid sodium reacts adversely with the coolant circuits, pumps, pipes, heat exchangers, and instruments.

Fuel for the LMFBR
The fuel that is used in the LMFBR is a combination of ^{235}U and ^{238}U. The ^{235}U in the central core is used to start the chain reaction. Around the core is a blanket of ^{238}U. Any neutrons escaping from the core are absorbed by the ^{238}U and instantly converted to ^{239}Pu, which is separated chemically and used in other fast neutron type reactors.

Operation of the LMFBR
A diagram of the liquid metal fast breeder reactor is shown in Figure 7-29. The thermal energy that is created by fission is transferred to liquid sodium in the primary sodium loop. Since sodium becomes radioactive in passing through the reactor core, an intermediate heat exchanger is used to transfer the thermal energy to the secondary sodium loop, which is nonradioactive. This liquid sodium then transfers the thermal energy into a water system through a second heat exchanger or steam generator. The steam is then used to operate a steam turbine as in other electrical generating power plants.

Advantages of the LMFBR
Sodium can be heated to higher temperatures than water or helium, resulting in improved heat-transfer efficiency. At such high temperatures, pressure can remain lower to permit the

Liquid Metal Fast Breeder Reactor (LMFBR)

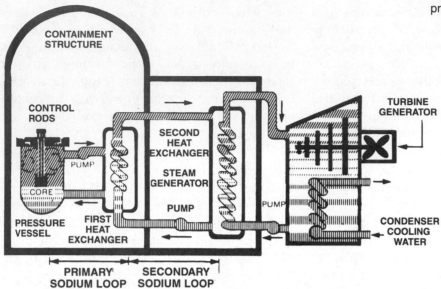

FIGURE 7-29 In the liquid metal fast breeder reactor, more fuel is produced than consumed.

use of low-pressure cooling circuits, which are less prone to failure. However, the major advantage of the LMFBR is that ^{238}U used in the core blanket is nearly 140 times more plentiful in nature than ^{235}U. Thus, using a LMFBR reduces the rate of exhausting the nonrenewable resource of ^{235}U and ensures a fuel for future nuclear power reactors.

Gas Cooled

Gas-Cooled Fast Breeder Reactor (GCFBR)

There are several technical options for the future in the nuclear energy field. One is the development of the gas-cooled fast breeder reactor. Although still in the experimental stages, this reactor is designed to produce a breeding ratio of 1:1.45. Gas-cooled fast breeder reactor development is currently supported by the General Atomics Company, the United States Department of Energy, and the GCFBR Utility Program Group (comprising over 50

utility companies, 70 rural electric cooperatives, and three European utility companies).

This type of reactor initially will use the built-up inventory of plutonium and depleted uranium discharged from light-water reactors. Therefore, current stockpiles of spent fuel rods will be retrievable. The goal is to develop a self-generated fuel cycle. The supplies of fuel made from breeding will be used for approximately three smaller-sized high-temperature gas reactors.

The reactor coolant is helium, which provides numerous advantages. The first is that helium does not slow down neutrons. It is also not as susceptible to radioactivity, eliminating the secondary heating loop used on the LMFBR. No corrosive problems exist, as with the use of liquid sodium in the liquid metal fast breeder reactor. Helium gas is also chemically inert, which means that it does not mix well with other chemicals nor is it affected by such mixing. Thus, only a small amount of dangerous chemical activity will take place. Consequently, reactor maintenance procedures are simplified and safety margins are improved with the gas-cooled fast breeder reactor.

If a rupture does occur in a helium pressure line, only the pressure on the helium will be reduced. Helium will not be lost, and circulation of low-pressure helium or even air within the reactor still provides adequate emergency core cooling.

The gas-cooled fast breeder reactor will be housed in a massive steel-lined, concrete vessel. The core and the entire reactor coolant system, including steam generators, will be contained within this structure. The steam and water for the turbine are the only components of the system that will enter and exit from the concrete vessel.

GCFBR Operation

The overall flow, temperature, and pressure sequence of the GCFBR are shown in Figure 7-30. As the helium is pushed downward through the reactor core by the helium circulator, it increases in temperature to approximately 1,022° F. This high-temperature helium is then sent directly into the upper part of the steam generator. This action increases the temperature of the water on the turbine side to nearly 900° F with a pressure of 1,178 psi. As the heated helium passes farther down the steam generator, a second heat exchanger captures steam to run the helium circulator, improving the overall plant efficiency. The heated steam is then sent out of the concrete reactor vessel and into a steam turbine. At the turbine, it is condensed and pumped back into the generator to repeat its cycle.

Advantages of the GCFBR

The economic advantages of the GCFBR are twofold. First, there is an increased efficiency over the LMFBR due to the higher operating temperatures of the GCFBR. Second, resulting from the use of helium, the system compo-

Gas-Cooled Fast Breeder Reactor (GCFBR)

FIGURE 7-30 This is a diagram of a gas-cooled fast breeder reactor showing circuits, reactor core, and the turbine and generator.

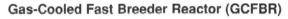

610°F 8996Kpa (1300 psi)

680°F 9272 Kpa (1344 psi)

RESUPER HEATER

1022°F

900°F 8128 Kpa (1178 psi)

TURBINE GENERATOR

890°F 19.992 Kpa (2900 PSI)

STEAM GENER-ATOR

410°F

CIRCULATOR TURBINE

HELIUM TURBINE

CONDENSER COOLING WATER

nents do not have to be designed to resist the corrosive effects of using sodium as in the LMFBR.

NUCLEAR WASTE

Overview of Nuclear Waste

During the operation of nuclear power plants, changes take place in the fuel during the fission process. Fragments of the fissioning process (pieces left over after the splitting of atoms) are left behind. These fragmented pieces are radioactive and, over time, reduce the efficiency of the chain reaction. Therefore, it is necessary to remove the oldest fuel assemblies, which have released their energy, and replace them with new fuel. About every 18 months in most commercial reactors, about one-quarter of the fuel assemblies must be replaced. These old assemblies (waste) must be disposed of in a safe manner to protect humans, other species, and the environment. The following section of this chapter presents issues related to the disposal of nuclear waste.

Types of Nuclear Waste

Nuclear waste is materials or products resulting from the nuclear process of fission. Examples are uranium, plutonium, and other materials (including cesium 137, krypton 85, strontium 90). These products fall into two categories as illustrated in Figure 7-31: low-level or high-level waste.

High Level

Volume of Waste from Refueling

A surprising aspect of nuclear waste is the small volume produced by a nuclear power plant. A typical plant produces about 30 tons of used fuel each year. This is about 3,000 tons of used

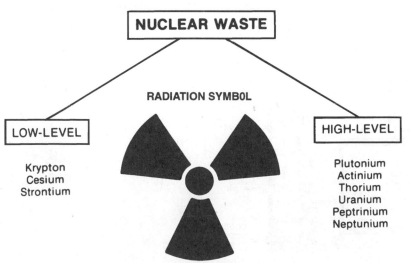

FIGURE 7-31 Two major categories of nuclear waste are low- and high-level wastes.

fuel every year for the approximately 100 plants in operation. This fuel is considered high-level waste. However, all the waste generated by nuclear power plants since their conception could cover an area the size of a football field to about five yards deep (this represents about 17,000 tons of spent fuel). By contrast, a 1,000 megawatt coal plant annually produces about 230,000 tons of coal ash along with tons of carbon dioxide and sulfur dioxides. The reason for the small volume of nuclear waste is that this energy form is so concentrated. However, it is important to point out that, although the volume of this material is not great, its toxicity is extremely high. One need only ingest a few micrograms of plutonium to cause death. Each year since 1980, commercial nuclear reactors have produced more than 26 million curies of high-level nuclear waste.

High-level waste includes such radioactive materials as plutonium, thorium, uranium, and others. Half-life for these can be as long as 25,000 years, and it may take 250,000 years before the radioactivity from these wastes decreases to a level safe to humans. High-level wastes are also a concern because of the heat generated by the radioactivity, which requires that it be stored and handled using unique methods. More will be presented on the storage of radioactive wastes later in this chapter.

The United States' inventory of spent fuel assemblies accounts for about 90 percent of the radioactivity of all the nuclear wastes generated. Unfortunately, the long-term effects of high-level waste contamination are not fully understood. However, research is continuing in this area.

Spent Fuel Assemblies
The spent fuel rods still contain usable uranium isotopes and must be managed correctly. Figure 7-32 illustrates some possibilities of

waste management of this high-level nuclear waste. Several choices for the disposal of spent fuel assemblies exist:

1. The fuel rods can be reprocessed to extract the uranium and plutonium still available. This collected fuel could then be used in the refabrication of new fuel assemblies.
2. The high-level wastes can be placed, temporarily, into a Monitored Retrievable Storage Facility.
3. The wastes can be placed directly into permanent waste storage.

If the spent fuel is put into temporary storage, it can either be reprocessed or put into permanent storage. Obviously, economic, political, and environmental considerations

Nuclear Waste Cycle

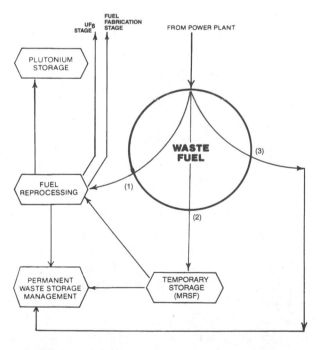

FIGURE 7-32 Several methods to manage nuclear waste are depicted in this illustration.

influence the decision where the waste will go. Both permanent waste disposal and reprocessing have many environmental implications. These aspects are presented in the following sections.

Reprocessing High-Level Wastes

In the reprocessing stage, the spent fuel assemblies are cut into pieces and dissolved in a pool of nitric acid. Once dissolved, this solution is fed to extraction systems to separate the plutonium from the uranium. This process may be repeated in numerous steps to remove 99 percent of the fissionable products. From the extracted uranium, both ^{235}U and ^{238}U are collected and put through a gaseous diffusion or other enrichment process. The collected plutonium is then mixed with the enriched uranium to make uranium and plutonium oxides, which can be used as a fuel for other uranium-powered reactors. Alternatively, the collected plutonium can be used to produce nuclear weapons or as a fuel for other reactors that use this type of nuclear fuel. The final option would be to place the reprocessed fuel directly in a waste storage facility.

The reprocessing of high-level wastes is done in several European and Asian countries. The United States has not advocated the reprocessing of commercial reactor fuel. In 1977, under the Carter Administration, a ban was placed on the reprocessing of spent fuel. The concern expressed by former President Jimmy Carter was that a proliferation of plutonium could lead to the availability of more nuclear weapons. Countries acquiring this plutonium, through either legal or illegal means, could fashion nuclear weapons or develop nuclear reactors to breed more plutonium. It takes less than 5 kilograms of plutonium to make a nuclear bomb. However, if the material is obtained as an oxide or from reprocessed commercial reactor fuel, larger quantities of plutonium are needed (less than 15 kilograms).

In 1981, former President Ronald Reagan lifted the Carter Administration ban on reprocessing spent fuel. This was done to encourage the nuclear power industry to reduce its demand for uranium by reprocessing the increasing volume of this high-level waste. (Figure 7-33 provides a graphic representation of the current and future quantities of spent nuclear fuel in thousands of cubic meters.) Because of the current large supplies of nuclear fuels, the nuclear power industry has not developed plans to establish a reprocessing facility. Without such a facility or a national storage site, the nuclear power industry has stored these wastes on-site.

On-site Storage of High-Level Wastes

Managing high-level waste safely requires it be put into temporary storage. Figure 7-34 shows several technicians storing high-level wastes in a fuel pool at a nuclear power plant. This storage of the spent fuel assemblies is necessary because, once removed from the reactor core, they are highly radioactive. Exposure to this material for only a few minutes could cause death to a person. The high radioactivity also makes the fuel assemblies hot. Therefore, it is important to allow this waste to cool in a pool of water for several years until the radioactivity has decayed to a manageable level. This cooling occurs in large concrete vaults lined with stainless steel. Once cooled, the high-level wastes can be placed in shielded casks and shipped to a reprocessing facility (if available) or a permanent-waste disposal site.

Besides spent fuel rods, nuclear power plants also generate high-level liquid wastes. Today, more than 80,000,000 gallons of radioactive solutions from nuclear reactors are stored in tanks at several Nuclear Regulatory

Quantity of Spent Nuclear Fuel (1985 with Projections to 2020)

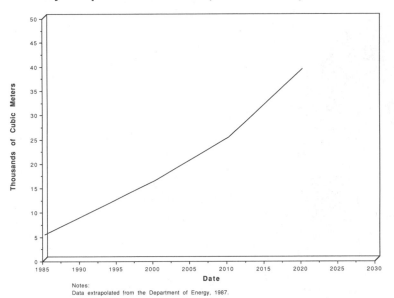

Notes:
Data extrapolated from the Department of Energy, 1987.

FIGURE 7-33 This graphic representation shows the current and anticipated quantities of spent nuclear fuel in thousands of cubic meters.

Commission sites. These tanks vary in size from 14,000 gallons to 1,000,000 gallons. Like the spent fuel rods, the liquid wastes must be stored in a safe manner to protect both plant and animal species from radioactivity. Research has gone on since 1955 to reduce the liquid wastes to dry, stable oxides—which reduces the

FIGURE 7-34 These technicians are storing spent fuel rods in a storage pool at the plant site. (John Coletti/Stock, Boston)

risk of spillage, makes the material safer to transport and handle, and provides for an overall reduction of material to be stored. A pilot study done at the Idaho Nuclear Engineering Laboratories successfully solidified more than 2.6 million gallons of liquid high-level wastes. Continued research in this area may provide further methods for the disposal of other radioactive wastes.

Low Level

Low-Level Waste
Low-level waste is not merely a product of the nuclear power industry. It is generated throughout the United States by more than 12,000 universities, the medical community, industrial manufacturers, electrical utilities, and the federal government. Figure 7-35 illustrates these various producers of nuclear waste with percentages representing each one's contribution (volume) to this waste stream.

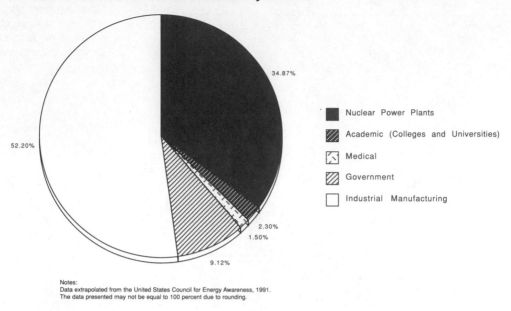

Sources of Low-Level Nuclear Waste by Volume

34.87%

52.20%

2.30%
1.50%

9.12%

■ Nuclear Power Plants

▨ Academic (Colleges and Universities)

▨ Medical

▨ Government

☐ Industrial Manufacturing

Notes:
Data extrapolated from the United States Council for Energy Awareness, 1991.
The data presented may not be equal to 100 percent due to rounding.

FIGURE 7-35 The largest producer of low-level wastes is industrial manufacturers followed by nuclear power plants, government, colleges and universities, and the medical community.

Industrial manufacturing is the largest producer of low-level nuclear wastes followed by nuclear power plants, government, colleges and universities, and the medical community. Each year in the United States, more than 80,000 cubic meters of low-level radioactive waste are produced from these sources. This volume of waste is expected to increase in coming years. Figure 7-36 illustrates the increase in low-level nuclear wastes from 1985 to 2020. Note the inset pie graph as part of this illustration, which shows the level of all the low-level radioactive wastes produced. The nuclear power industry contributes approximately 85 percent of this radioactivity.

Low-level radioactive waste products from nuclear power plants consists of used protective clothing, contaminated tools and equipment, used water-treatment resins and filters,

paper, and wood. A part of this waste is a result of daily operations, housekeeping, and maintenance activities within the power plant. Other wastes are generated when radioactive components are repaired or replaced, or when cooling water is treated to remove radioactive impurities. For example, if a valve is leaking radioactive water, the maintenance person may step in or touch a tool to the radioactive water. The shoes and tools are then considered a waste product. Figure 7-37 shows a technician preparing to enter a section of the reactor during standard maintenance procedures. If this technician is working on contaminated radioactive parts, then the clothing used and some tools may become part of the low-level radioactive waste stream.

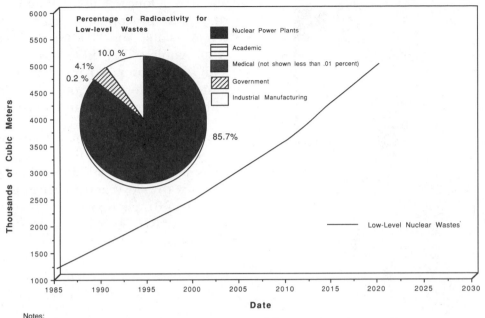

Quantity of Low-level Nuclear Wastes (1985 with Projections to 2020)

Percentage of Radioactivity for Low-level Wastes

- Nuclear Power Plants
- Academic
- Medical (not shown less than .01 percent)
- Government
- Industrial Manufacturing

10.0 %
4.1%
0.2 %
85.7%

Low-Level Nuclear Wastes

Thousands of Cubic Meters

Date

Notes:
Data extrapolated from the Department of Energy, 1987, and the United States Council for Energy Awareness, 1991.

FIGURE 7-36   Illustrated is the increase in low-level wastes from 1985 with projections to 2020. Approximately 85 percent of the radioactivity generated by low-level wastes is from the nuclear power industry, as shown in the pie chart inset.

Categories of Low-Level Wastes

In the United States, the Nuclear Regulatory Commission (NRC) is the federal agency responsible for assuring that nuclear materials are handled safely and protecting the public by overseeing the operations within the nuclear power industry. As defined by the NRC, all wastes are categorized as low-level except

1. Highly radioactive spent fuel rods from nuclear reactors.
2. Defense waste from the federal government.
3. Artificially produced elements (e.g. plutonium) that are radioactive.
4. Waste produced from the mining and milling of uranium.

This regulatory body divides low-level waste into three categories: Class A, Class B, and Class C. Each of these categories will denote the level of radioactivity of the wastes generated and what procedures are needed for their safe disposal.

Class A Wastes – This category contains materials that have the least radioactivity. Slightly more than 90 percent of all low-level wastes fall into this category. Class A wastes will lose their radioactivity in a very short amount of time (sometimes, a few days). This type of waste poses the least risk for plant and animal species and consists predominately of protective clothing and laboratory containers.

FIGURE 7-37 This technician is being readied to enter the reactor during normal maintenance procedures. (Courtesy of Nuclear Support Services)

To reduce the volume of Class A wastes, compactors or incineration may be used. With compactors, a 55 gallon waste drum can be reduced by as much as 90 percent. With incineration, 95 percent of the volume of material may be reduced. Filters on the smokestack of the incinerator limit the emissions of radioactivity to the environment.

Class B Wastes – This category of waste is more radioactive than Class A wastes and will remain so for a longer time. A portion of these wastes will generally lose their radioactivity in several weeks to several decades. However, some Class B wastes, such as the filters used in nuclear power plants, will remain radioactive for centuries.

Since Class B wastes are more radioactive and have a longer half-life than Class A wastes, they must be contained in a form that will remain stable for 300 years. Therefore, the NRC requires that these wastes be in a solid and stable form. To meet these guidelines, waste generators may mix their liquid waste with cement or chemical stabilizers.

Class C Wastes – Radioactive wastes in Class C are more radioactive than both Class A and Class B wastes. In addition, these wastes can take up to 500 years to decay to safe levels. Because of their toxicity to plant and animal species, Class C wastes require special shielding and disposal techniques. They are placed five meters beneath the surface of land with intruder barriers to protect from groundwater contamination. The barriers and placement must remain intact for 500 years. Class C wastes include such things as power plant parts from the nuclear reactor.

Disposal of Low-Level Radioactive Wastes

Some low-level waste products are disposed of at the plant site. In other cases, low-level waste is shipped to other states for burial. In 1980, the United States Congress passed the Low-Level Radioactive Waste Policy Act. This law requires that every state take responsibility for the waste created within its borders. As such, each state could choose to develop its own disposal site or make a compact with other states to dispose of their waste together. By 1985, no states had formed compacts or chosen disposal sites. It was, therefore, necessary for Congress to amend the 1980 Waste Policy Act to create deadlines and to impose penalties if the states failed to meet these. The amendments written into the original act further allowed the three states—South Carolina, Washington, and Nevada—with existing sites to refuse wastes produced outside their compact regions. Today, most states have developed their own storage facilities or have joined regional compacts for the disposal of low-level radioactive wastes. The decision to either join a compact or develop individual storage facilities is based upon economic, technical (i.e., meeting NRC guidelines for safe disposal), and political issues.

Regarding technical issues, the NRC has established strict guidelines for the safe disposal of nuclear wastes. These include choosing a location, operating the facility, and providing maintenance and security. The guidelines state that each established site must have geological formations that have remained stable for thousands of years. The mandates require that the disposal facility cannot be built on flood plains or in areas with excessive runoff from streams. The area must also have good drainage from groundwaters and must be located so that future community growth will not be affected. Finally, the storage facility must be a safe distance from economically recoverable resources, such as mineral or future oil or gas reserves.

Storing Nuclear Energy—Energy Issue 7-6

Low-level waste has a short half-life. Compare and contrast the advantages and disadvantages of storing these wastes directly on the nuclear power plant site. What are the concerns of the public regarding such storage?

NUCLEAR POWER PLANT LICENSING

Operation Period

Nuclear power plants in the United States are licensed to operate for 40 years, as specified by Congress in the Atomic Energy Act of 1954. Fashioned after the Communications Act of 1934, this act allows nuclear power plants to extend their licenses, as long as they meet specific criteria. Congress chose the 40-year time period because this was a typical amortization period for a power plant generating electricity. (An *amortization period* is the time for paying off the large capital outlay needed to build a power plant.) This 40-year license was not based upon safety, technical, or environmental factors.

Licensed nuclear power plants are given a set of requirements by the NRC based upon the type of plant. These requirements are called the plant's *licensing basis,* an evolving set of rules and regulations that plant operators must meet to maintain their operation. As technical advances are made and new research data become available, a plant's licensing basis may be changed—as mandated by the NRC. These new requirements then become part of the plant's licensing basis. This on-going examination by the NRC helps to ensure that a plant will operate safely throughout its license period.

Renewing a Nuclear Power Plant License

The decision to renew a nuclear power plant's license is fundamentally an economic one. This decision is made partly on estimates of future electricity demand in the utility's territory, partly on the cost to generate electricity by other methods (e.g., natural gas, wind power, and so on), and partly on the cost of continued operation of the nuclear power plant. Regarding the latter, this cost includes meeting any added NRC requirements based upon the plant's licensing basis.

In determining the need for a license renewal, the utility company is sure to base some of its decisions upon the fact that after a nuclear power plant's 40-year license, the initial capital costs for the plant would have been recovered. The recovery of these costs would allow the plant to operate very economically if a license renewal were granted. Based upon these considerations, most electrical utility companies are likely to wait until 5 to 10 years before their licenses expire before determining whether to apply for renewal.

The licenses for approximately 44 nuclear power plants will expire between 2002 and 2014. This represents about 32,500 megawatts of electrical generation. Each year between 2010 and 2030, an average of about 5,000 megawatts of nuclear generating capacity will be retired. Figure 7-38 illustrates this data.

Life Without Nuclear Energy—Energy Issue 7-7

A reduction in nuclear power plants (assuming none apply for and receive a license renewal) poses difficult choices for the future. (Review Figure 7-38.) What type of plants should be built to generate electricity in the future (e.g., nuclear, fossil fuel, and so on)? If no new nuclear power plants are built, how will the environment be affected if fossil fuels are used to generate electricity? What are the consequences of not building new power plants (consider both positive and negative consequences)?

Decommissioning

Eventually, all electric generating stations must be taken out of service, or *decommissioned*. This is true of both fossil-fuel and nuclear power plants. During the early economic planning of each plant, decommissioning was calculated into the overall cost. However, decommissioning processes have become substantially more expensive. Thus, some new costs will be passed

on to consumers. In the process of decommissioning, the nuclear power plant is placed into a condition where it presents no greater hazard to the public than it did while it operated.

Decommissioning of some existing power plants is scheduled to start in the first decade of the twenty-first century. Several decommissioning options exist:

1. *Mothballing* consists of placing the plant in protective storage and securing buildings to allow radiation to diminish to safe levels. The site would be decontaminated at a future date (some 25 to 50 years later) and the land restored to its original condition. Security and maintenance of the structure would continue until the plant site has been completely decommissioned.

2. *Encasement* consists of closing the radioactive part of the plant to allow the radiation to diminish to safe levels. Maintenance and security would continue until this safe level occurs and until the structure is removed.

3. *Entombment* is the process of using concrete to seal all highly radioactive components inside the reactor building. The structure would be completely sealed to prevent entry. The plant would remain in this condition to allow radioactivity to diminish to safe levels. In this method of decommissioning, no security or maintenance would be supplied.

4. *Disassembly* involves the removal of all radioactive materials and structures from the site. The land is then returned to its original condition.

Of the foregoing options, the NRC has endorsed disassembly or encasement without dismissing the possibility of entombment. As a plant readies for decommissioning, the NRC requires plant operators to demonstrate that adequate funds are available to complete the

Number of Nuclear Power Plant Licenses Expiring (2002 – 2030)

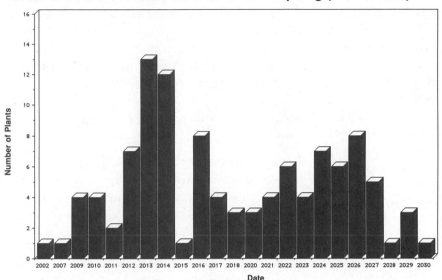

FIGURE 7-38 The licenses for approximately 44 nuclear power plants will expire between 2002 and 2014.

Notes:
Data extrapolated from the United States Council for Energy Awareness, 1993.

Decommissioning Costs of Nuclear Power Plants in Selected Countries

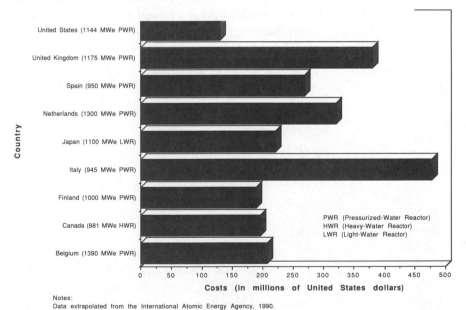

FIGURE 7-39 Projected costs to decommission various nuclear power plants in several countries are shown.

Notes:
Data extrapolated from the International Atomic Energy Agency, 1990.

task of shutting down the plant. The amount demonstrated must be at least $135 million for BWRs and $105 million for PWRs (in 1986 dollars), adjusted annually to account for inflation. Prior to decommissioning, an action plan with estimated costs must be submitted for closing the plant. Should the estimated costs exceed the amount established by the operator for decommissioning, the operator must demonstrate in the action plan how the difference is to be made up. Figure 7-39 illustrates the projected costs to decommission various nuclear power plants in several countries. Currently, the United States has the lowest costs associated with decommissioning a nuclear power plant. However, these costs are increasing and are expected to reach over $400 million by 2000.

ALTERNATIVE FUTURES

Transmutation of High-Level Wastes

One of the technical solutions to dispose of high-level nuclear waste is *transmutation*. This process may offer a solution to the disposal of large quantities of spent fuel assemblies. Researchers at the Los Alamos National Laboratory in New Mexico are major proponents of this solution.

In this method, a particle accelerator is used to accelerate neutrons into other radioactive atoms (i.e., spent fuel assemblies). When a waste atom absorbs the neutrons, it is converted into a different atom. Scientists hope that this method could be used to create less long-lived or even nonradioactive elements. Thus, spent fuel assemblies, which represent a major portion of nuclear wastes, could be converted into less harmful substances.

Although research is underway to make transmutation an economically feasible alter-native to storage, this goal may not be reached until well into the next decade.

SOCIAL AND ENVIRONMENTAL ASPECTS OF USING NUCLEAR ENERGY

Social and Environmental Benefits of Nuclear Power

Before illustrating some social and environmental costs of nuclear power, it is important to reiterate that nuclear power does not contribute to the production of greenhouse gases. Using coal-fired plants in place of nuclear power ones that have operated over the past 30 years would have produced additional emissions of 10,000 million tons of carbon dioxide and 200 million tons of acid gases, along with quantities of heavy metals. In addition, the use of nuclear power helps to reduce dependence on and depletion of fossil-fuel resources. As earlier noted, nuclear power supplies almost 22 percent of the United States' electrical needs and almost 17 percent of the world's electricity.

Social and Environmental Costs of Mining Uranium

The tailings left behind from mining uranium ores contain radioactive particles. These include radium, which decays to form radon gas. (Radon gas is believed by scientists to cause lung cancer.) Thus, as a preventative measure, tailing piles need to be covered to prevent excessive radioactivity in the surrounding air. In the early years of uranium mining, this practice was not common; therefore, high radon levels resulted in the vicinity of the tailing piles and radioactive dust was dispersed by the wind to surrounding communities. In years past, tailings were used in the construction of residential and commercial structures (in concrete) causing high concentrations of radon gas in these structures. Although such construction practices no longer occur, some buildings still retain these high concentrations of radon.

Social and Environmental Costs of Radiation Exposure to People

The NRC and the Environmental Protection Agency have set forth protection standards that take the form of (1) maximum permissible dose levels for on-site (workers) and off-site (public) populations; and, (2) power plant designs that set exposure limits that are "as low as reasonably achievable." Exposure of both the environment and workers is verified by sampling air, water, soil, and so on.

At a nuclear power plant, radiation exposures are measured by air samplings taken throughout the facility. The maximum allowed dose limits for workers ensure that the probability of harm is negligible for any one person. The dose limits pertain to single exposures, as well as to quarterly, annually, and occupational lifetime exposures. The NRC limit for workers is 5 rem per year, and the NRC must be notified whenever any worker receives 25 percent of the maximum allowed dose. On average, nuclear power plant workers receive between 300 and 600 millirem of radiation exposure beyond that of natural background levels.

Established standards permit the public to receive annually a maximum of 500 millirem above background levels. On average, nuclear power plants contribute much less to background levels than this established standard. Typical radiation exposures to the public that result from nuclear power plants is less than 0.1 millirem annually.

It is difficult to generalize about the health effects of low-level radiation (received either by working in a nuclear power plant or from living near power plants, reprocessing centers for fuel rods, radioactive waste disposal centers, or other such activities). The timing and severity of health effects from radiation are related to the level of exposure. High levels of exposure cause painful death within a few weeks. Intermediate doses have been shown to cause cancer and other health problems. The effects of low levels of exposure, particularly those received over a period of years, are still debated by the medical community. However, the prevailing view of scientists and the medical community is that no dose is harmless. Research evidence suggests that radiation—however small—presents a risk in the development of cancer and other illnesses. According to John Gofman, a medical doctor and physicist who worked on the Manhattan Project during World War II, the health dangers from low-level radiation cumulatively received are 6 to 30 times greater than what is currently believed to cause physical harm. Clearly, radioactive exposure is a serious concern in the generation of electricity from nuclear power and the disposal of nuclear wastes.

Social and Environmental Costs of a Nuclear Accident

The accidents that can occur in a nuclear reactor are grouped in classes of increasing severity. The International Nuclear Event Scale (INES), illustrated in Figure 7-40, categorizes events from level-0 (for one having no safety significance), to level-7 (for a major accident). A level-7 event would have widespread health and environmental consequences.

Today, 32 countries use this event scale to describe accidents that occur in nuclear instal-

International Nuclear Event Scale (INES) Levels

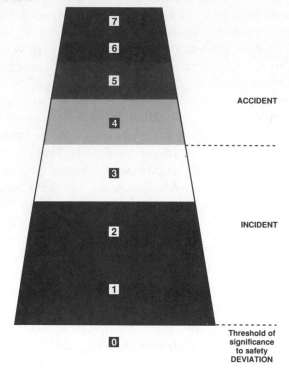

ACCIDENT

INCIDENT

Threshold of
significance
to safety
DEVIATION

INTERNATIONAL NUCLEAR EVENT SCALE (INES) LEVELS:

Level-7: **Major accident (widespread health & environmental effects)**
Level-6: **Serious accident (significant off-site effects)**
Level-5: **Accident with off-site risk**
Level-4: **Accident without significant off-site risk**
Level-3: **Serious incident**
Level-2: **Incident**
Level-1: **Anomaly**
Level-0: **Below scale (deviation; no safety significance)**

FIGURE 7-40 The International Nuclear Event Scale is used to describe various levels of accidents in the nuclear industry.

lations. The INES has been used chiefly for reporting events at nuclear power plants. However, the scale has been used to describe accidents in all aspects of the nuclear industry, including the transportation of radioactive wastes.

One of the more serious events that can occur in a nuclear power plant is a loss of coolant accident (LOCA). As previously noted, the function of the coolant in a nuclear power plant is to remove the heat from the reactor and transfer this heat through various systems to turn a turbine for electrical generation. In United States' light-water reactors, the coolant is water, which may also serve as a moderator to slow down the fissioning process. However, in a situation where there is a sudden loss of coolant, there is a danger of a core *meltdown*. This meltdown results from the decaying fission products generating excess heat.

In a worst-case scenario, there would be no cooling to remove the heat generated in the core. In such an instance, the fuel temperature would soar, causing the fuel rods to become a molten mass. It is conceivable, because of the large heat source, for this mass to melt through the reactor vessel and containment. This would create the possibility of a large-scale release of radioactivity in the environment.

As a safety mechanism, all nuclear power plants have numerous backup *emergency core-cooling systems* to extract the heat generated within the core should an LOCA occur. However, there is a probability that an LOCA will occur. Estimates by engineers and scientists experienced in high-pressure systems say that there is between .0001 and .000001 (or 10^{-4} to 10^{-6}) chance for each reactor, during a year's operation, to experience an LOCA. With hundreds of reactors in operation around the world, the probability of an LOCA would be a few percent per year or one every 20 to 40 years, using the high end of the chance range. Clearly, this figure should be considered unacceptably high and further safety measures and research are needed to mitigate the consequences of an LOCA. Further note here that, although tests have been performed on small experimental reactors with emergency core-cooling systems, no widescale test of the system has been undertaken to determine its usefulness in an LOCA.

The effects of a meltdown of a nuclear power plant could be devastating. The release of radioactive material from the containment structure could kill several thousand people with many more suffering future risks of cancer. The 1986 accident at Chernobyl in Russia (the worst nuclear accident to date) illustrates this point. Because of both poor design and numerous human errors, the reactor core at Chernobyl had a meltdown (level-7 on the INES) on April 26, 1986. The meltdown killed two people from the resulting explosions. An additional 29 firefighters and power plant workers later died from acute radiation poisoning. The accident caused the evacuation of over 140,000 people living within 20 miles of the reactor and contaminated thousands of square miles of agricultural land. This radioactive contamination was spread quickly to other countries by the wind. Following the accident, vegetables in many parts of Europe contained levels of radioactivity above the limits recommended by the medical community. Cows grazing on contaminated grass also produced milk with significant levels of radioactivity. For several months, millions of people had to alter their diets so as not to consume these substances. Current estimates of future cancer deaths range from 14,000 to 475,500 people worldwide, and the costs associated with cleaning are expected to reach $5 billion.

Social and Environmental Costs of Storing Spent Nuclear Fuel

Since the start of research with radioactive elements, scientists have proposed many plans to safely dispose of nuclear wastes. These proposals include burying the waste in Antarctica's ice cap, in the earth's surface, in deep ocean sediments; launching the waste into solar orbit; storing the waste in specially constructed buildings; reprocessing the waste; and, transmuting it (i.e., converting the waste to shorter-lived isotopes through neutron bombardment). Of these alternatives, geologic burial is advocated by most countries.

Geologic burial entails the use of deep underground repositories (approximately 1,000 feet below the earth's surface). The repository would be made up of several rooms drilled in granite, volcanic tuff, clay, or salt. The nuclear wastes would be placed in specially made containers to be deposited into the repository. These containers would be surrounded by clay to retard contact from groundwater and then sealed with cement. Once full, the repository would be sealed from the surface and the public would be warned for millennia of the dangers of the materials buried underneath.

Scientists in the nuclear industry have asserted that burying radioactive wastes underground is a safe solution to the problem of disposal. However, many environmentalists charge that geologic disposal is nothing more than a calculated risk. Changes in geology, land use, settlement, and changing patterns in groundwater flow all affect the ability to isolate nuclear waste safely. Konrad Krauskopf, a Stanford University geologist, has noted that no scientist or engineer can provide an absolute guarantee that radioactive waste will not someday leak from even the best of repositories.

The timeframe for building repositories is unclear. In 1975, the United States planned on having a waste disposal site in operation by 1985. This date was later changed to 1989, then to 1998, 2003, and now 2010. This continual change of operation dates reflects both the technical and political problems of finding a suitable location. Many other countries have also placed on hold the process of establishing geological repositories to 2020.

The costs associated with establishing and securing these repositories has risen 80 percent since 1983. A single site in the United States holding approximately 96,000 tons of high-level waste is projected at $36 billion. Without a mechanism for disposing of nuclear waste, the building of new nuclear power plants is clearly in jeopardy. To address the issue of wastes, some nuclear scientists have advocated reprocessing.

Social and Environmental Costs of Nuclear Proliferation

The issue of reprocessing high-level nuclear wastes is seen as a sound mechanism for both reducing waste and providing resources for the future. As earlier noted, reprocessing extracts the plutonium and unused uranium from spent fuel rods for future use. However, because it reduces the most potent components of the waste, it increases the production of intermedi-

ate and low-level isotopes. By industry estimates, reprocessing of spent fuel increases the quantity of intermediate wastes by nearly tenfold; and, low-level wastes are increased by an even greater amount. In total, reprocessing expands the volume of radioactive waste by 160 times the original irradiated fuel.

The general concern expressed with reprocessing is the added availability of nuclear material. While reprocessing does not make nuclear weapons, it can make the acquisition of nuclear material more available for this purpose. For example, in 1974 India became one of the countries able to produce a nuclear bomb, a direct result of acquiring plutonium from the country's reprocessing facility. Iraq has also made attempts to manufacture nuclear weapons. These attempts included the separation of plutonium from irradiated fuel. Note that Iraq could do this extraction of plutonium while in compliance with the Non-Proliferation Treaty and under the International Atomic Energy Agency safeguards. Reprocessing of nuclear waste may only make it easier for countries to obtain nuclear weapons and accelerate proliferation.

POINT/COUNTERPOINT 7-1

TOPIC:

Increasing or Decreasing the Use of Nuclear Energy in the United States for the Production of Electricity

Theme: In the past few years, the public has placed an emphasis on reducing the role of nuclear power to produce electricity. In addition, there has been a significant amount of misinformation regarding nuclear energy. Yet, society must have enough electricity to meet its energy demands. Attempt to answer the following questions in a classroom discussion or debate:

1. What has caused the public opinion of nuclear energy to change in the past few years?
2. What are some acceptable risks for nuclear energy based upon the number of power plants today?
3. If more nuclear power plants are built, is the populace willing to take the risks associated with the disposal and possible safety problems of handling nuclear wastes?
4. If more nuclear power plants are built, what will be done with the wastes generated?
5. Will the costs of electricity produced from nuclear energy remain competitive if more/fewer power plants are used in the future?

6. What environmental hazards will surface if more power plants are built?
7. If the usage of nuclear power is reduced, what will happen to the costs of producing electricity from other sources?
8. If the usage of nuclear power is reduced, what fuel could be used to make up the difference so that electrical demands could be met?
9. If nuclear energy is reduced and coal is its replacement, are the environmental risks more or less than those of nuclear reactors?
10. If nuclear energy is reduced, would the public accept the fact that fossil fuels will need to be exploited to a greater degree?
11. If it is accepted that newer type reactors are better, such as HTGRs and LMFBRs, who will pay for the additional research needed to make these power plants commercially available?
12. If nuclear energy is to be increased, who will pay for the added research and development to solve the waste problem?
13. What part should each of the following interest groups play in making the decision to increase/decrease nuclear production of electricity: government, power companies, environmentalists, and the public?
14. If enough money is directed into the nuclear industry, could many of the waste and safety concerns be eliminated? If so, from where would the money come?
15. Possibly the best way to allow for the reduction of electricity in society is to use fewer materials, goods, and services. How could society be persuaded to reduce its energy demands?

SUMMARY & REVIEW

Summary/Review Statements

1. Reasonably assured uranium resources are those that are considered recoverable with existing facilities, present technology, and at current costs and price levels.
2. Speculative reserves are those believed to exist in the judgment of geological engineers, but discoverability, availability, or economics associated with their extraction are not considered.
3. The largest uranium reserves can be found in the former Soviet Union and the Peoples Republic of China.
4. Nuclear energy was discovered because of the theory of relativity, which states that mass can be made into energy and energy can be made into mass.
5. Only five producers (Canada, the United States, Australia, Namibia, and France) produce over 77 percent of the world uranium outside of centrally planned economies.
6. Worldwide, there are more than 400 nuclear power plants operating in 25 countries around the world. These power plants supply almost 17 percent of the world's electricity.
7. The world annual growth rate for electricity produced by nuclear power is anticipated to be between 1 and 3 percent through the turn of the century.
8. Today, there are over 100 nuclear power plants in operation in the United States.
9. Nuclear power plants supply over 570 billion kilowatts of electricity to the U.S. This accounts for almost 22 percent of the electricity produced in the country.
10. Control rods in a nuclear reactor are used to slow down or speed up the reaction in the core.
11. The two types of nuclear energy processes are fusion (binding together to produce energy), and fission (splitting apart to produce energy).
12. When a ^{238}U element is hit by a neutron it momentarily turns into ^{239}U, then converts to ^{239}Pu, an artificial element.
13. Enrichment is a process in the fuel cycle in which the relative percentage of the ^{235}U element is increased to enhance the probability of fission.
14. The most common nuclear reactors used today to produce commercial electricity include the BWR (boiling-water reactor) and PWR (pressurized-water reactor).
15. The high-temperature gas reactor uses helium as a medium to transfer the thermal energy from the nuclear core.
16. The liquid metal fast breeder reactor uses sodium as a medium to transfer the thermal energy from the nuclear core.
17. There are two major categories of nuclear waste: low-level and high-level waste.
18. Isotopes consist of atoms that can vary in atomic weight while the atomic number remains constant.
19. Many natural and manufactured isotopes are radioactive. Radioactivity is the spontaneous disintegration or decay of the nuclei of atoms.
20. In disintegration, the emission of alpha, beta, and sometimes gamma particles occurs. The emission of these particles is often called alpha decay, beta decay, and gamma decay, respectively.
21. The most convenient method of describing or measuring the rate at which a material loses its radioactivity is in terms of its half-life. The term *half-life* refers to the time needed for half of the unstable atoms originally present in a

radioactive substance to decay into their new form.

22. Three radioactive units used to quantify the amount of radiation emitted by a radioactive element are the curie, the rad, and the rem.

23. Each year the average United States' citizen receives about 360 mrem of radiation exposure. This type of radiation is typically called natural radiation or background radiation.

24. Einstein's equation $E = mc^2$ states that energy (E) equals mass (m) multiplied by the square of the speed of light (c) in a vacuum. The equation suggests that mass (m) and energy (E) are merely two different forms of the same physical reality.

25. Materials that are fissionable are said to fissile. There are only a few fissionable elements. These include ^{235}U, ^{239}Pu, ^{233}Pu, and ^{233}U.

26. Light-water reactors have efficiencies of about 32 percent, as compared to 38 to 40 percent for modern fossil-fuel plants.

27. Approximately one-third of the reactors operating in the United States are BWRs.

28. Almost two-thirds of the commercial power reactors in the United States are PWRs.

29. High-level wastes include such radioactive materials as plutonium, thorium, uranium, and others. Half-life of these can be as high as 25,000 years. It would take almost 250,000 years before the radioactivity from some of these wastes decreases to a level safe for humans.

30. Licensed nuclear power plants are given a set of requirements by the NRC based on the type of plant designed. These requirements are called the plant's licensing basis.

31. The licenses for about 44 nuclear power plants will expire between 2002 and 2014.

32. Eventually, all electric generating stations must be taken out of service, or decommissioned. Several decommissioning options are mothballing, encasement, entombment, and disassembly.

Discussion Questions

1. What are three advantages and three disadvantages of using nuclear fission in our society today?

2. What would be the effects on our society if a governmental decision were made to cancel the operation of all nuclear power plants in one year? in five years? in ten years?

3. Conversely, what would be the effect if the government decided to encourage utility companies to build new nuclear-fueled power plants to meet the expanding demand for electricity?

4. Project a scenario of what might happen if a nuclear power plant had a meltdown. Identify both long- and short-term consequences.

III

Introduction to Nondepletable and Renewable Energy Resources

Section III is concerned with energy resources that have been categorized as nondepletable and renewable. Nondepletable resources are those that are not decreased as they are consumed. An example of a nondepletable resource is solar energy. Renewable resources can last (theoretically) forever only if they are managed correctly (i.e., renewed). The use of wood for heating is an example of a renewable resource. If managed correctly, wood could be used for an indefinite time period. However, with mismanagement, the resource could be depleted through deforestation.

There are numerous types of nondepletable and renewable energy resources used today. Research on others is conducted for possible large-scale uses in the future. Section III examines the availability of and technology associated with nondepletable or renewable energy resources. As in previous sections, these resources are presented in light of their social, economic, political, and environmental effects. This section has been divided into chapter 8, Solar Energy Resources, and chapter 9, Indirect Solar Energy Resources.

Chapter 8 introduces solar energy technologies in which sunlight is collected directly and immediately used as either electrical or thermal energy. This chapter includes the technology associated with the collection, storage, control, and distribution of both active and passive systems. In addition, this chapter provides detailed information on the development and applications of photovoltaic systems.

Chapter 9 provides an overview of indirect solar energy resources. These resources are derived from the sun's energy, but over time and through a natural process have been changed into a form that is not directly usable. Therefore, the energy form must be changed to make it viable. For example, wind energy (a form of solar energy, is in the form of mechanical energy) must be changed by a rotor and generator into electrical energy before being used. In this chapter various forms of indirect solar energy are addressed: wind, bioconversion, ocean thermal energy conversion, hydroelectric energy, tidal power, and geothermal energy.

chapter 8

Solar Energy Resources

PURPOSES

Goals of Chapter Eight

Industrialized societies power most of their machines with energy obtained by burning fossil fuels (e.g., coal, oil, and natural gas). However, there is clearly a limit to these resources, and the use of them has caused environmental and social harm. The combustion of oil and coal on a global scale makes breathing hazardous in many cities, generates acid rain, and is the principal contributor to global warming. Understanding these problems, many people are becoming aware of the need to live in a harmonious relationship with nature and to be aware of the consequences that technological development has on future generations. Confirmation of these statements can be found in the rapid development of organizations (both national and international) promoting the concepts of clean energy sources, efficiency, and the use of renewables as a practical solution to energy conservation and restoring the environment. Organizations such as the Alternative Energy Institute, American Solar Energy Society, Biomass Energy Institute, Development Centre of the Organization for Economic Co-Operation and Development, Geothermal Resources Council, National Renewable Energy Laboratory, Union

of Concerned Scientists, and WorldWatch Institute are all examples of groups committed to the restoration of the environment by the promotion of energy conservation and alternative energy sources.

In the United States, numerous polls have been conducted that indicate that people are willing to support new technologies that emphasize energy efficiency. According to several recent polls, 63 percent of people in the United States believe the U.S. can meet its future energy demands without building new power plants. These polls found that people wanted more funding and a priority placed on the development of energy-efficient technologies and the use of renewables. However, despite this public interest in such initiatives, the use of these energy sources has not been significantly addressed in the nation's energy policy.

Solar energy (a nondepletable energy form) can be used in the heating of buildings or producing electricity directly from the sun. There has been much research and development of this technology over the past two decades. According to the Solar Energy Industries Association, private outlays are currently over $8 billion. By the early part of the twenty-first century, solar energy expenditures are expected to exceed $40 billion.

The development of solar energy was stimulated by the 1973 oil embargo and by federal, state, and local governmental policies. Many U.S. industries, both public and private, began to realize that energy resources other than oil had to be studied, experimented with, and implemented. However, as fuel prices leveled off and tax incentives were removed during the 1980s, interest in solar energy diminished. Nevertheless, many companies continue their research. Solar energy retains its promise for commercialization, offers a means of restoring the environment from the burning of fossil fuels, and provides a long-term energy source for the future.

The purpose of this chapter is to explore the technology associated with solar energy systems—including its history, collection, storage systems, and useful applications. At the completion of this chapter, you will be able to

1. Investigate the history, abundance, and types of solar energy.
2. Define solar radiation characteristics.
3. Analyze flat-plate collectors and their efficiency.
4. Compare parabolic collectors to flat-plate collectors.
5. Determine geographic heating values for solar energy.
6. Examine residential solar space-heating systems.
7. Investigate residential solar hot-water systems.
8. Analyze various cost factors involved in implementing solar energy technology.
9. Analyze advantages and disadvantages of solar thermal storage systems.
10. Define the technology associated with photovoltaic solar cells, and understand how to apply solar energy systems to meeting future energy needs.

Terms to Know

By studying this chapter, the following terms can be identified, defined, and used:

Absorber
Absorptivity
Active solar systems
Ambient
Change-of-state storage
Design heat loss
Direct solar energy
Electromagnetic spectrum
Envelope design
Heliostats
High-temperature collectors
Hydronic
Incident solar radiation
Indirect solar energy
Insolation
Life-cycle costing
Parabolic dishes
Passive solar systems
Phase change
Photoelectric effect
Photovoltaic systems
Point focus collectors
Solar cells
Solar collector efficiency
Solar constant
Solar orientation

INTRODUCTION

History of Solar Energy

Solar energy is the radiant energy transmitted to the earth from the sun. It has provided, either directly or indirectly, almost all of the sources of energy on earth since the planet's formation. Analysis shows that most forms of energy used today originated from the sun except for nuclear and tidal energy. The importance of the sun in human endeavors cannot be understated. A study of either Western or Eastern philosophies reveals many myths and legends that illustrate both human awareness and debt to the sun in providing useful heat and the necessary mechanism for crop production. Solar energy has been used for thousands of years for the drying of food, for fuel, and for obtaining salt from brine. Through time, architecture has evolved to take advantage of the sun for heating or to take refuge from it for cooling.

Over 2,000 years ago, the ancient Greeks were among the earliest peoples to incorporate solar energy designs into their architecture. Socrates noted that houses faced south allowing the sun to penetrate the entrance of the structure in the winter. During the summer months, the sun was directly overhead; therefore roofs were built to shade the tops of buildings. These structures were further constructed with either thick adobe or stone walls to keep out the summer heat. With the deforestation of the Roman Empire (because of increased pressures of farming, building, and wood burning), access to the sun was protected legally with the Justinian Code of Law adopted in the sixth century. This code provided citizens with an energy source to heat their homes.

Other cultures throughout the world have also used a knowledge of solar energy for both heating and cooling their structures. The Chinese during the fourth through sixth centuries built their structures in courtyards to face south. In their system of building, the sunlight was diffused as it entered the structure by wood-lattice windows and rice paper during the summer months. This helped to keep the buildings cooler in the summer by reducing the heat generated in the home, and warmer in the winter by allowing the radiant energy to enter without the use of shading material.

A knowledge of solar energy in cooling a structure can be found throughout most of tropical Asia. The use of open-sided pole-and-thatch construction allows adequate ventilation and protection from incoming solar radiation. Thatch, which is an excellent insulator, has been used for thousands of years in Asian countries to help protect the inhabitants of structures from the oppressive heat during the summer months. In these areas, cooling towers have also been used to draw air into buildings (the warmer air rising to the top of the insulated thatch tower draws the surrounding air in as well), helping to cool the structure.

Early Experiments with Solar Energy

The use of solar energy has not been limited to the heating or cooling of buildings. Early experiments with this energy source helped in numerous scientific discoveries and supplied researchers with a power source for inventions. For example, in 1774, the British chemist Joseph Priestley found that, by concentrating rays of sunlight onto mercuric oxide, a gas would be given off. Priestley noted that a candle would burn more brightly when surrounded by this gas. Priestley thought that this gas was air in a greater perfection. With similar experiments concentrating light by glass lenses, the French chemist Antoine Lavoisier concluded, correctly, that the gas was oxygen.

In 1872 in Chile, a solar distilling operation was built to provide fresh water from saltwater. In this experiment slanting roofs of glass were placed over the saltwater. The solar energy transmitted through the glass caused the saltwater to increase in temperature. This increase in temperature caused the water, but not the salt, to vaporize and condense on the underside of the glass. The condensed (distilled) water flowed down with the aid of channels into freshwater collection bins. This experiment later turned into a profit-making operation providing almost 6,000 gallons of fresh water per day to the surrounding population.

Despite the historic evidence that demonstrates the successful use of this energy source, the application and use of solar energy has not been widespread. This is partly because of the intermittent nature of the energy source (sunshine only occurs for a portion of each day in most places) and partly because of the cultural preferences and convenience of using fossil fuels to meet industrial and economic needs. Yet, the promise of solar energy has not diminished. For many scientists, technologists, and social critics, using this energy source is tantalizing because it is plentiful, nonpolluting, and free to those who choose to exploit it.

Abundance of Solar Energy

Researchers at the United States Department of Energy have illustrated how much solar energy is available. They note that on a global scale, the amount of solar energy that arrives in a two-week period is equivalent to the fossil energy stored in all the known reserves of coal, oil, and natural gas. In fact, the total amount of solar energy striking the earth's atmosphere in a year is equivalent to 35,000 times the energy used annually by humans. At the outer limits of the atmosphere, the average intensity of incoming solar energy is equal to 1.36 kilowatts per square meter (measured on a plane perpendicular to its path). This number is called the *solar constant* and is received at this maximum rate only when the sun is directly overhead.

During other periods of the day, the energy received depends upon the angle at which the sunlight strikes the surface of the atmosphere. Other variables influencing the solar energy received include the latitude, the time of day, and the season of the year. Note also that, at any one time, about half of the earth is receiving no direct solar energy.

Imagine if today's technology could capture the quantity of solar energy available at the outer limits of the atmosphere. Energy shortages would cease. Countries could no longer be held hostage by other countries who possess a wealth of fossil fuel resources. This abundance of energy would allow developing and industrialized nations to compete more equitably in the global marketplace. (Remember, the use of inexpensive energy sources is a requirement in the production of goods and services.) This amount of energy obviously cannot be easily comprehended, but it reinforces the concept that solar energy is truly a nondepletable energy resource.

Much of this energy, however, cannot be captured. As it is diffused throughout the atmosphere, some of it is reflected back into space, or absorbed by plants and bodies of water. Considering these losses, the sun still provides a tremendous supply of energy. For example, approximately 13 percent of the energy at the earth's outer atmosphere reaches the ground. Suppose that this quantity of energy was converted to electricity at only 20 percent efficiency. There would be enough energy to supply all of the electrical power needs of a state. In practical terms, imagine if all solar energy that falls on a residential roof measuring 32 feet × 42 feet were utilized. It would be enough to supply all of the electrical power needed to run an average, air-conditioned, all-electric house.

Advantages of Solar Energy

One of the inherent advantages of using solar energy is that it is free from long-term costs. The consumer generally has only the initial investment of a systems technology and its maintenance. Over the life of a solar energy system, the cost is much lower than if solar energy were produced and sold by a power company. The concept of considering the cost of energy over a system's life span is referred to as *life-cycle costing.*

The sun's energy, for all practical purposes, will never run out. Solar energy is environmentally attractive because little waste is generated in its use. Furthermore, solar energy for residential and commercial applications does not need to be transported to the site of consumption. No other resource now in use can claim these advantages.

Disadvantages of Solar Energy

Although solar energy is abundant, there are certain disadvantages:

1. The sun's energy is diffused or spread very thinly. Assuming, for example, the solar radiation on a clear day is 100 percent, a hazy day would only allow 60 to 80 percent of the solar radiation to reach the earth's surface. On a cloudy day, only 5 to 50 percent solar radiation would reach the earth's surface. Thus, solar energy needs to be concentrated to make it useful.

2. The sun's energy is intermittent. It can be easily reduced by clouds. It is only present for an average of twelve hours per day (depending upon the geographic location and time of year). Thus, technology must be improved to store the solar energy until it is needed by the consumer. This calls for improved and more efficient storage systems.

Classifying Solar Energy

The technology for the utilization of solar energy can be characterized as either direct or indirect (as shown in Figure 8-1). These two terms are used to help classify different types of renewable and nondepletable energy resources. *Direct solar energy* means that the sun's radiation comes to the earth and then is immediately converted by technology into a usable form of energy. *Indirect solar energy* means that the sun's radiation has come to the earth and over time, through natural processes, has been changed into a form that cannot be immediately used. Therefore, it must be converted to a usable form. Only direct solar energy is discussed in this chapter.

Direct solar energy systems can also be subdivided into thermal and photovoltaic systems. Thermal systems are designed to convert the sun's radiation directly into thermal energy for use in the commercial or residential sectors. Photovoltaic solar systems are designed to convert the sun's radiation directly into electrical energy for a variety of uses.

Thermal solar systems can also be subdivided into active, passive, and focused technologies. *Active solar systems* are those that use pumps, fans, and an external energy source to transfer the heat collected. *Passive solar systems* do not use pumps, fans, and so on, but generally use natural convection currents to move the thermal energy through a structure. Focused systems are used to concentrate the sun's rays.

All active, passive, and focused solar energy systems are designed to include four common technologies:

1. *Collection*—All solar thermal systems must have some means of obtaining or collecting the energy. For example, active solar energy systems use collectors. Passive

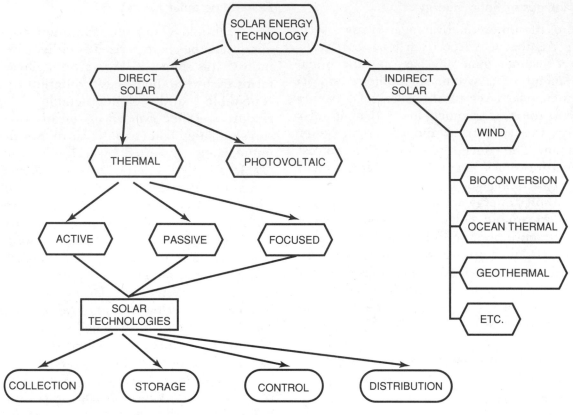

FIGURE 8-1 There are two major classifications of solar energy technologies, direct and indirect. This chart shows the various technologies of each class.

systems might use a greenhouse (floors, walls, and so on) as the collection device.

2. *Storage*—Solar thermal systems usually have some type of storage. Active systems use rocks or water for thermal storage. A passive system might use the walls, brick, concrete, and so on, as the thermal mass for storage.

3. *Control*—Solar energy systems usually use devices to direct the heat. Via this control, system components may be turned off and on or vents may be closed or opened to direct the heat source into a building. In active solar energy systems, control may be

achieved by a computer, various sensors, thermostats, and solenoids. In a passive system, the structure's owner usually controls the systems manually by doing things such as opening or closing shades.

4. *Distribution*—All solar energy systems require some type of distribution system. Distribution is used to move the thermal energy from the point of collection and/or storage to various parts of a structure. This is usually done in active systems by tubes, ducting, and the like. Passive systems use natural convection currents for distribution.

PRINCIPLES OF SOLAR ENERGY

Solar Radiation

Solar energy comes to earth in a form known as electromagnetic radiation. This wave form of energy is much like the waves created when a stone is dropped into water. Radiation is measured in terms of wavelength and frequency (waves per unit time). The shorter the wavelength, the higher the frequency, and vice versa. Visible light, sound waves, and radio waves are examples of electromagnetic radiation. Light waves travel from the sun to the earth at a speed of 186,000 miles (300,000 kilometers) per second.

Electromagnetic Spectrum

Figure 8-2 illustrates the frequency range of the *electromagnetic spectrum* of known energy waves. On this scale, the length of the wave is measured in centimeters (cm). Other units of measure are shown for reference—including the angstrom, micron, and meter. (The electromagnetic spectrum and these units of measure were previously presented in chapter 1 of this text.) Visible light is shown on the left center of the spectrum. Radio waves, sound waves, and other frequencies, such as gamma rays and X rays, are shown for reference only.

Of the light waves from the sunlight falling on the earth's surface, approximately 9 percent are in the ultraviolet range. These are light waves, starting beyond the violet end of the visible light spectrum. Sun tanning and sunburns are caused by ultraviolet waves.

Electromagnetic Spectrum

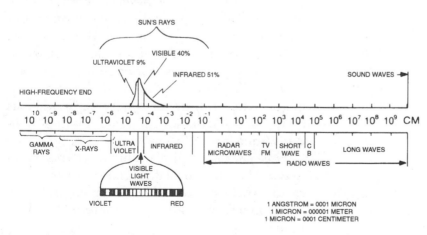

FIGURE 8-2 This diagram of the electromagnetic spectrum shows the location of visible, ultraviolet, and infrared light waves in relationship to all wavelengths.

The range of visible light transmitted to earth accounts for nearly 40 percent of the electromagnetic energy received from space. Each color in the visible spectrum has a different wavelength. Violet is the color on the high-frequency end of the spectrum, while red is the color on the low-frequency end.

The third frequency range of electromagnetic waves emitted from the sun is infrared radiation, which accounts for approximately 51 percent of the sun's total light waves. An object that is warm will emit infrared rays. In fact, any object that contains any thermal energy will emit infrared radiation. The infrared rays transmitted from the sun are absorbed by objects on the earth as radiant energy. These rays are then converted to thermal energy dispersed throughout the receiving object. Solar collectors utilize this infrared radiation as a means of collecting the energy given off by the sun.

Infrared Radiation—Energy Issue 8-2

Define **infrared radiation** by determining its location on the electromagnetic spectrum, and list four practical uses of this radiation.

Measuring Solar Energy

The terms solar constant, insolation, and langley are used to quantify or measure solar energy falling upon the earth's atmosphere and the surface of a solar collector. The exact term will depend upon how it is used. Refer to Appendix G for a detailed analysis of solar energy data.

Solar Constant

As noted earlier, the solar constant is the amount of energy in all the radiation reaching the outer edge of the earth's atmosphere. Figure 8-3 illustrates this concept. The solar constant is normally measured as Btu's striking a certain area in a specific time. The solar constant is equal to 4,290 Btu/square meter/hour or 1,354 W/m^2. Represented in yet another way, the solar constant is equal to 1.5 horsepower/square yard.

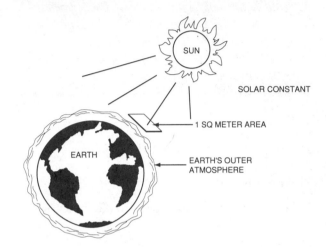

FIGURE 8-3 The solar constant is a measure of the sun's energy at the earth's outer atmosphere, upon a one square meter area.

Insolation

The second term used to define energy from the sun is *insolation*, or *incident solar radiation*. Insolation is solar radiation received per unit area for a given unit of time on the surface of the earth. While insolation can be measured in several units, the most common is Btu/square foot/hour. Figure 8-4 shows an insolation meter. When this meter is held perpendicular to the sun's radiation, it reads insolation in Btu's/square foot/hour.

Insolation is a more useful term than *solar constant* because insolation is a measure at the earth's surface, not at the outer limits of the atmosphere. Insolation is, therefore, useful when determining the efficiency of a solar collector. Note that insolation can vary according to air density, degree of cloud cover, geographic location, time of day, ambient temperature, and angle of the sun in reference to the earth's surface. In general, the insolation received at the earth's surface will range from 0 to about 360 Btu /ft^2/hr.

Insolation can also be used to compare the sun's radiation per unit area from one location to another. For example, when determining the placement of a solar collector, a comparison might be made of the solar energy striking a wall facing due south and a wall facing due west.

FIGURE 8-4 This meter is able to read insolation in Btu/ft^2/hr.

Solar Energy on a Roof—Math Interface 8-1

A large amount of solar energy strikes the roof of a building each day. Calculate how much energy strikes the roof by using the following data:

- Assume that the roof has a size of 28 ft × 40 ft.

- For the first 3 hours, an average of 110 Btu/ft^2/hr is received at the surface.

- Over the next 2 hours, an average of 160 Btu/ft^2/hr are received; and

- During the last 4 hours of daylight, an average of 150 Btu/ft^2/hr are received.

Total Btu's for 9 hours equals

_____

Langley

The third term used to measure solar energy is the langley. Langleys are measured by calories per centimeter squared. One langley is equivalent to one gram calorie per square meter of irradiated surface (221 Btu/ft^2/hr). Langleys are used in the study of weather patterns by the United States Weather Bureau. For example, if the daily average solar radiation were recorded for one year and then plotted on a map or graph, a view of the sun's intensity in langleys could be shown. The numbers in Figure 8-5 illustrate mean or average daily solar radiation in langleys over the entire year for the United States. There are also charts available for each month of the year for different geographic locations. Obviously, the months of November, December, January, and February will have much lower recordings and the summer months will have higher ones.

Often, it is necessary to convert these units into other solar energy units. (Also refer to Appendix H.) When converting from insolation in Btu/ft^2/hr to

New Unit	X	Multiple
Langleys/min	multiply by	0.00452
calories/cm^2/hr	multiply by	0.271
watts/m^2 (solar constant)	multiply by	3.152

Solar Energy Units—Math Interface 8-2

Several units are used to represent the amount of sun striking the earth. Calculate the following conversion problems:

If 241 Btu/ft^2/hr strike a surface, how many langleys/min, calories/cm^2/hr, and watts/m^2 strike the same surface?

1. Langleys/min would equal

   _____.

2. Calories/cm^2/hr would equal

   _____.

3. Watts/m^2 would equal

   _____.

Solar Collector Basics

Solar Collector Efficiency

Solar collector technology continues to improve. Many new designs have been developed to increase *solar collector efficiency*, which is the rate of useful heat collected in relation to solar radiation on the surface of the collector.

Ideally, if all solar insolation were captured, the solar collector would be 100 percent efficient. Obviously, this is not possible since some energy is reflected and so on. If a collector cap-

tures half of the insolation, 50 percent efficiency is achieved.

Determining Collector Efficiency

The calculation of collector efficiency can be complex because of several variables:

1. The design and materials of the collector—such as the glazing (i.e., the transparent cover used), amount of insulation, type of absorber surface to collect the incoming solar radiation (common absorber surfaces include copper, aluminum, tin), and so on.
2. The outside ambient temperature.
3. The wind velocity.
4. The inlet and outlet temperature difference of the fluid flowing through a collector.

Each factor is important. The more these factors are considered when designing, purchasing, building, or installing a solar system, the more efficient the collector will be.

Performance Charts, Operating Characteristics, and Efficiency

Many types of performance charts are used to represent the efficiency that can be expected from a solar collector. Some commercial charts illustrate differences in ambient temperature; some, insolation differences; and others, internal materials and their differences. Still, others represent efficiency in terms of temperature input versus temperature output. Although commercial manufacturers of solar collectors all report the efficiency of their units by a variety of means, ratings are meant to be used for comparative purposes only. Real output efficiency will vary depending on how and where the solar collector is installed.

One commercial collector performance chart, shown in Figure 8-6, compares the collector efficiency at different insolation quantities. In this chart, two lines are shown, one for

Mean Daily Solar Radiation (Langleys) Annual

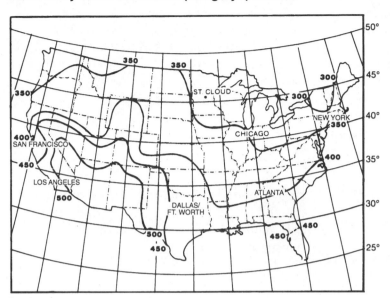

FIGURE 8-5 This chart shows the mean daily solar radiation in langleys for the year.

an insolation of 235 Btu/ft^2/hr and one for an insolation of 360 Btu/ft^2/hr. Efficiency of the collector is shown as a percentage on the vertical axis. The horizontal axis shows the outdoor (or ambient) temperature. To read the chart, use the following procedure:

1. Determine the insolation from the sun by using an insolation meter, say 235 Btu/ft^2/hr.
2. Select the ambient temperature on the horizontal axis, say 32° F.
3. Starting at the horizontal axis at 32° F, move up until the correct insolation line is crossed (235 Btu/ft^2/hr).
4. Now move to the left vertical axis and read the collector efficiency at these particular conditions (i.e., 47 percent).

Note that as the *ambient* or surrounding temperature decreases (moving to the right on the horizontal axis), collector efficiency decreases. Two conclusions can be drawn from this efficiency chart.

1. As ambient temperature increases, efficiency of the solar collector increases. This is shown by the fact that from 0° F to 60° F on the 235 Btu/ft^2/hr line, efficiency increased from approximately 35 percent to 58 percent (as represented by the dotted lines).
2. The higher the insolation, the higher the efficiency at ambient temperatures, below 60° F. This is represented by the fact the 360 Btu line is above the 235 Btu line (higher efficiency) to near 60° F.

A second collector efficiency chart is shown in Figure 8-7. This type of collector chart is produced from the manufacturer or by a solar collector testing firm. It is calculated for different ambient temperatures with specific insolation figures. Efficiency charts such as these are available from the solar collector manufacturers and are usually included with collector sales literature.

In this chart, the numbers on the horizontal axis (.1, .2, .3,...1.0) illustrate the difference between the outlet and inlet temperatures of the collector at various ambient temperatures throughout the day. The numbers are calculated by using a specific formula for the horizontal axis, shown in Figure 8-7. The formula is a relationship between the inlet and outlet temperatures of the collector, the ambient temperatures, and the insolation striking the solar collector.

Two efficiency lines are shown to illustrate different *absorber* temperatures within the collector. The absorber section of a solar collector is the inside surface area that absorbs the radiation after it passes through the glass or plastic glazing. The following is an example of how to calculate collector efficiency. Suppose the following data were applied to the horizontal axis formula:

Inlet temperature = 158° F
Outlet temperature = 194° F
Ambient temperature = 50° F
Insolation = 300 Btu/ft²/hr
Applying this data:

$$\left\{ \frac{\dfrac{194+158}{2} - 50}{300} \right\} = 0.42$$

If these figures are applied to the formula, the calculated number for the horizontal axis is equal to 0.42. Using 0.42 on the horizontal axis, and following to the 120° F (absorber temperature) line, the efficiency under these conditions is approximately 69 percent. (See the dotted lines.)

From this efficiency chart, several characteristics of solar collectors can be inferred:

Collector Efficiency

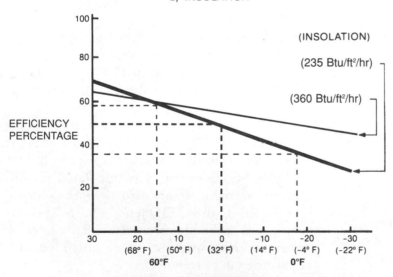

VARIABLES: 1) AMBIENT TEMPERATURE
2) INSOLATION

(INSOLATION)

(235 Btu/ft²/hr)

(360 Btu/ft²/hr)

EFFICIENCY PERCENTAGE

DEGREES CELSIUS (DEGREES FAHRENHEIT)
AMBIENT TEMPERATURE

FIGURE 8-6 This solar collector efficiency chart illustrates that as insolation increases, efficiency of the collector increases up to a certain point. (Adapted from Daystar Solar Products)

1. As the fluid (either water or air) is sent through the collector by the pump or fan, it absorbs thermal energy. The slower the fluid flows (lower speed pump or fan), the more thermal energy can be absorbed per given volume of fluid. Thus, the output temperature of the fluid increases. As the output temperature increases (keeping the inlet temperature the same) the efficiency will decrease. To illustrate this point, take two conditions, complete the formula using Fahrenheit scales, and apply them to the efficiency chart (using the 120° F absorber temperature) illustrated in Figure 8-7.

	Case # 1	Case # 2
Outlet Temperature	220	195
Inlet Temperature	158	158
Ambient Temperature	50	50
Insolation	300	300
Formula Answer	46	42
Efficiency (approximate)	64 percent	68 percent

By using these formulas and data, it can be seen that the speed of the fluid becomes important in determining the collector efficiency. As the temperatures of the output and stored fluid increase, the efficiency of the system will decrease.

2. The lower the absorber temperature, the greater the collector efficiency. Absorber temperatures vary with the amount of fluid passing through the solar collector and the insolation striking the absorber. As can be seen in Figure 8-7, if the absorber temperature is at 221° F, the efficiency will be lower than if it is held at 120° F. The major rea-

Collector Efficiency

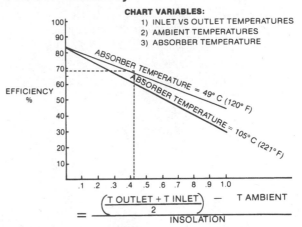

FIGURE 8-7 This collector efficiency chart shows that the higher the absorber temperature, the lower the efficiency of the collector. (Courtesy of Daystar Solar Products)

son for this is that as more thermal energy is held within the collector, more of the incoming radiation is reflected back into the atmosphere by the increased heat from the absorber. Under the latter conditions, the collector is saturated from the amassed heat. Thus, it cannot absorb increased amounts of radiant energy.

Solar Energy Efficiency—Energy Issue 8-3

Research has shown that cooler solar collectors are more efficient. Why is this?

Solar Energy for Space Heating—Math Interface 8-3

Solar energy can contribute significantly to heating needs. Calculate the following sequential problems dealing with solar energy:

1. A structure has a total of 30 active collectors. Each collector has 21 square feet of absorber space. After determining the total square feet of collector space, determine how many Btu's can be obtained from the collectors in 24 hours. For this problem assume that the collectors are rated at 67 percent efficiency and that the insolation is rated throughout a 24-hour period at 85 Btu/ft²/hr.

   Btu's collected during a 24-hour period equal _____.

2. If the building (from problem 1) needed 7,000 Btu's per hour for heating purposes, how many hours could the solar energy collected supply heat to the structure?

   Hours of solar energy supplied to the building equal _____.

SOLAR COLLECTORS

Flat-Plate Solar Collector

Design of a Flat-Plate Collector

Many types of active flat-plate solar collectors have been designed, developed, and installed in structures in the United States. The majority of these have been placed in residential dwellings. Two major types have been adopted: the hot-air collector, and the hydronic (water) collector. The major distinction between them is the medium or fluid that carries the heat from the collector absorber plate. The air type uses forced air or air moved by convection currents to remove the captured energy from the solar collector. Forced-air collectors are nor-mally used with forced-air heating systems for homes or commercial buildings, but they can also be used to heat water.

Hydronic collectors utilize a liquid, usually water, with or without antifreeze. Small tubes within the collector, on the absorber surface, carry the liquid. The antifreeze keeps the fluid from freezing in cold weather and acts as a lubricant for the pump. There are also drain-down and drain-back systems to evacuate the fluid during freezing conditions if antifreeze is not used in the system.

Hydronic collectors can be used to heat water for a building. The heated water (without antifreeze) can either be used for itself or for space heating. Regardless of whether an air or hydronic system is used, the collector's efficiency characteristics and basic operating principles remain essentially the same.

Parts of a Flat-Plate Collector

Figure 8-8 illustrates some of the basic components of a flat-plate collector. Collectors are housed in a rigid frame made of plastic, wood, metal, or other suitable material for support and weathering. The back section of the frame is insulated so that a minimum amount of thermal energy will be lost through conduction.

Absorber

Directly above the insulation is an area called the absorber. As noted earlier, the absorber has the function of collecting as much of the sun's radiation as possible. It then converts the radiant energy into thermal energy that can then be transferred out of the collector. Absorbers have many designs, depending upon the manufacturer and the application. For example, in a hydronic system the absorber will be made of a high thermal conductivity type of tubing for transporting the liquid. Copper is

Flat Plate Solar Collector

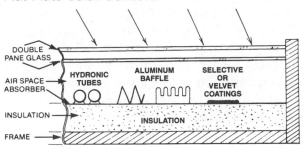

FIGURE 8-8 The design of flat-plate collectors varies with the application. Many types of absorbers are used to collect and trap the sun's radiation.

one of the more common types of tubing used. Copper has high thermal conductivity, enabling the heat to be transferred quickly through the tubing and into the liquid. In some absorbers the tubes are cast directly in the absorber plate.

Another type of absorber that has been tested with a hydronic system is a thin-walled transparent plastic channel. The channel will carry the heat-absorbing liquid throughout the solar panel. With the use of these transparent plastic channels, the liquid must be dyed a dark color to absorb the sun's thermal radiation.

Use of Baffles   Various other solar collector designs have been tried. One of these using baffles captures the short- and long-wave radiation of the sun. If the baffle is honeycombed, it is cut with an angle that reflects the short- and long-wave radiation of the sun to the absorbent surface to achieve a higher efficiency. This honeycomb baffle additionally will allow the collector to be installed vertically against a wall. However, the main purpose of baffles is to control the negative effects of convection currents. As the temperature of the medium within the collector increases, convection currents often

develop within the collector. These currents cause the thermal energy to move randomly throughout the collector, thus cooling the absorber. With the use of baffles the convection currents are reduced and efficiency is improved.

A second purpose of using baffles is to ensure that as short- and long-wave radiation strikes the baffles, the rays do not reflect directly back into space, but into a second or third absorber surface on the baffle. This added mechanism for absorption aids in improving the overall efficiency of this type of collector.

Selective Coatings   Velvet cloth coatings, special paints, and other selective surfaces (such as electroplating with nickel and zinc sulfide) have also been tried on the absorber. *Selectivity* means that certain materials are capable of absorbing more of the incoming sun's radiation than others. With the use of selective coatings, the radiation is conducted into the material, causing more energy to be collected. Absorbers are designed to have high **absorptivity** and low emissivity. This means that the absorber can take in radiant energy and not emit it back to the surrounding air.

Another important characteristic of an absorber is the material's ability to remain stable under excessive heat. Absorbers are designed to resist structural changes (i.e., elongation) as a result of the absorbed heat. A collector temperature may reach as high as 250° F. This temperature could have certain detrimental effects on the material or the structural integrity of the solar collector.

Absorber designs continue to undergo research and development using many new and highly efficient materials. During their development, consideration is given to the insolation available, economics of design, materials, conductivity, selectivity, and reflecting characteristics.

Absorber Color The way in which the surface of the collector absorbs solar energy is a factor of its efficiency. The color of the absorber is important since high absorption and quick transfer of heat is desired. Several colors have been tested and applied on the absorber section of flat-plate collectors. When solar radiation falls on a darkened surface, the radiant energy is absorbed and converted to thermal energy. Any color that is present within the absorber reflects that color's frequency back to space. Thus, it is not absorbed. Because of this, black, which is the absence of color, is most commonly used on the absorber surface. (Collectors will use a matte black paint to reduce reflection.) Other colors have been tested, including several shades of blue-green, but black remains the most popular because of its superior ability to absorb thermal energy.

With selective paint coatings, a black metal oxide is applied to the absorber. Although these coatings provide high thermal transfer of incoming solar radiation, they are costly and require complex application processes. Therefore, many laypersons experimenting with solar collector designs use commercially available matte black paints.

Glazing

Another design feature of a flat-plate collector is the protective glazing or transparent cover. The glazing is made of various materials including Mylar®, Tedlar®, polyethylene, and glass. Glazing has several purposes. First, the cover serves as a protection against dirt and moisture. Second, and more importantly, the glazing provides protection against heat loss by convection and thermal radiation.

Most of the sun's radiant energy is in the form of very short wavelengths. These include ultraviolet, visible, and infrared wavelengths. When the sun's radiant energy falls upon the collector, it easily passes through the glazing. It is then captured by the absorber plate. Figure 8-9 shows the warmed interior of the absorber and the thermal losses incurred due to radiation and convection. Notice how the glazing redirects some wavelengths back to the absorbent surface.

Although a single-glazing layer provides this protection, double- and triple-glazing layers have also been used in solar collectors. With the use of multiple glazings, heat energy that has been captured will remain within the collector with less conduction loss to the outside. A standard 0.1875-inch-thick glass cover will transmit about 85 percent of the solar radiation into the collector. Glazings often designated as lead free or clear glass will transmit up to 95 percent of the solar radiation into the collector.

The loss of heat from the collector also depends upon the size of the air gap between the glazing and the absorber. An air gap ranging between 0.375 and 0.5 inches will provide higher efficiencies than those with larger spaces. This is because the larger air gap will allow greater thermal loss by convection.

Collector Flow

Often on an active hydronic system, multiple collectors are used. Figure 8-10 shows how the fluid would typically flow through three collectors used for heating water in a structure. Note that the collectors in this figure are connected in a parallel design. One manifold is used on the top and one on the bottom of the system to allow the fluid to enter and exit the collectors.

On active air systems, multiple collectors are also used to capture the radiant energy.

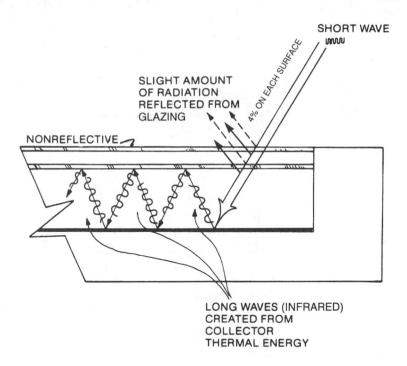

SHORT WAVE

SLIGHT AMOUNT
OF RADIATION
REFLECTED FROM
GLAZING

4% ON EACH SURFACE

NONREFLECTIVE

LONG WAVES (INFRARED)
CREATED FROM
COLLECTOR
THERMAL ENERGY

FIGURE 8-9 As radiation comes to a collector, it is taken in by the absorber and converted into long-wave radiation, which cannot pass through the glazing surfaces as easily.

Generally, the air in the absorber should be held in the collector long enough to gain sufficient temperature. Yet, it must move fast enough to gain maximum efficiency. Typically, multiple air collectors are placed so that the air is channeled from the top to the bottom through all collectors. Figure 8-11 illustrates these flow patterns. Manifolds are used on the top and bottom of the collectors to allow the air to enter and exit appropriately.

Focused Collectors

The use of focused collectors can produce much higher temperatures than flat-plate collectors. Since focused collectors can produce steam, they can be used in industrial applications, such as electric utilities and any manufacturing enterprise requiring temperatures greater than 200° F. (Focused collectors can produce temperatures to 4,000° F.) This is due

to collector performance. Appendix F shows a comparison of different collector configurations, with tracking being the highest yield.

Focused collectors can be divided into four basic types—parabolic dishes, parabolic troughs, central receivers, and solar ponds. The concepts used in focused collectors are similar to those of solar heating by flat-plate collectors. That is, these collectors capture the short- and long-wave radiation of the sun and transfer this into usable thermal energy.

Parabolic Dishes
Parabolic dishes, also called *high-temperature collectors* or *point focus collectors*, are used in a variety of applications today. Parabolic collectors use a reflecting surface to concentrate the sun's parallel rays to an absorber (also known as a receiver or boiler in these applications) that is at a focal point in front of the dish. This concentration allows these collectors to trans-

Hypertonic Collector Flow

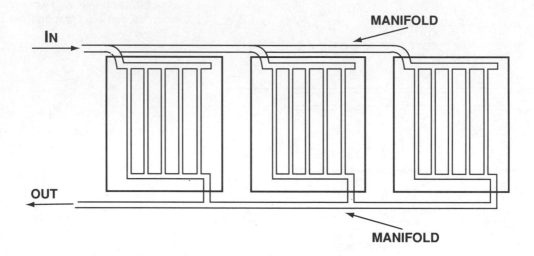

FIGURE 8-10 When several hydronic collectors are used, manifolds are used on the top and bottom of the collector system to transfer the fluid to and from the structure.

Air Flow

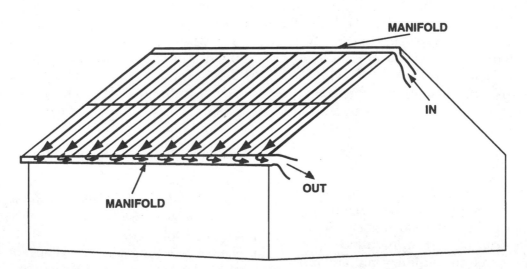

FIGURE 8-11 Air is channeled from the top to the bottom of the collectors in an air system. Manifolds are used on the top and bottom of the collectors to transfer the air in and out of the system.

fer extremely high temperatures to the fluid contained within the absorber.

In the design of parabolic dishes, the focusing of the sun's light may be accomplished by one of several reflective surfaces. The dish may use curved mirrors of glass, metal, or plastic. If metal or plastic is used, these are usually coated with aluminum to increase reflectivity to the absorber. Note that parabolic dishes may be made in a variety of configurations that will influence the overall efficiencies obtained. Examples of variables that influence efficiency include the placement of the absorber, type of reflective surface, and shape of the circular focusing collector. Figure 8-12 illustrates a parabolic dish with its focal point, tracking mechanism, and reflective surface indicated.

Although flat-plate collectors can obtain temperatures of 200 to 250° F, parabolic collectors can attain temperatures up to 4,000° F by following, or tracking, the movements of the sun. The high temperatures obtained from these devices allow them to be used for space

Parabolic Dish

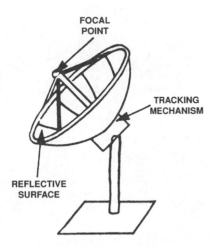

FIGURE 8-12 A parabolic collector is illustrated here with its component parts called out.

heating, refrigeration units, steam creation, and the generation of electricity. Experimental uses of parabolic dishes have also been applied to decompose toxic wastes (e.g., dioxins and PCBs). In these experimental uses, efficiency levels of over 30 percent have been achieved.

Parabolic Troughs

Parabolic troughs are similar to parabolic dishes since they also focus sunlight to a focal point to concentrate the sun's thermal energy. However, in this design the absorber is a vacuum-enclosed glass or metal tube that extends the length of the focal point of the trough. Figure 8-13 illustrates a parabolic trough. The tube is filled with water or oil; when heated, it can obtain temperatures in excess of 740° F.

Like parabolic dishes, the trough is usually tracked with the movements of the sun to obtain greater operating efficiencies. In a system, there may be multiple trough collectors connected to capture enough energy for a specific application. One example of this type of modular design is the parabolic troughs operated by Luz International in southern California. By connecting a series of parabolic troughs, Luz produces 354 megawatts of electric power for over 165,000 homes. The consumer cost for this electricity is as low as $0.08 per kilowatt hour.

Central Receivers

Central receivers are solar collectors that use many large mirrors to focus the sun's light onto an absorber on a high tower. The absorber in this type of a solar collector is called the receiver, and the large mirrors and their tracking systems are **heliostats**. Circulating through the receiver tower is a fluid that can be heated to about 2,600° F by the focused energy. To

Parabolic Trough

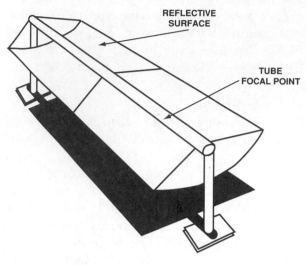

REFLECTIVE
SURFACE

TUBE
FOCAL POINT

FIGURE 8-13 A parabolic trough is illustrated here with its component parts called out.

increase the efficiency of the central receiver, computers track the mirrors to the movements of the sun. Like other focusing collectors, central receivers can be used to generate electricity or steam for industrial processing because of the high temperatures achieved. Figure 8-14 shows an example of a central receiver power plant in operation.

Solar Ponds

Solar ponds are bodies of shallow water that trap heat at the bottom. In these ponds the water may have several layers (usually two to three) of varying degrees of salinity. In the building of solar ponds, the black bottom of the pond is filled with a concentrated brine solution. Over this brine is a layer of ordinary water (or several layers of varying degrees of brine). Normally when heated by the sun, warm water will rise, but the brine layers in the pond stop this action. This is because of the dif-

ferent densities between the brine water at the bottom of the pond and the ordinary water at the upper level. Therefore, the warm water is trapped at the bottom of the pond and increases in temperature.

The temperature at the bottom of the pond will rise to near its boiling point after several days of heating by the sun. This heat can then be extracted to produce a low-pressure steam for operating an electric turbine. To increase the efficiency of solar ponds, these bodies of water may be covered by plastic to reduce the vaporization of water at the surface and the heat loss that would result.

One example of the use of solar ponds can be found in Israel. The government there supported the development of a solar pond in 1979 that produced more than 200 kilowatts of electricity, and a current project is producing up to 5 megawatts of electricity at a cost of $0.10 per kilowatt hour. Worldwide, several countries are planning to develop the use of solar ponds to supply electricity or heat for industrial processing. These include Australia, Israel, Italy, Japan, and Mexico.

FIGURE 8-14 This power plant uses mirrors to focus the sun's energy to the central receiver. The steam that is produced is used to turn a turbine to generate electricity. (Courtesy of Southern California Edison Co.)

Cost Advantages of Focused Collectors

The use of focused collectors offers several cost advantages over traditional fossil fuel for generating electricity in power plants. The construction of focused collector systems can be placed on site ready for operation in as little as nine months, as compared to the six to twelve years it takes to build a conventional power plant. Due to their modular design, focused collectors can also be designed to be added over time, which allows power plants to meet the changing population or energy needs of a particular area.

The energy produced by focused collectors has become more economically competitive with that produced by fossil-fuel power plants. Figure 8-15 illustrates the cost of energy created by focused collectors since 1980 with projections to 2000. The cost is expected to decrease to $0.05 per kilowatt hour by the turn of the century. Already, some focused collector systems applied on a large scale (like the parabolic trough system operated by Luz International in California) are competing economically with traditional fossil-fuel electrical generating plants. This economic competitiveness is a good reason for further research and development of solar energy technologies.

> **Parabolic Collectors—Energy Issue 8-4**
> Parabolic collectors can be designed to be either a dish or a trough. What are some advantages and disadvantages of using a parabolic collector instead of a flat-plate collector?

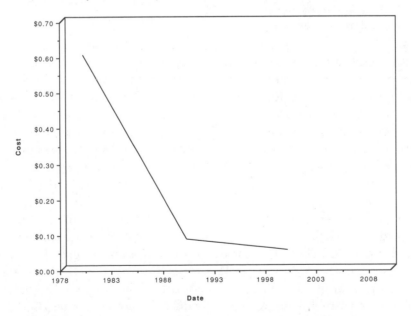

Estimated Costs of Energy Produced by Focused Collectors (1980–2000)

FIGURE 8-15 This figure illustrates the cost of energy produced by focused collectors since 1980. By the turn of the century these technologies are expected to decrease to $0.05 per kilowatt-hour.

SOLAR ORIENTATION

Structure or Collector Placement

Solar orientation is the study of how a structure or a solar energy system is placed (north, south, east, and west) in relationship to the direction of the sun. Solar orientation is applicable to both active and passive solar systems. Installers of active systems use solar orientation to help determine the exact position of the collectors in relationship to the movement of the sun. Installers of passive solar energy systems are especially concerned with the direction from which the sun's radiation strikes south-facing windows within a structure. In fact, solar orientation should be considered when designing and building any structure. To use solar orientation with solar energy systems, knowledge of the sun's position throughout the year becomes important.

The sun rises and sets at different positions each day of the year. During the winter months the sun is lower in the sky than during the summer months—as shown in Figure 8-16. The degree to which this happens depends upon the latitude of the observer. For example, in Figure 8-16 the sun's track is shown for a latitude of 40 degrees north for both December 21 and June 21. In the winter the sun sets much closer to the horizon than in the summer months. The design of the building including overhang lengths, position of windows, patios, overall shape of the structure, and collector positions, all play an important role in maximizing the sun's energy.

When studying solar orientation, two angles are often used. These are the azimuth and zenith angles. The azimuth angle is defined as the angle the sun makes with due south. An example is if a person were to look straight south and then look at the sun. The angle between the south and the sun is the azimuth angle. The zenith angle is defined as the angle the sun makes with the horizon. Often this is called the altitude angle. In Figure 8-16 the zenith angle is 26 degrees on December 21, and 73-½ degrees on June 21.

Solar Orientation—Energy Issue 8-5

Solar orientation is often used when building a structure to maximize the use of the sun's energy. Determine the advantages of maximizing the solar orientation of a structure. Afterward, take a drive through a new housing development to determine if residential structures are designed to maximize the benefits of solar orientation. If these structures are not, examine the zoning laws for the city or town to determine if they are the cause of contractors not using the benefits of solar orientation.

Collector Angle

Because of the varying positions of the sun during each day and each season of the year, fixed collectors for active systems must be set at the most efficient angle. The most efficient angle is the angle at which the sun's radiation is perpendicular to the collector surface. Since the sun continuously changes position, it is impractical to keep changing the position of the collector. Therefore, active thermal collectors used for heating air should be set at the angle most efficient for the coldest month of use. This coldest month is normally January in the northern hemisphere. For January, the optimum altitude angle for heating air collectors is found by adding 15 degrees to the latitude where the collectors are used. For example, the optimum altitude collector angle for a solar collector in a latitude of 45 degrees would be

Position of Sun at Noon for Latitude 40°N.

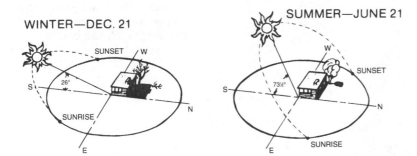

WINTER—DEC. 21

SUMMER—JUNE 21

FIGURE 8-16 The position of the sun during winter and summer is considerably different. Passive and active solar energy designs take advantage of the seasonal changes with respect to the sun's position throughout the day and the year.

60 degrees for the horizon, facing the south (in the northern hemisphere).

When determining the collector angle for heating water, it must be remembered that solar energy is needed the entire year. Therefore, a middle angle between the lowest and highest zenith angle of the sun should be selected.

SOLAR HEATING SYSTEMS

Energy Use in Residential and Commercial Structures

Some more practical applications of solar energy include residential and commercial air and water heating. As described earlier, the residential and commercial sectors of society utilize approximately 18 percent of all energy consumed within the United States. A significant proportion of this energy is supplied by natural gas and petroleum, with an increasing proportion coming from electrical energy generation. The major uses of energy in the residential and commercial sectors are heating, air-conditioning, and water heating. The remaining energy is used in applications, such as lighting and appliances.

The potential for solar energy use within the residential and commercial sectors is immense when one considers that these sectors currently use annually the equivalent of over 5 billion barrels of oil for heating, cooling, and lighting. The solar energy devices now in commercial production are capable of supplying all of the energy needs in a residence and a major portion of the energy needs in a commercial structure. However, some of these devices are too costly to justify their use (notably the meeting of electrical needs). Nonetheless, solar energy could readily be applied to the commercial and residential sectors in many areas for air and water heating needs.

Figure 8-17 shows average percentages of energy used by various consumption devices in a standard home. For example, central heating consumes most of the energy used within the home. If solar energy supplied the needed heat for a structure, an immediate and significant reduction in the consumption of hydrocarbon fuels and in individual fuel costs would be realized. There are, noted earlier, two major types of solar systems used in the residential and commercial sectors—passive and active. Before presenting information on these systems, a brief review of insulation and design heat loss is presented.

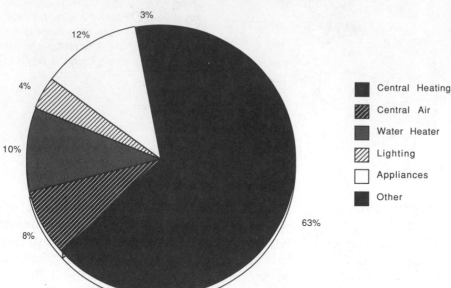

Energy Consumption in the Residential Sector (in percent)

3%
12%
4%
10%
8%
63%

Central Heating
Central Air
Water Heater
Lighting
Appliances
Other

FIGURE 8-17 Energy is used in various ways in a residential structure. This chart shows the relative percentages of each type of consumption device.

Solar Energy and Insulation

One important concept to consider when designing any building is the need to insulate the structure adequately. Maximum amounts of insulation should first be planned into the structure before a solar energy system is installed. It would be cost inefficient to spend money on any energy system and then lose the energy because of inadequate insulation. This is true of both passive and active solar energy systems. Estimates by heating and cooling experts note that a $400 to $600 investment in added insulation can provide a yearly savings of $125 to $150 (or more if the structure is air-conditioned). For comparative purposes, the importance of insulation can be realized by studying the concept of design heat loss.

Design Heat Loss

Design heat loss is a mathematical computation used by building contractors to estimate the maximum hourly energy consumption of a structure. This computation is usually performed for the severest heating or cooling conditions expected for a given building. Contractors will use design heat loss calculations to help determine a structure's heating and cooling requirements. Although it is unimportant for laypersons to be able to perform these calculations, the concept of design heat loss can be used to illustrate the importance of adequate insulation. In a poorly insulated structure in a cold climate, the building can have a design heat loss of about 65 $Btu/ft^2/hr$ in the living space. However, the same structure in the same climate with adequate insulation can have a design heat loss of 18 $Btu/ft^2/hr$ in the living space. Thus, insulation can have a direct effect on one's annual energy savings and on the ability to conserve finite resources.

Passive Solar Energy Systems

Passive solar energy systems incorporate natural convection, conduction, and radiation principles to transfer the solar thermal energy throughout a building. One of the first systems designed to test passive solar concepts used large barrels filled with water to absorb and transfer heat, as shown in Figure 8-18. These barrels were placed near a glass wall, which was protected at night by a large cover that was raised into place. When the sun was shining and the cover was lowered, the inner surface of this wall acted as a reflector. Heat from the sun was absorbed through the glass by the black surface on the outside of the barrel faces. This heat was then transferred to the room by conduction, convection, and radiation.

The outer end of each barrel was painted black for absorption; the inner end was painted white for appearance. Systems such as this had to incorporate a supplementary heating system for backup. The barrel system had the four technologies of any solar system: collection (the metal barrels), storage (the water in the barrels), control (operator control), and distribution (convection currents in the room).

During the summer months the system operated in reverse. The outside cover was opened during the night hours and closed during the daylight hours. The cool air at night absorbed the heat stored within the barrels during the day, thus getting the effect of a simple air-conditioning system. Today, some homes incorporate storage masses to hold or retain thermal energy during the day and release it at night.

Envelope Design

A schematic of another type of passive solar energy heating system is shown in Figure 8-19. This construction technique is called an *envelope design*. The left side of the structure is mostly made of glass, and faces south. The building itself is the collector. Storage is within the masses of the material used in the structure and in the rock below. Control and distribution occur naturally from convection currents passing through the walls of the building. In operation, as the sun's radiation heats up the left side of the building, a convective loop begins

Passive System

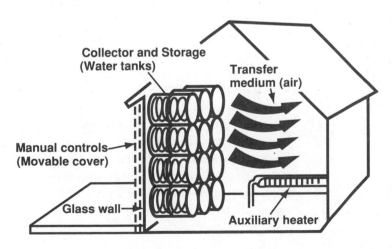

FIGURE 8-18 One of the first passive solar systems used barrels of water to act as the collector and absorber. The thermal energy was convected into the living space at night.

to develop. The warm air naturally rises and circulates through an envelope built around the structure. Based upon these principles, many other passive solar energy designs have also been tested and built. To be competitive, today's passive solar buildings must use solar orientation, convection, conduction, radiation, and storage masses effectively.

Comparing Cost of Passive Solar Energy— Energy Issue 8-6

Passive solar energy designs are used on many buildings today. What relationship exists between the costs of designing passive solar heating systems and the costs of energy saved by using these technologies? What factors can make passive solar energy more cost effective in the future?

Envelope Home

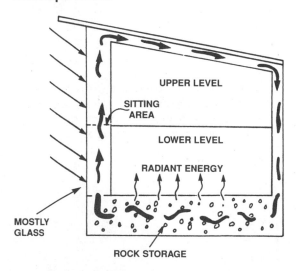

FIGURE 8-19 An envelope home is considered a passive solar heating design. As the air is heated, convection currents pass the warm air around the home, eventually heating the rock storage area.

Active Solar Energy Systems

Active space heating for residential buildings can be designed with either *hydronic* (antifreeze and water) or air solar collectors. The difference is the medium used to absorb the energy in the collector. Air systems are most commonly used with residential dwellings that have forced-air heating systems. Hydronic systems are mostly used for hot-water heating. Depending upon the economics, either a supplemental or primary system can be installed with any level of complexity.

There are distinct advantages and disadvantages of using either air systems or hydronic systems:

1. Active air solar energy systems allow the heat to be directed into existing ducts for either storage or for room heat. This helps to eliminate installation costs and improve the versatility of the heating and/or cooling system.
2. Leakage in an air system is not as severe as in a hydronic system. With a hydronic system, leakage could damage the walls and ceiling of the structure. If leakage occurs in an air system, efficiency is the only loss.
3. Air systems require less maintenance because they have less corrosion than hydronic systems. This helps keep the system maintenance-free. Hydronic systems generally build up corrosion within the pipes in the form of silicate and calcium, which causes flow-rate problems over long periods of time.
4. Collector design in air systems is usually less complex and not as costly as in hydronic systems. However, more collectors are needed than in hydronic systems. Generally, it is easier to design and build a collector to move air than water.

5. Air space collector systems can be operated at somewhat lower temperatures than can hydronic hot-water systems. Normally, solar air type heating systems operate at around 100° F. Or, if water is used for space heating, it can operate near 100° F. However, domestic hot-water systems need to operate the collectors at higher temperatures, near 140° F.

6. More room is required for air system ductwork than for the hydronic system. This disadvantage can become an important consideration if the solar heating system is installed after the structure was built.

7. Air systems can only be used with forced-air heating systems, unless heat exchangers are used for hot water, in which case, there will be additional losses in efficiency.

8. The efficiency of an air system is lower than that of water. Heat can be extracted more quickly from water than from air, which causes an air system to be less positive in terms of heat transfer. This difference is because of the differences in thermal mass and conductivity of the media.

9. Air systems are not recommended for the homes of people who are extremely sensitive to mold or dust. Warm, moist air entering a cool rock storage bin will cause condensation to occur. This condensation can cause mold and mildew to form on the rocks. Should this occur, the mold and mildew could emit spores to the air being circulated to heat the home.

The foregoing are only some of the advantages and disadvantages of using air and hydronic systems for air and hot-water heating within a structure. Each should be carefully considered when deciding which system to use. Other specific advantages and disadvantages will be presented later in this chapter.

Heat Exchangers

Although heat exchangers are not necessarily a part of an active solar heating system, many designers use these devices to transfer the heat collected. Generally, heat exchangers are used if the heat gathered by a solar collector must be transferred or exchanged to other sections of the heating system.

A heat exchanger is a unit within any heating or cooling system that transfers the thermal energy from one medium to another. For example, the thermal energy produced within a gasoline engine cooling system is transferred from the liquid coolant to the air by a heat exchanger called a radiator. Thermal energy is also transferred out of a refrigerator by a heat exchanger. In fact, all industrial or commercial air-conditioners and heaters use some type of heat exchanger.

In an earlier chapter three methods of transferring heat were presented: convection, conduction, and radiation. By applying these to heat exchangers, thermal energy can be transferred by various means. The most common types of heat exchangers used to transfer energy within a solar energy system are

1. liquid-to-air/air-to-liquid;
2. liquid-to-liquid; and,
3. air-to-air.

Liquid-to-Air/Air-to-Liquid Heat Exchangers

To accomplish the needed thermal exchange, two types of heat exchangers are normally designed in heating and cooling systems. By combining these exchangers, any type of heat transfer can be obtained. Hydronic collectors require thermal energy to be transferred from a liquid-to-air. In operation, the heated liquid medium from the collector is pumped through the piping into the exchanger as shown in Figure 8-20(a).

Heat Exchangers

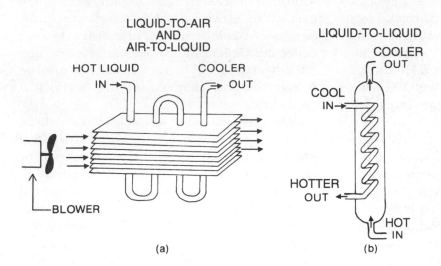

FIGURE 8-20 Thermal energy can be transferred from one medium to another by the use of heat exchangers.

The thermal energy from the liquid is then conducted or transferred into the surrounding pipe and baffle plates. This transfer occurs naturally because heat always moves from a hotter to a colder material. The heat is then transferred from the baffles when forced air is blown through them (heat transfer is by convection). The larger the surface area of the baffles, the greater the heat dissipation. Also, the greater the temperature differential across the two inputs (liquid and air), the greater the heat energy transfer within a given time period.

This entire process can easily be reversed so that thermal energy can be transferred from air-to-liquid. In this case, the air must be hotter than the liquid. A typical application for this type of heat exchanger is an air type of solar collector, transferring energy to the hot-water system of a structure. Although possible, this method of heat transfer is difficult because the density of air is far less than that of water, which causes poor heat transfer capability in this direction. Therefore, liquid-to-liquid systems are preferred.

Liquid-to-Liquid Heat Exchangers

The second type of heat exchanger is a liquid-to-liquid one, as illustrated in Figure 8-20(b). Here, thermal energy is transferred from one liquid to a second liquid. A typical example would be to transfer thermal energy out of antifreeze and water, into water used for domestic hot-water or space heating.

Although many designs of liquid-to-liquid heat exchangers exist, the more common ones use coils of tubing placed directly within the second liquid. The heat from the first liquid is conducted into the metal tube and then conducted into the second liquid.

Another method of liquid-to-liquid heat exchange has the tubes of each system physically contacting each other. Through conduction, the heat transfers from the hotter liquid to the colder liquid.

Air-to-Air Heat Exchangers

Air-to-air heat transfer is somewhat different. In this system, a series of input metal baffles are interchanged and interconnected with a second set of output metal baffles. The system allows for a primary and secondary air flow passage. The two air passages do not interconnect or mix the air. As warm air passes through the primary passage and the set of baffles, thermal energy is transferred through the baffles to the secondary air passage. As the secondary cool air passes through the set of baffles, it becomes warmer. Air-to-air heat exchangers are often used in very tight buildings (super insulated) to exchange air without losing thermal energy. This is done to allow fresh air to enter the building and minimize stale air within the structure. Figure 8-21 shows how the primary and secondary air flows through an air-to-air heat exchanger.

Forced-Air Supplementary System

One of the least costly ways to incorporate solar energy into a building is shown in Figure 8-22. The gas manifold and furnace are centrally located. Heated air from the combustion of natural gas is transferred through an air-to-air heat exchanger into the ductwork. The heated air is then forced into the lower part of the crawl space and then upward into the rooms. Two return vents are located in the ceiling to allow the air to return to the furnace, to be reheated by gas.

In this system, solar energy is added via a series of ten or more air collectors placed on the roof. The input of the solar collectors is taken from a cooled-air return to the furnace. The output of the collector is then sent directly into the return of the existing furnace. The connection into the furnace has an electronically controlled damper to close off the solar collector to the furnace. A second damper is closed at night to prevent cold air from entering the building from the collector.

Within the solar collector a temperature indicator senses the temperature of the air to determine if the dampers should be opened or closed. If the temperature is high enough, the

Air-To-Air Heat Exchanger

FIGURE 8-21 An air-to-air heat exchanger transfers thermal energy from the primary air flow to a secondary air flow.

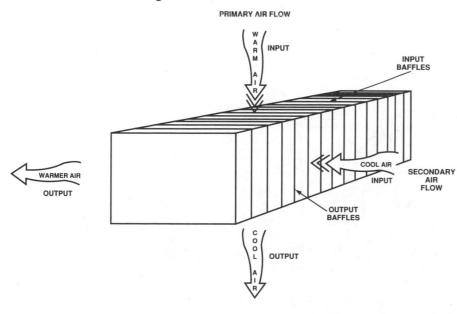

dampers will open allowing the warm air to flow into the structure.

When the thermostat within the building signals for an increase in heat, the furnace blower will turn on and the fuel will be ignited. As the furnace draws the cooled air from the return ducts, it will also draw warm solar-heated air if the collectors are warm. As this process continues, the energy captured from the solar collector is put into the existing forced-air system. If more energy is put into the system, the temperature within the building will rise quickly, shutting the thermostat off and, thereby, conserving fuel.

As the sun's radiation is reduced toward evening, the temperature indicator will close off the solar collector ductwork. The existing system will then operate normally, without the use of supplementary solar heating.

In this combined system, the collectors are used to obtain the sun's energy. The storage is composed of the materials in the structure.

The control is the existing thermostat, temperature indicators, solenoids, dampers, and so on. The distribution is the ductwork in the building.

A more complex active air solar system is illustrated in Figure 8-23. In this solar energy system three modes of operation are available: heating the structure directly from the collector, storing the thermal energy for later use, and heating from the storage. This system uses approximately 30 forced-air collectors, a rock storage area, an auxiliary heater, one blower, and several solenoid-operated dampers.

During the storage mode of operation, all solar energy collected goes directly into the storage area. See diagram 1 of Figure 8-23. Damper A, leading into the auxiliary heater and rooms, and damper B, the exit from the rooms, are closed. Damper D is also closed to shut off a bypass line. Having these dampers closed forces all the solar-collected thermal energy directly into the storage area, thus heat-

Forced Air Supplementary Solar Heating

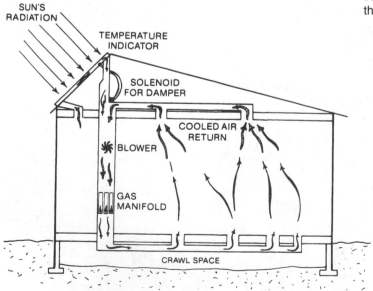

FIGURE 8-22 Solar collectors on the roof of a structure allow thermal energy to be directed into the existing forced-air system.

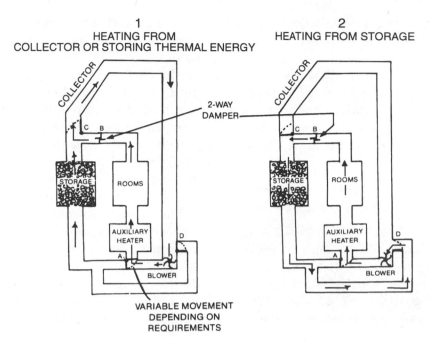

FIGURE 8-23 This schematic shows how a solar forced-air heating system operates in three different modes.

ing the rock and storing thermal energy. Should the temperature of the collector drop below a predetermined level (because of clouds or darkness), the blower will automatically shut off.

If the thermostat calls for thermal energy and the temperature of the collector is sufficiently high, thermal energy from the solar collector can be pumped directly into the rooms. During this mode of operation (see diagram 1), damper A closes off the storage area, opening the ductwork to the rooms. Damper B is also opened to allow air to return to the collectors. During this time the storage area is closed off to the circulation of air from the collectors.

A more complex system incorporates a variable-setting at damper A. Here, air needed within the dwelling will control the degree of opening of damper A to the rooms. For example, if a great deal of thermal energy were needed in the rooms, damper A would have a maximum opening to the rooms. If only a small amount of thermal energy were needed, damper A would open slightly to the rooms, allowing the remaining thermal energy to be put into storage. The automobile passenger compartment heating system works in much the same way. As the damper is opened or closed, more or less thermal energy is admitted into the passenger compartment.

During the third mode of operation, as shown in diagram 2, thermal energy is removed from storage and sent through the auxiliary heater to the various rooms. Damper C is closed to the collector. Damper A is open to the bypass duct so that thermal energy taken from storage can be connected to the suction side of the blower. Damper A is positioned so that thermal energy from the blower can be circulated through the rooms.

With this design, the auxiliary heater will supply energy for heating if the storage is depleted. In fact, any thermal energy created from the auxiliary heater can be recirculated out of the rooms and sent into storage to keep the energy within the structure.

Active Solar Systems—Energy Issue 8-7

Active solar energy air systems are not as popular today as in the past. What economic, technical, and political factors have contributed to this decline? What role can education play in dispelling many myths associated with solar energy?

Domestic Hot-Water Systems

Figure 8-24 shows a simple solar hydronic hot-water system. The three solar collectors located on the roof of the structure are used to collect the thermal energy. Because of the possibility of subfreezing temperatures, the fluid is a mixture of antifreeze and water, no greater than 60 percent antifreeze and 40 percent water. The antifreeze solution is kept separated from the existing hot-water system through a closed loop and a liquid-to-liquid heat exchanger.

The heated antifreeze solution from the collector is sent into an expansion tank. The purpose of this is to allow space as the hydronic solution is heated and expands; pressure from this expansion could otherwise cause a rupture in a line. The antifreeze solution is then pumped through a one-way flow check valve

Hydronic Hot-Water System

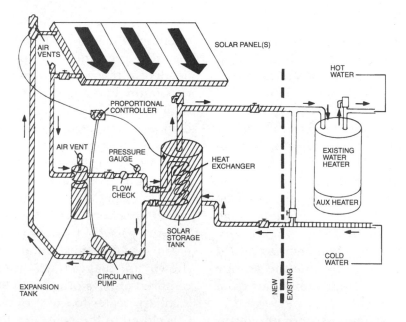

FIGURE 8-24 This domestic hot-water solar system is used to heat hot water in a house for cooking, bathing, and so on. (Courtesy of Times Mirror Magazines, Inc.)

and through a liquid-to-liquid heat exchanger. After the heat energy is extracted by the exchanger within the storage tank, the antifreeze solution is pumped back into the solar collectors for reheating. Within this closed-loop system, temperature controls on the collector and storage tank determine when the pump will turn on and off.

As the temperature difference between the collector and storage tank increases, the pump will start to circulate the fluid slowly in the collector. As the difference between the two temperatures continues to increase, the pump circulates the fluid faster, until eventually the pump is on continuously.

The storage tank, just as with any existing water heater, can store heat. As hot water is used for bathing, dish- or clothes washing, the hot water is taken directly from the existing tank. As the water is depleted, hot water from the solar storage tank replaces the hot water used. In this system the solar storage tank is in series with the incoming cold-water system. It acts as a preheater to the hot-water system.

If the solar energy system tank and the existing water heater do not maintain sufficient temperature (about 110–140° F), the auxiliary heater in the existing tank turns on and supplies the necessary energy for hot water.

Hydronic Solar Systems—Math Interface 8-4

Solve the following domestic hot-water problems:

1. A residential dwelling uses a hydronic solar energy system to heat its water. The insolation for a 7-hour period is 125 Btu/ft²/hr. The collector surface area is 68 ft². The collector operates at 53 percent efficiency. How many Btu's were captured in the 60-gallon water storage tank in 7 hours?

   Btu's stored in the tank equal

   _____.

2. Considering the temperature rise (in degrees) of the 60-gallon storage tank, what is the final temperature of the storage tank (see Problem 1) if the original or starting temperature was 55° F? Note: One Btu is the thermal energy necessary to raise one pound of water one degree Fahrenheit; one gallon of water weighs 8 pounds.

   Final temperature of the storage tank equals _____.

Total Hydronic System

A more complex hydronic system is shown in Figure 8-25. In this system both hot water and space heating can be obtained. Only the schematic and air flow is shown. The dampers, valves, expansion tank, and pressure and temperature sensors are not illustrated in this figure.

This collector system also uses a closed-loop design, keeping the antifreeze solution separate from the water. The heated solution from the solar collectors is sent into the hot-water storage tank. With a liquid-to-liquid heat exchanger, the heat is transferred into the hot-water storage.

A second liquid-to-liquid heat exchanger is used to transfer the thermal energy from the hot-water storage to the conventional hot-water heater. This part of the system is similar to any domestic hot-water system. Several variations are possible, including a second heat exchanger within the hot water heater and the use of a second pump.

Hydronic Solar System

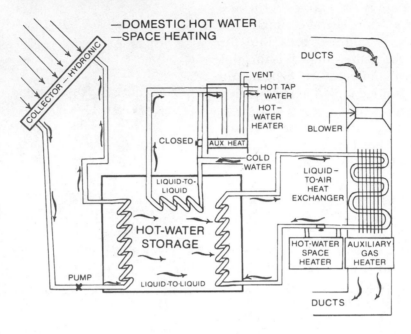

FIGURE 8-25 With this hydronic solar energy system, both domestic hot water and space heating needs can be met.

A third liquid-to-liquid heat exchanger extracts heat from the hot-water storage tank to be sent to the existing hot-air furnace. Within the furnace there is a liquid-to-air heat exchanger, which extracts the heat from the liquid to be used by the forced-air heating system.

Two auxiliary or supplementary heating systems are possible. If natural gas is used, the auxiliary heater would be located below and in series with the forced hot air. If a hot-water system is used as auxiliary heat, the heater would be located in parallel with the water pipes from the hot storage tank.

> **Building a House with Solar Energy—Energy Issue 8-8**
>
> *Suppose that you are going to build a new house for your family. To what degree would you incorporate the following technologies into your design or construction phases? Consider the costs to build, the environmental advantages/disadvantages, and the technology available for*
>
> *1. active air systems;*
>
> *2. passive air systems;*
>
> *3. hydronic hot-water systems;*
>
> *4. parabolic systems; and,*
>
> *5. more insulation.*

THERMAL STORAGE FOR SOLAR ENERGY

Collecting Energy for Later Use

A major part of any solar heating system is the thermal storage system. In the previous section of this chapter, several storage systems were mentioned as an integral part of solar heating. Thermal storage becomes a necessity if solar heating is to be available on a continuous basis. A great deal of solar radiation falls on a collector, especially when this energy is not needed for heating. With the use of thermal storage systems, much of this extra thermal energy can be collected and used later. Such systems allow energy storage during sunny days, for use at night and in overcast weather.

Thermal storage can be used in either hydronic or air systems, both of which have advantages and disadvantages. For further reference, Appendix E shows the specific heat and heat capacity of various materials.

Thermal Water Storage

Water thermal storage systems are designed so that the thermal energy from a hydronic collector is transferred to water in a large insulated container usually found in the basement of a structure. Water can absorb and store a great amount of thermal energy for its volume. Moreover, the flexibility of using hydronic collectors makes water storage popular in today's solar energy market.

If water thermal storage is used, there are certain aspects that should be considered. The water storage tank must be protected from freezing temperatures. Tanks must also be protected from corrosion—either by using corrosion inhibitors or by building the tanks with corrosion-resistant materials.

The tanks must also be insulated so that any heat stored will not escape. Sprayed polyurethane foam, mineral wool insulation, and fiberglass may all be used for storage tank insulation.

Storage tanks generally hold between 600 and 1,200 gallons of water for space heating. For domestic hot-water systems, usually a 60- to 150-gallon storage tank is used. The size depends upon the planned number of days without sunshine and the number of people in the residence. Some systems plan a two-day storage capability while others plan five days of storage capability. The usual rule of thumb for 2 to 3 days of water storage is to incorporate approximately 1.5 gallons of water storage for each square foot of collector. For example, if a building has 30 collectors, each having 21 square feet, there would be 630 square feet of collectors. Multiplying 630 times 1.5 would mean that about 945 gallons of water storage is used.

Thermal Rock Storage

For air type solar energy systems, rocks or pebbles are often used because they have the ability to absorb and store thermal energy effectively. Hot air from the collector is directed through the rock to heat it. Generally, the greater the surface area of rock exposed to the air, the greater the thermal energy storage. Because of this principle, rocks 0.5 inches to 1 inch in size are most often used. If larger rocks are used, there will be less total surface area. Smaller rocks will cause a restriction of the air flow; the collector will operate at a higher temperature, which will lead to a lower operating efficiency for the heating system.

Rock storage systems have several disadvantages. The biggest disadvantage is that rock storage requires more area than water storage.

In fact, rock storage needs approximately 2.5 times as much space as water storage.

This also means that the storage container will weigh more. The rule of thumb for sizing rock storage is to use 60 pounds (0.5 cubic feet) of rock per square foot of collector space.

Another disadvantage of rock storage is the possibility of getting dust and mold spores into the air of the building. Since mold grows in warm dark places, the warm humid air that flows through the storage area may cause problems for persons with asthma or other upper respiratory problems.

Change-of-State Storage For Thermal Energy

Change-of-State Operation

The third type of storage, thought by some experts to be the most effective, is called *change-of-state storage*. This method uses materials such as eutectic salts for the medium to store heat. In certain salts, such as sodium sulfate decahydrate, thermal energy is stored in the latent form. Latent heat is referred to as hidden heat.

Eutectic salts are typically contained in a closed structure or container. Within the container, the salts change from a solid to a liquid form as they increase in temperature. The change permits the storage of more thermal energy per pound than with other forms of storage. When eutectic salts reach temperatures of near 90° F or above (from a collector or passive gain system), the salts turn to a liquid. If cool air from a structure is blown over the container, it transfers the stored thermal energy to the structure. During cooling, the liquid changes back to a solid state again, giving up heat in the process.

A chart used to illustrate the concept of latent heat and change-of-state storage is shown in Figure 8-26. The horizontal axis rep-resents time; the vertical axis represents temperature. When placed in the sun, eutectic salts increase in temperature. As they warm, the salts remain in the solid state until their *phase change* temperature is reached (i.e., 90° F).

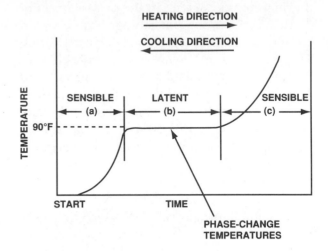

FIGURE 8-26 As thermal energy is applied to eutectic salts, both sensible and latent heat is stored. During cooling, both sensible and latent heat is given off.

At 90° F the salts turn into a liquid. The thermal energy collected up to this point is called sensible heat. Eventually, all the eutectic salts turn into a liquid above that temperature. The thermal energy stored at this temperature is called latent heat. Once all have turned to a liquid, the temperature of the liquid again begins to rise and sensible heat is again stored.

As cooler air is passed over the salt containers, the temperature of the salts begins to drop. Both sensible and latent heat are given off until all the salts turn back into a solid.

Advantages and Disadvantages of Eutectic Salts

There are advantages and disadvantages of using eutectic salts. One advantage is the large amount of heat that can be stored in a structure with limited storage space. For example, to store thermal energy from an active hot-air system, rock storage needs about 20 times as much volume in cubic feet as eutectic salts. Water requires about 2.5 times as much capacity as that of eutectic salts.

A second advantage is that heat energy can be stored with low temperature inputs from the solar collector. Anytime 90° F can be obtained from a collector, heat can be stored. Thus, collecting some thermal energy on partly cloudy or hazy days is possible.

The storage can be maintained at low temperatures as well. A 90° F temperature is all that is required to keep the salts in a liquid state. The solar collectors can then be operated at cooler temperatures, improving their efficiency. In addition, lower storage temperature will also have a slower rate of heat loss because of the smaller temperature differential between the storage area and the surrounding area.

Salt storage, however, has some disadvantages. It is expensive because of the materials and containerization, and its expected service life is approximately 20 years.

There is research ongoing in the technology thermal storage systems. Although the future is somewhat unpredictable, these advantages and disadvantages will help determine what type of thermal storage will be used in the future.

PHOTOVOLTAICS

Solar Photovoltaic Systems

Solar energy can be used to convert the sun's energy directly into electricity by using solar *photovoltaic systems*. Figure 8-27 shows a photovoltaic system operated by Virginia Power. This process is accomplished by photovoltaic cells, also called *solar cells*. The name *solar cell* is often

FIGURE 8-27 This photovoltaic solar cell system is designed to convert the sun's energy directly into electricity. (Courtesy of Virginia Power)

applied to a variety of energy-converting devices or systems. It may also be applied to devices that produce electricity from solar radiation. However, the term usually refers to devices that produce an electric current when light falls upon them. This phenomenon of converting sunlight into electricity is more properly termed the *photoelectric effect*.

Discovery of the Photoelectric Effect and Its Uses

Solar cells are devices that can convert radiant energy directly into electricity. The first photoelectric effect (i.e., conversion of radiant energy into electricity) was discovered in 1839 by a French physicist, Edmund Becquerel. However, no useful application of this effect occurred until the late nineteenth century. During this period, the photovoltaic effect was applied to photography and photocopiers. Later, in 1954, Bell Laboratories in their research on transistors and semiconductors made silicon photovoltaic cells that converted sunlight to electricity with efficiencies of about 6 percent.

The most useful early application of solar cells was for the space program. In fact, since 1958, most of the electronic components on space vehicles have been powered by solar cells. This technology is now applied in remote areas throughout the world in both industrial and agricultural societies. Applications consist of water pumping, lighting and power, refrigeration, and communication systems. Figure 8-28 shows a photovoltaic system used on oil production platforms in the Gulf of Mexico. Although small photovoltaic systems are now in use in many applications, future large-scale use is very promising.

FIGURE 8-28 Solar cells are used on this oil production rig, providing electricity for part of the electrical needs. (Courtesy of Conoco Incorporated)

Use of Photovoltaic Cells Worldwide

Shipments of photovoltaic cells worldwide have steadily increased over the last 20 years. While this technology had not been used commercially until the 1970s, it now supplies over 56 megawatts of energy worldwide. Figure 8-29 provides an overview of photovoltaic usage worldwide since 1971 with projections to 2005. Since 1982, photovoltaic systems have averaged a 15 percent annual growth rate. Should this rate of growth continue, by the year 2005 over 356 megawatts of photovoltaic cells will be in place worldwide, as illustrated in Figure 8-29. This remarkable rate of growth is a result of both cost reductions and technical improvements in solar cell construction.

Technological Improvements in Solar Cell Construction

Recent technical improvements in solar cell design, materials, and manufacturing have increased the operating efficiency and reduced the cost of solar photovoltaic systems. Efficiencies of these devices are now above 20 percent, and some experimental designs have reached efficiencies greater than 27 percent using concentrated sunlight (e.g., mirrors, and so on, to focus the light onto the cells). These improvements are a direct result of increased research funding and advances in solid-state circuitry. Since their introduction, photovoltaic systems have dropped in price by 20-fold to about $4/watt. This translates to electricity costs as low as 30 cents per kilowatt hour over their 20- to 30-year operation period. It is expected, however, that newer manufacturing processes associated with photovoltaic technology will easily reduce the cost of these devices to between $1.15 and $1.25/watt. These advances, a result of automated manufacturing processes, would reduce the cost to approximately 4 to 6 cents per kilowatt hour. Figure 8-30 illustrates the projected cost for photovoltaic systems. These systems are forecasted to decrease significantly by 2030.

Uses of Solar Photovoltaic Energy

To date the largest market for photovoltaic systems is in locations that are not served by a util-

World Photovoltaic Usage (megawatts), 1971 with Projections to 2005

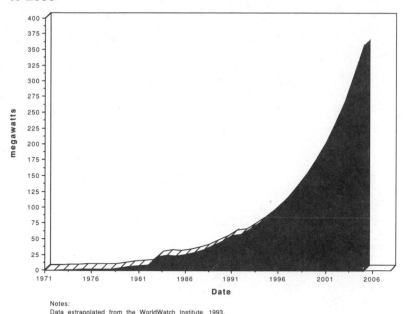

FIGURE 8-29 Shipments of photovoltaic cells worldwide have steadily increased over the last 20 years. By 2005, over 356 megawatts of photovoltaic cells are forecasted to be in place.

Notes:
Data extrapolated from the WorldWatch Institute, 1993.

ity grid. In such places the costs of purchasing, installing, and using these solar energy systems is considerably cheaper than extending electrical power from a utility company.

Energy Use and Photovoltaic Energy Systems in Developing Countries

Many developing countries face vast difficulties in supplying the needed energy to raise the standard of living for their populations. The use of inexpensive energy has long been recognized as a prerequisite for economic prosperity and industrialization. International funding institutions appear never to question the assumption that an expanding energy supply is the necessary foundation for expanding industries, providing jobs, and increasing the standard of living in developing nations.

Because of this assumption, developing countries have more than quadrupled their energy use since 1960 and per-capita use has more than doubled. The World Bank estimates that developing countries spend more than $49 billion a year for energy subsidies. Unfortunately, such initiatives have left these countries with extremely large foreign debt (from financing the burning of fossil fuels or construction of nuclear power plants), environmental damage, and still critical energy shortages. To improve this situation, developing countries are now promoting the use of photovoltaic energy systems in rural areas not served by a utility grid.

During the last 10 years, some 60,000 photovoltaic systems have been installed in developing countries. Algeria, Brazil, Colombia, the Dominican Republic, Kenya, Mexico, Sri Lanka, Thailand, and Venezuela have made major investments in these technologies. The countries of Africa have also experienced an

Average Prices for Photovoltaic Modules with Projections to 2030

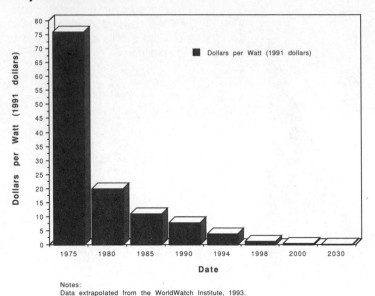

FIGURE 8-30 The projected cost for photovoltaic systems is forecasted to be less than 1 dollar per watt by 2030.

Notes:
Data extrapolated from the WorldWatch Institute, 1993.

unprecedented boom in photovoltaic power systems since the mid-1980s. In 1991, Botswana installed over 100 kilowatts of photovoltaic power. In Zaire, a republic in central Africa, photovoltaics were used at the Bulape Hospital to power over eighteen buildings without backup systems. Also, medical clinics in rural areas of Rwanda, Tanzania, and Zambia have used photovoltaics for over twenty years to supply energy for refrigerators that store vaccines.

Photovoltaic Energy Use in the United States
The largest user of photovoltaics in the United States is the Coast Guard. Having some 10,000 navigational devices with photovoltaic systems to supply their power requirements, the Coast Guard has been able to reduce taxpayer costs for maintenance. The Coast Guard saves taxpayers approximately $5,000 per device by eliminating the need to

remove, replace, and dispose of batteries from buoys and lighthouses.

Although using photovoltaics to power residential and commercial structures is not currently cost-competitive unless these structures are not connected to a utility grid, their use is increasing. This increase is a result of a utility company augmenting its rates during peak times of the day (i.e., when energy demand is high, utility companies may charge more for electricity). In such a situation the use of photovoltaics may be competitive with or even cheaper than utility-supplied electricity. The average costs for electricity in the United States are between $0.04 and $0.20 per kilowatt-hour. Peak utility rates may be as high as $0.25 per kilowatt-hour.

Another reason that photovoltaics are becoming more popular is that excess energy generated by the solar cells can be sent back to the utility company. Should the solar cells placed on top of a structure generate more energy than can be used, this energy is sent back to the utility company to be used by other consumers. This typically results in a utility rebate for the owner of the solar cells.

Producing Solar Photovoltaic Energy

As noted earlier, the basic component of a photovoltaic system is the solar cell. These cells can convert sunlight into electricity (i.e., the photoelectric effect). Many metals and some nonmetals exhibit the photoelectric effect. When light strikes the atoms contained within these substances, energy is imparted to the outer loosely-bound electrons. If the imparted energy is capable of moving the electrons away from the atoms and if a load is connected in a circuit with the electron-emitting substance, an electric current will flow.

Figure 8-31 illustrates graphically how solar energy is converted into electricity. As the sun shines on the solar cell, the light is absorbed into the cell and knocks electrons loose from the atoms it strikes. To allow this to happen, solar cells are made of a semiconductor material. Semiconductors are materials that allow the flow of electricity at a rate between that of metals (high conductivity) and insulators (no conductivity). Most semiconductors, even those used in transistors, have two materials that constitute a junction. For example, phosphorus-silicon, considered a negative material (−) is frequently paired with a mixture of boron-silicon, considered a positive material (+). The negative material is transparent and very thin. Electrons in the material absorb energy from the sunlight that bombards it. This causes a voltage to be set up across the junction. The voltage can then be used to power electrical circuits.

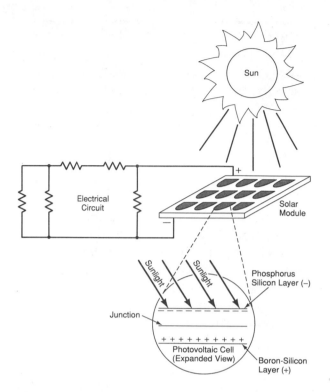

FIGURE 8-31 When sunlight strikes a photovoltaic cell, voltage is produced.

Once the solar cells are manufactured, they are connected to obtain a module. Today, a 4-foot by 4-foot solar module is capable of producing between 12 and 48 volts. Depending upon manufacturing design, modules are often available with 35, 36, or 40 cells. For example, a 4-foot by 6-foot module can be purchased in 12, 24, 36, or 48 volt versions. For large applications, modules can be connected to form a solar array. Various voltages and current capabilities can be obtained depending upon the connections in the solar array. Thus, any power output can be supplied by combining modules in various arrays.

Concentrating Photovoltaic Systems

Photovoltaic cells are made more efficient by
increasing the sun's energy on the cell. Various
types of reflectors and concentrators are under
investigation to increase the insolation on the
solar cell per unit area. To do this, materials
within solar cells must withstand greater temperatures—a difficulty at present.

The photovoltaic system shown in Figure
8-32 uses mirrors to concentrate sunlight onto
the solar cells. This type of system is referred to
as Concentrating Photovoltaic Cells or CPVC.
Note that when concentrating collectors are
used, the unit has to track the sun, which
improves both efficiency and electrical output.
Experimental designs with these collector systems have efficiencies greater than 27 percent
with 300 kilowatt outputs.

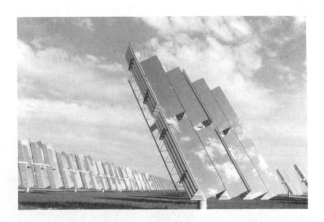

FIGURE 8-32 These platforms and mirrors are used to
concentrate the sun's radiation onto the solar cells in the
center to improve efficiency. (Courtesy of Arco Solar
Inc.)

Several other solar cell facilities are also
under consideration by the government and
private industry. One of the more promising
plans is shown in Figure 8-33. Here, a platform
of concentrating collectors reflects the sun's
energy onto solar cells located beneath the
center strip on the top and the bottom sections
of the platform. In operation there may be as
many as 300 to 400 platforms, all producing
electricity. It is estimated that approximately 30
megawatts could be produced with this many
platforms. Small facilities are also being
designed for use on university campuses, in
small cities, and so on.

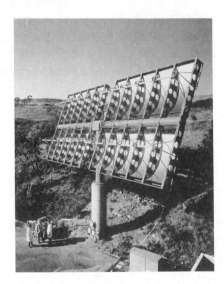

FIGURE 8-33 This solar concentrating platform uses
trough-type mirrors to direct the sun's radiation onto
solar cells located beneath the two center strips.
(Courtesy of Alpha Solorco)

Another futuristic idea is to design paint and
other colored materials that behave as a solar
array. Potential applications include automobile, boat, train, or airplane finishes. The
electrical energy collected could charge a
group of DC batteries used to operate motors

for propulsion. This is already being studied by several automotive manufacturers. Although highly experimental, the idea does have interesting implications for the implementation of solar energy in the transportation sector.

Another possibility is to manufacture a roof shingle that has solar cells built into its structure. NASA has already developed and tested such a device. Such shingles are placed onto a collection strip, developing almost any combination of voltage and current necessary for the structure.

ALTERNATIVE FUTURES

Electric Vehicles

Emissions from the transportation sector are the leading cause of urban smog. Despite federal legislation to improve the smog controls and fuel efficiency of new cars, over 90 United States cities fail to meet federal clean-air standards. Additionally, more than half of all people in the U.S. live in areas where smog levels exceed federal standards; and, these polution levels are still on the rise.

Efforts to reduce vehicle exhausts are stymied by the necessary growth in the country's use of automobiles. According to the Union of Concerned Scientists, the air quality is expected to worsen in the coming years as a direct result of vehicle exhausts. By 2010, emissions of pollutants are projected to increase by about 40 percent because of increased driving in congested conditions.

In response to these forecasts, there has been increased federal and state legislation to better the air quality in the country by reducing harmful vehicle emissions. Besides attempting to clean up tailpipe emissions, specific states have taken even stricter approaches to vehicle exhausts. California has become the first state to require 2 percent of all new cars to emit no pollutants starting in 1998. After 2003, 10 percent of all new cars sold in California must meet zero-emission standards. Yet, the question remains: are zero emissions possible?

The answer to the above question is yes—zero emissions are possible, but not feasible using fossil fuels. There is only one practical and economic solution to meeting California's requirements of zero tailpipe emissions: to require the use of electric vehicles.

A joint study by the Electric Power Research Institute and United States Department of Energy determined that switching to electric vehicles could almost eliminate nonmethane organic gases—such as hydrocarbons, carbon monoxide, and particulates. Additionally, electric vehicles would allow for deep reductions in nitrogen oxides almost immediately. According to this study, electric vehicles would provide a 30 percent or more reduction in greenhouse gases. Given these projections, electric vehicles provide a reasonable action plan for combating global warming.

Speed and distance have always been the holdup in electric-car production. Today, the automobile industry is working to overcome these drawbacks in order to develop electric vehicles to reduce global warming and meet California's strict emission standards. One example of an electric vehicle is General Motors Corporation's Impact test car. The Impact is a two-seater sports coupe that can accelerate from 0 to 60 mph in 8 seconds and has a top speed of 75 mph. This electric vehicle can travel 120 miles before it requires recharging from a standard residential wall outlet. This automobile has fewer moving parts than a conventional vehicle and is expected to require less maintenance over its useful operation period.

Overall, the potential market for electric vehicles is extremely large. Cities such as Baltimore, Chicago, Hartford, Houston, Los Angeles, Milwaukee, New York, Philadelphia, and San Diego have been identified as not meeting federal clean-air standards. The introduction of electric vehicles into these areas could prove to be financially beneficial to automakers as well as environmentally beneficial to citizens. Just over half of all United States' households have more than one car. The second car could be replaced with an exhaust-free electric vehicle for commuting and short trips—thus helping to reduce global warming, acid rain, and improving urban air quality.

SOCIAL AND ENVIRONMENTAL ASPECTS OF USING SOLAR ENERGY

Social and Environmental Benefits of Solar Energy

Solar energy technologies have numerous benefits over the use of traditional fossil fuels or nuclear energy. Solar technologies are already a cost-effective way for societies to save the finite resources of oil and natural gas. The capital cost for solar hot water are, on average, nearly 25 percent lower than those for hot water using electric energy from power plants. The energy source is both nondepletable and free, and solar technology is the least destructive to the environment. The use of these technologies do not contribute to greenhouse gases or the production of environmentally harmful by-products.

Social and Environmental Costs of Active and Passive Systems

The building of the equipment used to manufacture solar energy systems is likely to require the burning of fossil fuels or production of nuclear energy. In this aspect, solar energy technologies require the depletion of nonrenewable energy sources for their construction and installation (from both the energy sources and the materials used). Solar energy systems may also use fluids that are toxic (producing corrosion or pollution) and have to be changed over time. These substances, like other hazardous wastes, must be disposed of in a proper manner to assure protection to the environment, humans, and other species.

Social and Environmental Costs of Photovoltaic Energy Systems

Again, it needs to be repeated here that the production of electricity from photovoltaics generates no air pollution, and the production of silicon solar cells contributes to very little waste. However, an estimated 97 percent of the environmental risks of photovoltaic energy systems are a result of the construction of these devices. In the manufacture of these systems, particularly those using newer technologies, photovoltaic materials can be potentially toxic if not handled and disposed of properly. The problem is a result of solar cells developed with materials like gallium arsenide or cadmium sulfide. The production of these newer solar cells creates some toxic waste that must be disposed of or recycled in a safe manner.

POINT/COUNTER POINT 8-1

TOPIC:

Funding Research and Development for Solar Energy

Theme: In the past few years the emphasis on solar energy research has remained rather stable, and the use of solar energy has been applied primarily to the residential sector. There is also very little incentive (financial or social) to install solar energy systems in other sectors of society. Research funding needs to be available for both the promotion of solar energy and to develop newer systems with increased efficiency. This research funding could come from a variety of sources, including tax incentives from the federal and/or state governments, from private industries, user fees (i.e., taxes) placed on energy resources, or increased taxes placed on goods or services requiring the use of energy-intensive industries. In a discussion or debate, attempt to answer the following questions:

1. What consequences would be experienced by taxpayers if the government provided tax incentives for solar energy research or installation?
2. How can solar energy be used in enough applications to make it economically advantageous for industry to provide continued research?
3. How can the price of solar energy systems be reduced to make them more competitive with fossil fuel systems?
4. What social mechanisms are necessary to educate consumers about the need to continue research and development of solar energy systems, even if this may mean increased taxes or costs of consumer goods?
5. What relationship exists between the projected development of solar energy and the existing price structure of fossil fuels?
6. To what type of applications could solar energy systems be applied in the various energy-using sectors of society?
7. What are the environmental and social implications of large-scale development of solar energy systems?

SUMMARY & REVIEW

Summary/Review Statements

1. Solar energy is the radiant energy transmitted to the earth from the sun. It has provided, either directly or indirectly, almost all of the sources of energy for the earth since its formation.

2. Despite the wealth of historic evidence on the successful use of solar energy, it has not been employed on a broad basis.

3. The amount of solar energy that arrives in a two-week period is equivalent to the fossil energy stored in all the earth's known reserves of coal, oil, and natural gas.

4. At the outer limits of the atmosphere, the average intensity of incoming solar energy is equal to 1.36 kilowatts per square meter. This number is called the solar constant.

5. Solar energy can be characterized as either direct or indirect. Direct solar energy means that the sun's radiation is converted by technology directly into a usable form of energy. Indirect solar energy means that the sun's radiation is first converted to other forms of energy before being used.

6. Thermal systems are designed to convert the sun's radiation directly into thermal energy for use in the commercial or residential sectors.

7. Solar energy is abundant and could be more widely adopted and developed for the residential and commercial sectors.

8. Solar energy comes to the earth in a form known as electromagnetic radiation. Visible light, sound waves, and radio waves are all examples of electromagnetic radiation.

9. Insolation is solar radiation received per unit area for a given unit of time on the surface of the earth. While insolation can be measured in several units, the most common is $Btu/ft^2/hr$.

10. Solar collector efficiency is the rate of useful heat collected to total solar radiation on the surface of the collector.

11. Of the installed flat-plate collectors, the majority have been placed in residential dwellings. Two major types have been adopted: the hot-air collector and the hydronic (water) collector.

12. Focused collectors can produce much higher temperatures than flat-plate collectors. Since focused collectors can produce steam, they can be used in industrial applications—such as electric utilities and any manufacturing enterprise requiring temperatures greater than 200° F. (Focused collectors can produce temperatures to 4,000° F.)

13. Focused collectors can be divided into four basic types—parabolic dishes, parabolic troughs, central receivers, and solar ponds.

14. Parabolic dishes concentrate radiant energy by using a reflecting surface, which concentrates the sun's parallel rays to an absorber that is at a focal point in front of the dish.

15. Parabolic troughs also focus sunlight to a focal point to concentrate the sun's thermal energy. In this design the absorber is a vacuum-enclosed glass or metal tube that extends the length of the focal point of the trough.

16. Central receivers are solar collectors that use many large mirrors to focus the sun's light onto an absorber stationed on a high tower. The absorber in this type of a solar collector is called the receiver, and the large mirrors are called heliostats.

17. Solar ponds are bodies of shallow water that trap heat at the bottom.

18. Solar orientation is the study of how a structure or a solar collector is placed in relationship to the direction of the sun.
19. Although heat exchangers are not necessarily a part of an active solar heating system, many designers use these devices to transfer the heat collected. Generally, heat exchanges are used if the heat gathered by a solar collector must be transferred or exchanged to other sections of the heating system.
20. Thermal storage becomes a necessity if solar heating is to be available on a continuous basis. Thermal storage systems allow energy storage during sunny days for use at night and in overcast weather.
21. Change-of-state storage uses materials, such as eutectic salts, for the medium to store heat. In certain salts, such as sodium sulfate deca hydrate, thermal energy is stored in the latent form. Latent heat is referred to as hidden heat.
22. Solar energy can be used to convert the sun's energy directly into electrical energy by using solar photovoltaic systems.
23. Converting sunlight into electricity is termed the photoelectric effect.
24. Recent technological improvements in solar cell design have reduced the cost of solar photovoltaic systems. Efficiencies of these devices are now above 20 percent, and some experimental designs have reached efficiencies greater than 27 percent using concentrated sunlight.
25. To date, the largest market for photovoltaic systems is in locations that are not served by a utility grid. In these locations, the costs for purchasing, installing, and using these solar energy systems is considerably lower than extending electrical power from a utility company.

Discussion Questions

1. What factors could be changed in an active solar collector to improve efficiency of operation?
2. How can solar energy be used in various energy-using sectors of society? What technologies must be improved to accomplish this task? What role can education play in accomplishing this task?
3. Estimate a time period by which solar energy may constitute more than 60 percent of the total energy used in the residential sector. List some constraints that influenced your selected time period.
4. Identify three disadvantages and three advantages of using active solar space heating systems.
5. What are the advantages of solar energy over fossil fuels?
6. Why are utility companies reluctant to adopt solar energy technologies in the production of electricity?

chapter 9

Indirect Solar Energy Resources

PURPOSES

Goals of Chapter Nine

While the direct use of solar energy is researched and implemented, indirect solar energy resources are also developed and utilized. As energy demand continues to expand and hydrocarbon fuel supplies decline, more energy will be needed to supply the global society. Of particular need are energy resources that are reliable, renewable, or non-depletable, and as environmentally benign as possible.

Clearly petroleum has supplied the energy for industrialization for the past 50 years. However, oil cannot be considered a reliable energy source for the indefinite future. During the past 20 years, there have been three periods where oil prices have risen sharply. Oil supplies are finite and limited to certain geographic areas. About three-fourths of the world's reserves are located in the politically unstable Persian Gulf region. Additionally, the burning of petroleum (and other fossil fuels) has contributed to global warming. Scientists have noted that the combustion of all the world's remaining fossil fuels would raise the concentration of carbon dioxide in the earth's atmosphere by as much as tenfold. Such a rise would cause unprecedented destruction to the earth's environment and place the continued existence of humans in danger.

Today, the only known recourse to reverse the effects of global warming is to place limits on the combustion of fossil fuels. This prospect, however, is difficult for both developing and industrialized societies. With the growing world population and pressures to provide economic growth, countries are becoming more dependent upon the exploitation of energy. Therefore, to stabilize global warming, emphasis has been placed on increasing the energy efficiencies of devices that burn fossil fuels and the development of renewable and nondepletable energy sources.

The overall purpose of this chapter is to explore various types of indirect solar energy currently used or under research. The status, availability, significance, and environmental advantages and disadvantages of indirect solar energy are presented. At the completion of this chapter, you will be able to

1. Investigate the use of wind as a possible energy resource.
2. Examine the use of bioconversion as an energy resource.
3. Evaluate the technology associated with hydroelectric energy.
4. Determine the advantages and disadvantages of using tidal power as a possible energy resource.
5. Define and delineate the associated technology of geothermal energy.
6. Identify the process of extracting energy through ocean thermal energy conversion (OTEC).

Terms to Know

By studying this chapter, the following terms can be identified, defined, and used:

Aerobic
Anaerobic
Bioconversion
Biomass
Darrieus vertical axis turbine
Dry steam hydrothermal reservoirs
Energy winds
Geothermal energy
Hot-dry rock
Hydraulic fracturing
Hydro energy
Hydrologic cycle
Hydrothermal reservoirs
Indirect solar energy
Multivane
Normal thermal gradient
Ocean Thermal Energy Conversion (OTEC)
Penstocks
Power coefficient
Prevalent winds
Thermal gradient
Tip speed ratio
Wet steam reservoirs
Wind farms
Yaw

INTRODUCTION

Today, many types of energy resources are being studied for future use. One category, called *indirect solar energy* resources, are those that come from solar energy, but are not in the correct form. These resources must be converted to another form before being used. For example, wind energy, although considered a form of solar energy, must be converted to mechanical energy or electricity before it can be used. Thus, it is not direct, but indirect solar energy.

Another form of indirect solar energy is that of hydroelectric energy. The sun's energy causes the rain cycle to occur. So, water that is held back by a dam can be considered a form of solar energy. However, it is not in the correct form for use by our society; thus, it is considered a form of indirect solar energy.

This chapter presents various forms of indirect solar energy. Several are already being used. However, all of these indirect energy resources will become part of the energy supply in the future.

WIND ENERGY

History of Wind Energy

Harnessing the wind's energy for useful purposes is not a new idea. Since the earliest of recorded times (some five thousand years ago) humans have used the wind to power their sailing ships. In addition, since the seventh century, wind power has been harnessed to pump water for crop irrigation and livestock, for turning millstones to grind flour from grain, and for the sawing of wood. The popularity of wind power grew most rapidly until hydrocarbon fuels and the internal combustion engine became prominent in the early 1900s. In fact, by the end of the nineteenth century, several hundred thousand windmills were in place throughout the world. In the United States from 1850 to 1970, some 6 million wind machines were installed. These machines produced about 3 kilowatts of power in a 15 mile per hour wind and helped farmers and ranchers to meet their daily water needs.

Harnessing the wind's energy for useful purposes is not a new idea.

Eventually, wind power fell out of favor as a power source with the introduction of low-cost and reliable fossil fuels. This was a result of the introduction of the steam engine. Recently, however, because of dwindling fossil energy resources and steady increases in energy demand, wind energy has been identified as a means for producing electricity.

Availability of Wind Energy

Wind is produced because of sunlight. As the sunlight falls on different areas of the earth,

the atmosphere is unevenly heated. Since warm air weighs less than cool air and rises as the temperature increases, air currents are created. On a global scale, air currents consist of cool polar air drawn toward the tropics to replace the lighter warmer air that rises and moves to the poles. Even within this natural flow pattern, consistent winds develop because of the uneven heating of the earth.

Of the solar energy that reaches the earth from the sun, only 2 percent is converted into wind. Yet this small percentage contains more energy than humanity uses in one year. Of the 2 percent, about 30 percent is generated below the 3,120 foot level of atmosphere, which allows it to be used as an energy source. Within the continental United States, the wind energy available is about 9 times the total amount of energy now being used.

Use of Wind Energy

Worldwide Use

During the past 15 years, tremendous progress has been made in the technology used to convert wind into electrical energy. These technical advancements have allowed wind power to expand significantly worldwide. Figure 9-1 illustrates the dynamic increase in wind power since 1981 with projections to 2005. As can be seen by Figure 9-1, by the early twenty-first century, wind energy could produce more than 17,700 megawatts of generating capacity. This increase in the growth of wind power is expected to be concentrated in the United States (California) and spread throughout Europe as governments encourage investment in this energy source.

The potential for generating electricity with wind power is enormous. In Europe, the use of wind power could eventually satisfy all the continent's electrical needs. The European

World Wind Energy Generating Capacity (Megawatts), 1991 with Projections to 2005

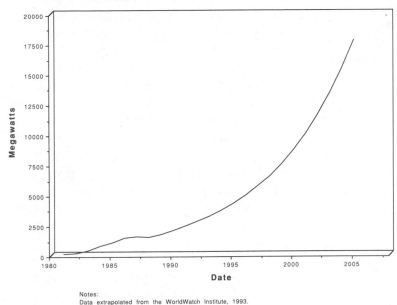

Notes:
Data extrapolated from the WorldWatch Institute, 1993.

FIGURE 9-1 This chart illustrates the increase in wind power since 1981 with projections to 2005. Wind energy could produce in excess of 17,700 megawatts of generating capacity by the early twenty-first century.

Community is now actively pursuing wind energy as a means of satisfying its energy requirements. This body has adopted a program to obtain some 8,000 megawatts of electricity by the year 2005. This ambitious goal would satisfy about 1 percent of the electrical needs for the region.

To date, the largest share of electricity generated with wind energy can be found in Denmark. In this country, wind generators are distributed throughout the country and supply 3 percent of the region's electricity. Other countries taking major initiatives in wind power include Germany, the Netherlands, and the United States.

Use in the United States
In the United States, wind energy currently supplies over 2.5 billion kilowatt-hours of electricity. In theory, the wind generated in the Great Plains could meet all of the nation's elec-

tricity requirements several times over. The growth of wind generation in the United States is forecasted to increase substantially over the next 10 years. This expected increase is a result of governmental initiatives that provide a tax credit of 0.015 cents per kilowatt-hour of electricity generated from wind. Plans now exist in several states to capitalize on this tax credit. For example, Iowa, Maine, Minnesota, North Dakota, South Dakota, and Washington are expected to develop wind farms over the next decade. *Wind farms* consist of a group of wind generators or turbines installed in a central location to produce electricity for sale to a local utility.

The world leader in the development of wind energy is the state of California with more than 16,000 wind generators currently in use. California's growth in the progression of wind energy started over a decade ago. At that time, California, like many other states, was dependent on fossil fuels for the generation of electricity. In response to high petroleum costs during the mid-1980s, the federal government offered investment and energy tax credits for the development and installation of renewable or nondepletable energy sources. These tax credits, which equaled 25 percent of the total investment, helped to create the proper incentives for wind development. From 1982 to 1986, wind power generation increased in California from 6 million kilowatt-hours to over 1.2 billion kilowatt-hours.

While many of the energy tax credits implemented during the 1980s have been removed, the use of wind energy has continued throughout the United States. Utility companies and entrepreneurs in California have continued to install wind farms to meet the increasing need for electrical generation. Today, more than 16,000 wind generators are operating in California. These machines have the capability of producing more than 2.9 billion kilowatt-hours of electricity—enough electricity to meet the needs of more than 1.2 million people. More wind farms are planned for the future.

The attractiveness of wind energy in the United States and worldwide is a result of several unique attributes of this resource. First, the potential of wind energy in meeting the world's electrical needs is huge. According to energy experts, if the world's wind resources were developed to their maximum potential, about 210 trillion kilowatt-hours of electricity could be produced annually. Note that this figure far exceeds current global electricity production. Second, electrical generation by wind power produces no air pollution. Third, it produces no acid rain or carbon monoxide and carbon dioxide emissions. Finally, it is the least costly of all the newer technologies in its extraction and conversion to useful energy.

Classification of Wind

Wind can be classified into two types: energy wind and prevalent wind. Before studying these two classifications, note that wind speed is dependent upon geographic location. Stronger energy winds are found mainly in the temperate and polar regions; weak or prevalent winds are in the tropics. In addition, winds are stronger in oceanic and coastal areas than in continental regions. Winds generally blow stronger in mountain areas as well. Figure 9-2 shows the location of wind energy throughout the United States. To show the effect of local topography, Figure 9-3 lists the average ground wind speeds, in miles per hour, for several cities within the United States.

Energy winds range in speed from 10 to 25 miles per hour, and blow at this rate about two to three days per week. *Prevalent winds* are slower—averaging 5 to 15 miles per hour—and blow between three to five days each week. Energy winds are the more effective type of wind for use in electricity production because of their greater speeds. Prevalent winds are less efficient for conversion to electricity and are, therefore, less economically attractive.

Scientists at the Pacific Northwest Laboratory in Richland, Washington, have developed another categorizing scheme to describe the characteristics of wind. In this scheme, wind is classified in seven different ways. Class 1 winds have an average speed of less than 12.5 miles per hour. Class 7 winds are equal to or are greater than 19.7 miles per hour. Note that most electrical generation from wind machines occurs in areas with Class 5 winds (i.e., those that average about 17 miles per hour), but improvements in windmill design are expected to allow electrical generation with wind speeds as low as 14 miles per hour (Class 3 winds).

Wind Energy Locations

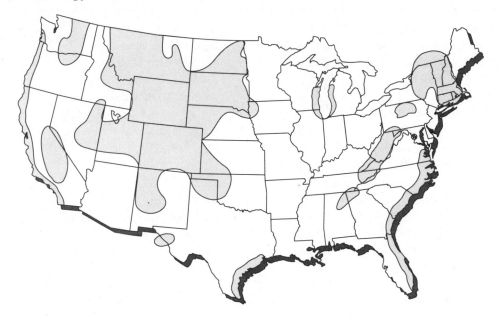

FIGURE 9-2 This map shows the areas within the United States that have high enough wind speeds for wind turbine/generator operation.

Factors Affecting Wind Energy Output

Three ingredients determine the amount of energy that can be obtained from the wind. These are the altitude of the wind turbine, the size of the wind turbine rotor, and the wind speed. Figure 9-4 shows that the higher the altitude the greater the speed of the wind. For example, at 100 feet in the air, the wind speed would be about 30 miles per hour. If the wind turbine rotor were higher, say 200 feet, the speed of the wind striking the rotor at this level would be 40 miles per hour.

The rotor size of the wind turbine also affects the energy produced. Figure 9-5 shows the relationship between the average wind speed in miles per hour, rotor size, and yearly energy production. Various rotor diameters are graphed. Referring to the figure, you can see that when the speed of the wind increases or the size of the rotor increases, annual energy production also increases. Thus, to get the most energy from the wind, the wind turbine rotor should be as high as practically and economically possible, the rotor should be as large as possible, and the wind generator should be placed in the highest practical wind speed.

Wind Turbine Configuration

There are two basic configurations of wind turbines: the horizontal axis and the vertical axis. With horizontal axis turbines, the blades are attached to a shaft that is parallel to the ground. Vertical axis turbines use a shaft that is perpendicular to the ground. These two configurations have been used for centuries. With

Average Ground Wind Speeds in Major Cities (Miles Per Hour)

City	Speed	City	Speed
FARGO, North Dakota	14.5	DETROIT, Michigan	10.5
WICHITA, Kansas	13.5	DENVER, Colorado	9.9
BOSTON, Massachusetts	13.0	KANSAS CITY, Missouri	9.7
NEW YORK, New York	12.7	ATLANTA, Georgia	9.6
FORT WORTH, Texas	12.4	WASHINGTON, D.C.	9.6
DES MOINES, Iowa	12.0	PHILADELPHIA, Pennsylvania	9.5
HONOLULU, Hawaii	12.0	PORTLAND, Maine	9.5
MILWAUKEE, Wisconsin	12.0	NEW ORLEANS, Louisiana	8.9
CHICAGO, Illinois	11.5	LITTLE ROCK, Arkansas	8.8
MINNEAPOLIS, Minnesota	11.5	SALT LAKE CITY, Utah	8.8
CLEVELAND, Ohio	11.0	MIAMI, Florida	8.9
INDIANAPOLIS, Indiana	10.7	ALBUQUERQUE, New Mexico	8.6
PROVIDENCE, Rhode Island	10.6	TUCSON, Arizona	8.0
SEATTLE-TACOMA, Washington	10.6	BIRMINGHAM, Alabama	7.8
SAN FRANCISCO, California	10.5	ANCHORAGE, Alaska	6.7
BALTIMORE, Maryland	10.5	LOS ANGELES, California	6.7

FIGURE 9-3 Average ground wind speeds (in miles per hour) in major cities are illustrated in this table.

Altitude vs Wind Speed

FIGURE 9-4 The higher the wind turbine is placed on a tower, the higher the average wind speed.

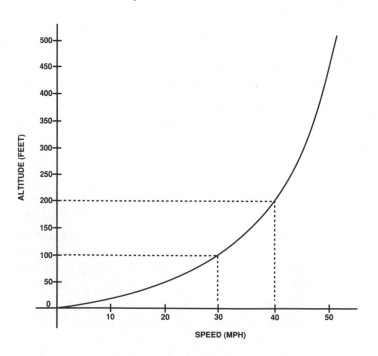

either type, the wind provides mechanical energy to turn a turbine—either on a horizontal or vertical shaft. The mechanical energy is then used to pump water or to turn an electrical generator.

Both direct current and alternating current generators are used in wind machinery. Direct current (DC) units can be used for smaller and more remote locations. They can also be connected into the public utility grid after the DC has been converted to AC voltages via an inverter. A typical DC generating system operates when its turbine blades are rotated by the wind to turn a generator. The energy from the generator is then used to charge a battery system, which feeds the electric demand. See Figure 9-6.

As the electrical energy is generated by the turbine and generator, the voltage is regulated before it goes into the battery storage bank.

Batteries are used to provide DC storage. Usually 12, 24, 110, or 220 volts DC can be produced by using a specific generator. All electrical energy including electricity produced during periods of low demand is stored in the batteries. If AC voltage is needed, an inverter will change the DC to AC before being sent to the local utility.

In the operation of a wind turbine, there are advantages and disadvantages to both AC and DC systems. The majority of all domestic electrical equipment uses AC. This is because AC is transmitted more efficiently over long distances through the utility grid. Direct current could be used in long-distance transmission, but much of the energy would be dissipated as heat. Thus, such a transmission scheme would be less efficient than using AC. The major advantage of DC systems is that they can store energy in batteries.

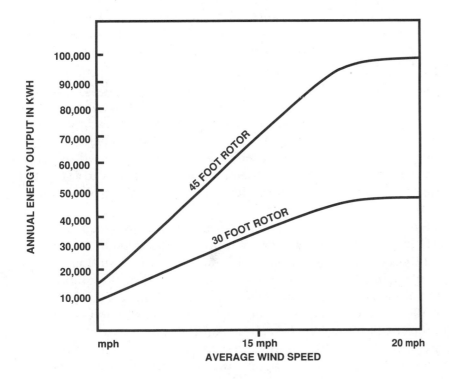

FIGURE 9-5 The larger the wind turbine rotor and the higher the wind speed, the greater the wind turbine output. (Courtesy of Northern States Power)

turbine and rotor. In practice, most wind turbines can extract, overall, about 35 percent of
the wind's energy. This efficiency represents an
overall efficiency rate (a combination of both
the rotor and the generator). Note that a 35
percent efficiency rating is comparable to the
efficiency of many conventional fossil-fuel
power plants.

Horizontal
Wind Turbine

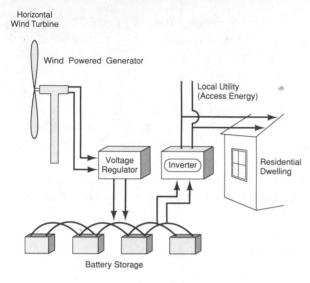

FIGURE 9-6 When a wind turbine is used in a single
structure, additional components—such as battery storage, voltage regulators, and inverters—most often are
part of the system.

Power Coefficient

Each type of wind machine has a specific *power
coefficient*, which is a measure of the energy that
can be extracted from the wind by the turbine.
Because of the dynamics of wind and the manner in which wind causes a propeller to spin,
theoretically only 59.3 percent of the energy in
wind can be extracted. If that percentage of the
wind energy has been extracted, the machine is
said to have a power coefficient of 100 percent.
If a wind turbine has a power coefficient of 80
percent, this means that only 80 percent of the
59.3 percent maximum can be extracted by the

Tip Speed Ratio

Tip speed ratio is an important characteristic
of wind generators because it helps to determine the best speed of a wind turbine rotor.
Tip speed ratio is the ratio of the tip of the turbine rotor blade to the wind speed. A tip speed
ratio of 6 to 1 means that the tip of the wind
turbine blade is moving 6 times as fast as the
wind speed. Each type of wind machine operates most efficiently at a definite tip speed
ratio. Tip speed ratios range from 1 to 1, to 6 to
1 or more, depending upon the type of windmill used.

Figure 9-7 illustrates the tip speed ratio of
three types of wind machines. Each has a maximum tip speed ratio and the resulting power
coefficient. The vertical axis of the graph represents the power coefficient, or the percentage of energy that can be extracted from the
59.3 percent maximum. The horizontal axis on
the graph shows the tip speed ratio. Referring
to the figure, the *multivane* fan wind turbine,

Ratio of Tip Speed to Power Coefficient

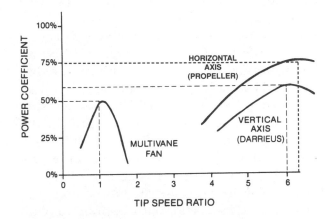

FIGURE 9-7 Each type of wind generator has an optimum power coefficient determined by the tip speed ratio.

often used for pumping water on farms, operates most efficiently at a tip speed ratio of 1 to 1. This means that the tip speed of the turbine and the wind are moving at the same speed. Under these conditions the power coefficient would be about 50 percent. As the speed of the multivane windmill increases beyond this point, the power coefficient drops. The multivane wind turbine should, therefore, be operated at a speed to produce the tip speed ratio of 1 to 1. Since this is a slow speed, this design is not suitable to generate electricity.

In the vertical axis turbine, the tip speed ratio of 6 to 1 produces a power coefficient of about 58 percent. However, the power coefficient decreases as the tip speed ratio goes beyond 6 to 1. The power coefficient of the horizontal axis (propeller) wind turbine rotor is highest (about 75 percent) at a tip speed ratio of 6.25 to 1. This design of wind turbine is often used to produce electricity on a large scale. Each wind-powered generator should be operated at a predetermined (optimal) speed to produce maximum efficiency. Since the

maximum efficiency of a windmill occurs at a definite ratio between wind speed and tip speed, wind turbine controls are added to keep the blades at a constant and efficient number of revolutions per minute (rpm).

Types of Wind Turbines

As noted earlier, most of the wind turbines designed and used today fall into two basic categories: horizontal axis and vertical axis. Almost all of the large federally funded projects undertaken in the past 10 years have used the horizontal axis type. There are, however, other types of wind turbines that are considered practical from both economic and technological standpoints. Figure 9-8 illustrates various profiles of different wind turbines. The first wind turbine is called the ***Darrieus vertical axis turbine***. The other profiles shown are all horizontal axis wind turbines.

Vertical Axis Wind Turbines

Various types of vertical axis wind turbines have been designed and tested. However, the three

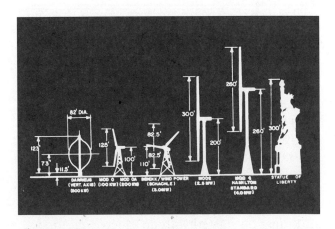

FIGURE 9-8 There are various types of wind turbine systems tested and used today. This is a comparative profile of different wind turbines. (Courtesy of Southern California Edison Co.)

FIGURE 9-9 A Darrieus wind turbine can accept wind from any direction easily. This vertical wind turbine/generator is capable of supplying enough electricity for 50 houses. (Courtesy of American Petroleum Institute)

basic styles that have emerged are the Darrieus wind turbine, the Savonius rotor, and the Giromill.

The Darrieus wind turbine shown in Figure 9-9 has airfoil blades that can accept wind from any direction. It is named after the French engineer G. J. M. Darrieus, who conceptualized this design in the 1920s. This unique configuration does not need to redirect itself as the wind direction changes. The Darrieus wind turbine resembles the lower section of a kitchen egg beater. The electrical generator is connected through gearing to a vertical shaft, which in turn provides the electrical output. Note that the Darrieus wind turbine is not self-starting and must be started with an electric motor.

In the Savonius wind machine, the output torque is controlled by an adjustment of air flow, as illustrated in Figure 9-10. This design consists of a cylindrical shell, split in half and mounted to rotate between circular top and bottom plates. The two halves separate so that wind may flow between them. When the center passage is closed, shown in Figure 9-10(a), air circulation is not allowed between the two blades. This causes low pressure to develop on the back side of the blade. The low pressure has a slowing effect on the rotor. If the blades are adjusted, as shown in Figure 9-10(b), then the area that was a vacuum now has a pressure because of the wind passage through the blades. This would cause the torque output to be increased by a factor of three or more. With this design, turbine speed can be controlled.

The third type of vertical axis wind turbine is the Giromill, illustrated in Figure 9-11. The rotor consists of a set of blades attached to the axis with support arms. As the rotor turns, the position or pitch of the blades changes to capture the maximum force of the wind.

Savonius

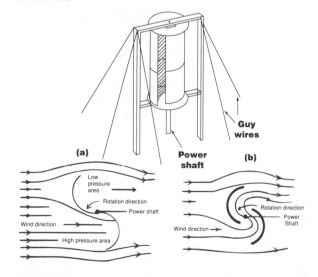

FIGURE 9-10 This Savonius wind turbine has adjustable baffles so that a maximum tip speed ratio can be obtained.

Giromill

FIGURE 9-11 This vertical wind turbine is called the Giromill. It has adjustable outside foils to control turbine speed.

Horizontal Axis Wind Turbines

The horizontal axis wind turbine is capable of generating large quantities of electrical power. This turbine can either be designed in an upwind or downwind configuration. Upwind means that the wind hits the turbine blade before the generator; a tail vane is needed to keep the assembly pointed into the wind. Downwind means the wind hits the generator first, then the turbine blades. The major advantage of the downwind configuration is that it adjusts to the directional wind changes much more easily without a tail vane. Figure 9-12 illustrates both configurations.

Figure 9-13 illustrates the typical parts of a horizontal tower-mounted wind-powered generator. Key parts include the following:

1. Pitch actuator—used to change the pitch or angle of the rotor blades.
2. Gear box—used to keep the wind turbine at the correct speed for producing electricity.
3. Alternator—used to make electricity (AC current) from the rotating shaft.
4. *Yaw* drive—used to turn the entire assembly about its center axis.
5. Nacelle—the shroud used to cover the entire assembly.
6. Disk and yaw brakes—used to stop rotational motion.

Research Developments in Wind Energy

The United States Department of Energy and the Lewis Research Center of the National Aeronautics and Space Administration (NASA) have completed several major projects designed to test the feasibility of using wind energy as an alternative resource for producing electricity. The wind generators were called the MOD Projects. Included were the MOD O at Sandusky, Ohio, the MOD OA at Culebra (near

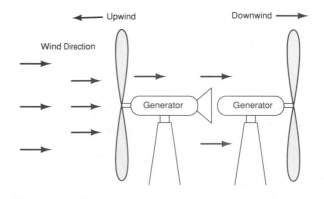

FIGURE 9-12 Horizontal wind turbines can either be designed in upwind or downwind configurations.

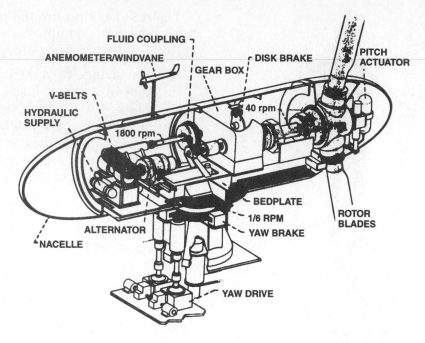

FLUID COUPLING
ANEMOMETER/WINDVANE
GEAR BOX
DISK BRAKE
PITCH ACTUATOR
V-BELTS
HYDRAULIC SUPPLY
40 rpm
1800 rpm
BEDPLATE
1/6 RPM
YAW BRAKE
ROTOR BLADES
ALTERNATOR
NACELLE
YAW DRIVE

FIGURE 9-13 The major parts of a horizontal wind turbine are shown in this figure. (Courtesy of NASA)

Puerto Rico), and MOD 1 and MOD 2 projects in North Carolina and Washington, respectively. Each project tested the feasibility of different wind turbine generator designs.

The first MOD project was at Sandusky, Ohio. This project, shown in Figure 9-14, was coordinated between the Department of Energy and NASA. The turbine and generator system was mounted on a 100-foot open-truss tower and produced about 100 kilowatts of power in an 18 mph wind. The rotor was connected to a low speed (40 rpm) shaft that drove a gear box. The gear box increased the turbine shaft drive to 1,800 rpm, the speed needed to run a 60 hertz (cycles per second) generator. A wind sensor, a servomotor, and a gear-bearing assembly provided the rotor with directional orientation into the wind.

Since the Sandusky project, several larger wind generator projects have been completed. The second project, MOD OA, included the development of four wind turbines. One such turbine—shown in Figure 9-15—was built near Puerto Rico. Others were placed on Block Island, Rhode Island, in Clayton, New Mexico, and on the island of Oahu. Most of these projects were designed to connect the electricity directly into larger utility networks. These wind machines were able to produce up to 200 kilowatts of AC electrical power.

The MOD 1 project was placed in operation along with the Blue Ridge Electric Membership Corporation at Howard's Knob, near Boone, North Carolina. The rotor was 200 feet in diameter. This system could produce up to 2 megawatts of electricity in a 25 mph wind.

FIGURE 9-14 One of the first MOD experimental wind turbine systems was capable of producing 100 kilowatts. (Courtesy of NASA)

FIGURE 9-15 This MOD OA wind turbine operates in Culebra and is capable of producing 200 kilowatts of power in 18 to 34 mile per hour winds. (Courtesy of NASA)

The MOD 2 wind turbine project consisted of three wind turbines similar to the one shown in Figure 9-16. These were located at Goodnoe Hills, Washington. Each wind turbine generator unit was capable of producing 2.5 megawatts of power. Each had a rotor diameter of 300 feet. The electricity was used in the Bonneville Power Administration electrical energy grid system.

These systems have led the way to another generation of MOD projects (MOD 5 and MOD 5B)—which are now being developed, tested, and analyzed for cost comparisons to fossil-fuel generator systems. These units can produce 5 to 7 megawatts of electrical power.

Wind Energy Farms

Wind energy farms, shown in Figure 9-17, are clusters of wind turbine generators used to produce electrical energy. Several wind energy farms are currently in full operation and connected directly into public utility electrical grid systems. The wind turbines are spaced close together in a staggered pattern. In these wind farms, costs are lower than those associated with fossil-fuel plants because the wind turbine units are mass produced. A 25-unit wind energy farm can produce about 62.5 megawatts. Spaced across the country, these wind energy farms could produce a significant reduction in the use of fossil fuels for the production of electricity.

A great deal of new wind energy technology is developed each year. It is reasonable to expect that wind energy will supply an even greater amount of the United States' electrical power in the next decade. Since 1985, over 1.5 billion kilowatt-hours of wind power electrical generation have come on-line, and the prospects for continued development is even greater when one considers the economic advantage of wind turbines.

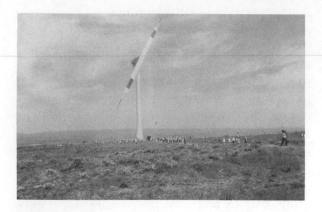

FIGURE 9-16 This MOD 2 wind turbine generator is capable of producing 2.5 megawatts of electrical power in winds of 25 miles per hour. (Courtesy of the American Petroleum Institute)

FIGURE 9-17 A wind energy farm consists of many wind turbines placed in one location. The output of each is added together. (Courtesy of Southern California Edison Co.)

Economic Advantage of Wind Turbines

The technical advances made to produce wind turbines have had a tremendous effect on the economic competitiveness of these devices. According to the California Energy Commission and the American Wind Energy Association, wind farms in California can generate electricity at a per kilowatt-hour cost of between $0.047 to $0.072. At this rate, wind energy is not only competitive with other forms of electrical generation plants, it is one of the least costly energy sources. In addition, the costs associated with operation and maintenance of wind energy range from a low of $0.008 to a high of $0.02 per kilowatt-hour (average costs are $0.012 per kilowatt-hour). Compared to nuclear and coal-fired electrical plants, wind energy is cheaper to operate and maintain. (Nuclear power plants and coal-fired plants have operation and maintenance costs of $0.02 and $0.022, respectively.)

There are also other advantages of using wind energy. With wind energy there are no waste problems from the disposal of spent fuel assemblies (nuclear) or ash (coal). Also, there is no cost associated with the purchasing of the fuel—wind is free. Unlike nuclear power or fossil fuels that become depleted and increase in cost, wind is nondepletable. It will always remain at its current cost since it is free. With these benefits, the nondepletable energy source of wind power offers a long-term alternative to the problems of finite energy resources.

Cost Benefits of Wind Energy—
Energy Issue 9-2

An electric power company would like to produce some of its electricity from a wind farm. The utility company is proposing a 5 percent rate increase to pay for the initial cost of the wind turbines. As a taxpayer, would you agree with this rate increase? Before attempting to answer this question, analyze the benefits of this energy source from both a short-term and long-term perspective (i.e., benefits from social and environmental perspectives).

**Considering a Wind Farm Proposal—
Energy Issue 9-3**

*A wind energy farm is planned in an area near
your home. The utility company would like the
wind farm to produce 400 megawatts of power:
the equivalent of a medium-sized nuclear power
plant. The power company indicated that 25 wind
turbine generator units can produce 60 megawatts of power. After determining how many wind
turbine units would be needed to produce the 400
megawatts of energy, provide a list of social and
environmental concerns for both the positive and
negative consequences of this proposal. Does the
proposal have merit, and would you support this
proposal?*

BIOCONVERSION

History of Bioconversion

Bioconversion is the process of producing usable
energy from organic matter, such as plant and
organic wastes. Any organic material (from
algae to wood), alive or decaying—including
various types of industrial and residential
waste—contains chemical energy that can be
released for use as a fuel. These types of materials are referred to as *biomass.* Bioconversion
takes the chemical waste and organic materials
and converts them into thermal energy.

Humans have used bioconversion processes
for centuries, mainly for the burning of wood.
It was the first energy source used by humans
to aid in their survival. In fact, the burning of
wood was the dominant energy source for
humankind until the last century. In the 1800s,
technical processes advanced and allowed
humans to exploit the energy in fossil fuels.

While biomass fuels form only a small percentage of the energy used worldwide (between
12 and 15 percent), they remain the main energy source for approximately half of the
world's population. According to the World
Bank, the developing nations in Asia and Africa
are heavy users of biomass energy sources.
About 65 percent of the energy consumed in
Asia is from biomass, and as much as 90 percent
supplies Africa with its needed energy.

In industrialized societies, biomass is once
again being considered as a viable energy
source. This is because biomass can be converted into gaseous, solid, and liquid fuels, or
converted by burning to produce electricity.
Also, the burning of biomass creates no net
gain in carbon dioxide in the atmosphere
because, if biomass is grown in large quantities,
the carbon dioxide released during combustion is offset by the carbon dioxide extracted
from the atmosphere during photosynthesis.

Bioconversion Processes

Bioconversion processes may be divided into
four major types, depending upon the source
of the organic materials:
1. Urban and industrial wastes;
2. Agricultural and forestry wastes;
3. Land and freshwater energy farming; and,
4. Ocean farming.

Chemical energy derived from organic
materials is similar to that obtained from fossil
fuels. However, organic materials can produce
the chemical energy within a much shorter
time. During photosynthesis (the primary bioconversion process), radiant energy from the
sun is captured by plants and converted into
chemical energy. This energy, in turn, is converted into oxygen and energy-rich carbohydrates (carbon and hydrogen) and remains
stored in the plant form. This energy may also
be stored after being eaten by humans or animals, in waste material. This stored energy is
released as heat, methane gas, sugar, alcohol,

or other useful chemical compounds when the biomass is burned, allowed to decay, or processed chemically.

Prospects for the Future

Energy experts estimate that early in the twenty-first century, 10 to 50 percent of the world's energy requirements could be met through bioconversion processes. This energy will come from energy crops (terrestrial and marine) and from organic waste (urban and agricultural waste). The latter is of major interest to industrialized societies. Each year the world produces billions of tons of urban waste. These wastes can be burned as a fuel or converted to other gaseous, liquid, or solid fuels by various conversion processes. Additionally, biomass can be found in any geographic area: agricultural areas, arid climates, wetlands, or forest regions.

Since biomass can be used to generate energy in many forms (including gases, heat, steam, electricity, and feedstocks for the chemical industry), this resource may be used in a variety of energy sectors.

In the United States the energy derived from burning wood accounts for almost 50 percent of biomass conversion for industrial and commercial uses. About two-thirds of this energy is used to generate steam to meet electrical energy needs. Florida, Maine, Hawaii, and California use biomass for meeting baseload requirements for electrical generation. Of these states, Maine is the largest user of biomass electrical generation—meeting about 24 percent of its electrical needs with this resource.

Many states, such as California, Florida, Maryland, and New York, use biomass derived fuels. Often, urban wastes are called municipal solid wastes. These wastes account for nearly 16 percent of the biomass fuels used in the United States.

Development of Bioconversion Technology

There are many current bioconversion projects producing data and giving scientists experience. A number of these are already producing energy for our society. Bioconversion processes will become an important energy form in the future. The following projects by public and private organizations are listed and grouped according to the bioconversion source:

Urban and Industrial Wastes
- Electricity by direct burning of shredded wastes mixed with coal in Iowa, Maryland, Massachusetts, Minnesota, and Missouri.
- Prolysis (destructive distillation) of wastes to produce gases, oils, and fuels for use as boiler, home-heating, and motor fuels in Baltimore, Maryland; Charleston, West Virginia; and San Diego, California.
- Collection of methane gas from sewage plants and sanitary landfills in Los Angeles and Palos Verdes, California.
- Fermentation (anaerobic digestion) of waste to produce methane gas in Pompano Beach, Florida.
- Enzyme hydrolysis of cellulosic wastes to produce industrial sugars and alcohol fuels in Natick, Massachusetts.

Agricultural and Forestry Wastes
- Inventorying of forest, field-crop, and animal wastes to estimate nationwide quantities potentially available for bioconversion.
- Investigation of technical, economic, and environmental factors in large-scale conversion of forest, field-crop, and feedlot residues to fuels and related chemicals.
- Development of small-scale farm digesters for use in converting wastes to methane gas.

Land and Freshwater Energy Farming
- Analysis of the technology and economics involved in large-scale production of forest and field crops, specifically for energy.

- Investigation of potential energy sources from algae cultures and lake plants (hyacinths).

Ocean Farming
- Testing of open-ocean rafts for large-volume growing of seaweeds to be converted into methane gas, foods, and chemicals.

Urban and Industrial Waste

Estimates indicate that the United States disposes of more than 250 million tons of municipal solid waste per year. Today, many cities have run out or are running out of room to dispose of their garbage. Disposal in landfills has become a national concern; estimates show that most of these waste sites will be closed by the turn of the century. The energy potential of this solid waste is equivalent to about 150 million barrels of oil per year. As the cost of petroleum increases, bioconversion of municipal solid waste could develop a substantial economic advantage.

Since much of the waste is scattered, the costs to collect it increases the overall cost of the bioconversion process. However, as the cost of fossil fuels increases, bioconversion becomes more economically competitive each year. The types and amounts of organic waste are presented in Figure 9-18. If each ton of this organic waste were converted to nine million Btu's, about 2 to 3 percent of the energy used today could be produced from bioconversion.

Waste Incineration—Energy Issue 9-4

Assume a large city is proposing to build a waste incinerator within the city limits of your hometown. The city planners report that the long-term prospects of this waste reduction plan are less expensive and less hazardous than landfilling. This plan does not propose the incineration of waste to produce electricity. After researching the advantages and disadvantages of this proposal, develop a list of questions or concerns that should be addressed by the city planners.

Urban and Industrial Waste Processes

The most inexpensive method of producing energy from organic waste is by combustion (i.e., the waste is burned). Before being sent to the boiler for burning, the organic waste is shredded. Glass and metals—sorted before or, in some cases, after the burning process—can then be recycled. In some systems, diesel fuel is added to the combustion process to increase the temperature in order to reduce emissions. Figure 9-19 illustrates one common waste combustion process.

The first step in the process is to collect the waste and dump it into a receiving pit. An operator working an overhead crane removes the waste from the pit and places it on a conveyor belt. The waste is then sent to a shredder. After shredding, the waste is sent to a boiler to be burned. In this case, water is heated for industrial processes. The glass and metal are then removed for recycling. The remaining ash (10

Dry Weight of Organic Waste

	Millions of Tons	
	Readily Available (estimate)	Total Amount (estimate)
Urban Refuse	60	225-250
Manure	26	260-290
Wood Manufacturing	5	55-65
Agricultural Crops	12	385-405
Industrial Waste	9	155-175
Municipal Sewage	3	16-19
Miscellaneous	2	55-60

FIGURE 9-18 Various types of organic waste can be used as fuel.

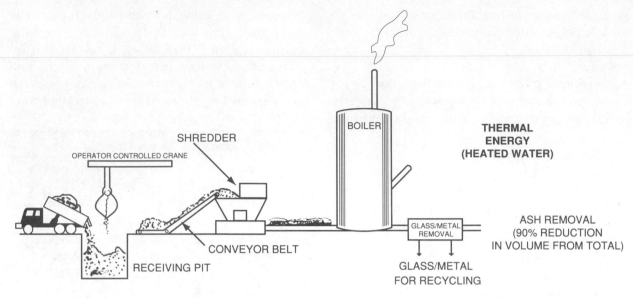

FIGURE 9-19 Organic waste is typically burned in order to produce thermal energy. Glass and metal must be removed during the process.

percent of the original volume) is reprocessed or placed into a landfill.

A second means of releasing the energy stored in organic waste material is by allowing the waste to decay. Any organic matter can be broken down by either *aerobic* (with oxygen) or *anaerobic* (without oxygen) processes. Methane gas, the basic ingredient in natural gas, is one of the by-products of anaerobic decay. Other by-products may include glucose (sugar), oil, and heavy tars. The type of by-product will depend upon the type of organic waste and method of decay. Glucose can easily be converted to alcohol, yielding a high-quality fuel.

Various experiments indicate that about 2.2 pounds of dry organic waste can produce between 6 and 11 cubic feet of gas with a 55 percent methane content. The remaining solids from the process can be heated and burned to produce electricity or steam for industrial processing.

Burning Versus Recycling Waste— Energy Issue 9-5

A great deal of waste exists within society that can be converted into usable fuels. However, there is also a strong incentive for recycling consumer and industrial waste. Identify the advantages and disadvantages of burning waste instead of recycling. Then, identify your position on this issue based on data that was collected.

Calculating Energy in Urban Waste—Math Interface 9-2

Solve the following sequential math problems concerning the use of urban waste:

1. Assorted urban waste can yield about 9,000,000 Btu's per ton. If a city produces 150 million tons of assorted

urban waste annually, how many Btu's are available for bioconversion?

Total Btu's available equal

_____.

2. If a barrel of oil has 5,600,000 Btu's (see Appendix L), how many barrels of oil equivalent are in the assorted urban waste described in the previous problem?

Barrels of oil equivalent equal

_____.

Energy Crops for the Future

Producing energy by growing various crops on land and in water, called organic farming, is under serious consideration. The United States Department of Energy (DOE) is planning to participate in several joint ventures designed to combine crop production with energy conversion facilities. The projects designed by the DOE would test energy crops like elephant grass, eucalyptus, bagasse (a by-product of sugar-cane refining and of ethanol production), poplar, sorghum, sugar cane, and sycamore. The planting of these energy crops also benefits the environment, as well as supplying useful biomass. These crops help to extract carbon dioxide from the atmosphere, reduce soil erosion, and rejuvenate unproductive soils. Grasses are particularly helpful. Switchgrass, for example, is a tall prairie grass that was responsible for creating the fertile soils of the United States Midwest.

The DOE expects farmers to be able to grow tree crops or perennial grasses on partially eroded or marginal agricultural lands. With the proper choice of tree crops and perennial grasses, the energy crops grown take far less energy to farm than food crops. These energy crops would also require less fertilizer, pesti-

cide, and irrigation than traditional food crops. Trees and perennial grasses, therefore, produce about 10 times more usable energy than farmers used to grow them.

Compared to other fuel sources, crops from organic farming form a substantial energy resource. One ton of dry biomass, heated in the absence of oxygen, produces the energy equivalent of about 1.25 barrels of oil, 1,322 cubic feet of medium Btu gas, or 825 pounds of coal-like solid residues. It has been estimated that if an area the size of the state of Texas were producing crops specifically for energy, all the United States' energy needs could be met.

Ethanol and Transportation—Energy Issue 9-6

The use of ethanol in transportation fuels has several advantages over gasoline. First, ethanol fuels are cleaner burning when used in internal combustion engines. Second, when engines are optimized for operation on alcohol fuels, they are about 20 percent more energy efficient than when operated on gasoline. Third, ethanol can be produced from renewable sources and is a product of bioconversion. In the United States, corn crops have been used to produce ethanol. Although the ethanol produced represents less than 1 percent of the transportation fuels used by people in the United States each year, the industry is well established in the United States and numerous other countries (Brazil, for example, has the largest alcohol-fuel program in the world. This country produces enough ethanol to meet about 20 percent of its transportation fuel requirements). After reviewing the social and environmental benefits of ethanol use in the transportation sector, identify inhibiting factors to the widespread use of ethanol-produced fuel (for the United States and for developing nations). How might these factors be overcome? What are the negative consequences of diverting crops to fuel?

Calculating Wood Heating Costs—
Math Interface 9-3

Wood is often used in the residential sector for heating purposes. Given the following data, determine if the use of wood to heat a home is cost effective for a monthly period. Assume a residential structure requires 15,000 Btu's per hour for heating in January (31 days). It would cost the homeowner $150 to heat the structure with energy provided by a utility company.

1. How many Btu's are needed for the entire month of January?

   Btu's required in January equal

   _____.

2. How many Btu's can be converted from burning a cord of air-dried hickory if the wood burning stove is 15 percent efficient?

   (Note: air-dried hickory has about 25,000,000 Btu's per cord.)

   Btu's converted from burning a cord of hickory equal _____.

3. How many cords of wood are needed to heat the home in January?

   Cords in January equal

   _____.

4. If each cord of wood costs $70, how much would it cost the homeowner to heat the home using wood as its primary energy source?

   Cost to heat with wood equals

   _____.

HYDROELECTRIC ENERGY

Introduction to Hydroelectric Energy

One of the oldest methods used to extract energy from the earth's resources is known as *hydro energy*. Hydropower is really a form of kinetic energy, using the gravitational forces of falling water to produce mechanical energy. Today, the mechanical energy often turns turbines that generate electricity. Hydropower is a result of the *hydrologic cycle*.

Hydropower is a form of indirect solar energy since it depends upon the hydrologic cycle. In this cycle the sun evaporates water, which is carried to the atmosphere, then to the land in the form of rain or snow, then back to the oceans by rivers and streams, and finally back into the atmosphere. This continuous cycle provides the world with its weather systems and the development of rivers and streams.

As the water in rivers and streams moves to the oceans, its kinetic energy can be converted to mechanical energy by utilizing the gravitational forces of falling or moving water. One such conversion occurs by the water turning turbines to generate electricity. Figure 9-20 shows two views of a hydroelectric dam in Quebec, Canada.

History of Hydropower

Waterpower is one of the oldest sources used by humans to aid in their survival. References to the use of hydropower in Western civilization can be found in the writings of the Greek poet Antipater, who noted that waterpowered gristmills had reduced the labor of women—who had hither to ground grain by hand.

The water mill was introduced in Greek civilization about 50 years before the Common Era. Although no written documents illustrate

FIGURE 9-20 (a) and (b) This hydroelectric power plant, located in Quebec, Canada, is used to convert the mechanical energy of falling water into electricity. (Courtesy of Hydro Quebec)

its development, it probably had its roots in the waterwheel of this time: a device consisting of a large wheel with buckets fixed to the circumference. This waterwheel was set into a stream and turned by human or animal power. The wheel would pick up water in the buckets and carry it to the top of the radius. Once at the top, the water was dumped into a reservoir or pipe for later use. It eventually became evident that by adding paddles to the buckets, the wheel could be turned by the flowing water itself.

The next development of the waterwheel was the incorporation of a gear on the axle of the paddle wheel. This allowed the mill to turn a vertical axle that turned a millstone. The first gears used round wooden pieces i.e., teeth, fitted at right angles. Figure 9-21 illustrates this design.

With these improvements, the water mill spread quickly throughout Western civilization. Mills later appeared throughout the Roman Empire and were applied to a variety of uses including the milling of grain, the forging of metal, and the cutting of wood and marble.

Having proved its worth in saving human labor, the waterwheel was applied on a larger scale worldwide. By the sixteenth century, water power had been applied to the industries of textile manufacturing, wood cutting, and mining. The writer and physician Georg Bauer wrote of the application of water power to

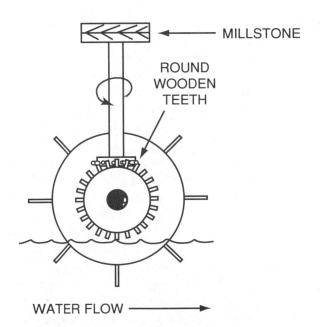

FIGURE 9-21 An early water mill is illustrated in this figure. Such mills used round, wooden dowels for the gear teeth.

machine tools. Bauer noted that water-powered machines (e.g., those using the water mill) could be applied almost anywhere.

Bauer also described the use of water mills in mining and metallurgy. He noted that water mills had been applied to pumping water, hoisting materials, ventilating mines, crushing and grinding ores, and moving bellows for smelting furnaces. Note here that the waterwheel in use during this time had remained essentially unchanged from the original designs implemented by the Greeks and Romans. Thus, with the exceptions of having tappets or cranks applied to the waterwheel, the designs remained unchanged until the development of the water turbine.

The Water Turbine

The water turbine or reaction turbine was highly influential in the development of power technology and for use in the generation of electricity in later years. In the reaction turbine, a wheel is propelled by the force of water coming out (instead of in, as with the waterwheel) through blades or vanes on its circumference.

The development of this device is credited to Bénoit Fourneyron in 1820. The water turbine had many advantages over the popular waterwheel. It was compact, submersible, and captured more of the energy in moving water; it also could be installed as a horizontal wheel with a vertical shaft. By 1837, Fourneyron had further refined the device by determining the optimal setting of the blades and dimensions for the wheels. This advanced design had a wheel diameter of 1 foot, weighed 40 pounds, and was capable of producing 60 horsepower at 80 percent efficiency. With further refinements, the water turbine was eventually applied to the generation of electricity.

The World's First Hydroelectric Power Plant

The world's first hydroelectric power plant, which used the water turbine, was established in Appleton, Wisconsin, in 1882. This plant was capable of producing 12 kilowatts of power—enough power to generate electricity for about 250 lights. The Wisconsin plant provided entrepreneurs with the technical background to establish other hydroelectric power plants.

The Potential of Hydropower Worldwide

The world has continued to increase its dependency on hydropower. Since 1973, hydroelectric power has almost doubled its production. Hydroelectric power is the largest nondepletable energy source in use today. It supplies almost 27 percent of the world's energy needs.

As a source of energy, the technology used in hydropower predates fossil fuels, and (with more efficient designs) continues to provide improvements. Hydro power has been proven to be competitive with fossil fuels. To date, only about one quarter of the world's hydropower potential has been utilized. However, experts note that it is currently impossible to develop every one of the remaining undeveloped resources because of the social and environmental problems of building large hydroelectric dams. These concerns will be examined in detail later in this chapter.

In recent years, hydropower has become more important to developing countries. Mexico, North Korea, and Venezuela have all increased their use of hydroelectric power generation since 1980. Additionally, several countries throughout Africa have begun to exploit this resource. Figure 9-22 illustrates world hydroelectric power generation for selected regions and countries. As can be seen in this figure, the combined areas of North, Central, and South America produce the bulk of hydroelectric power.

Worldwide, Canada produces the greatest amount of hydroelectric power, followed by the United States, Brazil, and the Former Soviet Union. This data is presented in Figure 9-23. Norway and Sweden have also invested greatly in this technology. Today, Norway obtains over 90 percent of its electricity from falling water; Sweden, about 50 percent.

Location of Hydroelectric Energy Resources

Hydroelectric energy resources are found throughout the world, often near mountainous or hilly regions where rivers can easily be dammed to provide a reservoir for storage water. In the United States, many larger sites are located in mountainous regions and the Pacific Northwest. Figure 9-24 shows a map of the United States with the distribution of large developed hydroelectric resources. A number are located around the mid-southern states (primarily in the Tennessee Valley and along the Appalachian mountain range) and in the northwest.

As prices of fossil fuels have increased, many smaller dams have been built or refurbished to provide hydroelectric power. However, in the United States, fewer than 3 percent of the existing dams produce electricity. Although there are over 65,000 dams in operation throughout the country, the majority are used for flood control. Many flood-control dams could be converted to produce electricity at a cost estimated at less than that of building a new coal- or oil-fired power plant.

Design Considerations for Hydroelectric Dams

Any hydroelectric dam for producing commercial electrical energy must have a minimum

World Hydroelectric Power Generation (Billion Kilowatt-Hours)

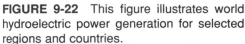
FIGURE 9-22 This figure illustrates world hydroelectric power generation for selected regions and countries.

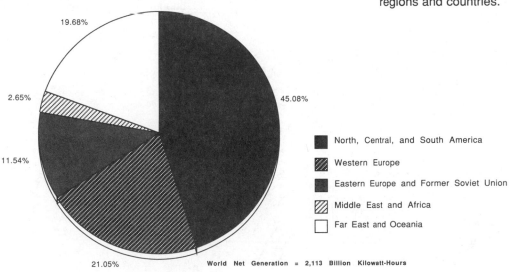

19.68%

2.65%

11.54%

45.08%

21.05%

■ North, Central, and South America

▨ Western Europe

■ Eastern Europe and Former Soviet Union

▨ Middle East and Africa

□ Far East and Oceania

World Net Generation = 2,113 Billion Kilowatt-Hours

Notes:
Data extrapolated from the Energy Information Administration, Annual Energy Review, 1991.

Hydroelectric Power Generation
(Selected Countries in Billion Kilowatt-Hours)

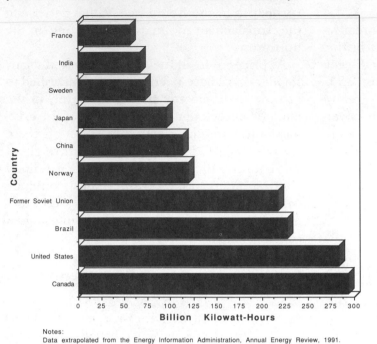

Notes:
Data extrapolated from the Energy Information Administration, Annual Energy Review, 1991.

FIGURE 9-23 As can be seen in this figure, Canada is the world leader in hydroelectric power generation with nearly 300 billion kilowatt-hours produced.

Distribution of Developed U.S. Hydroelectric Resources

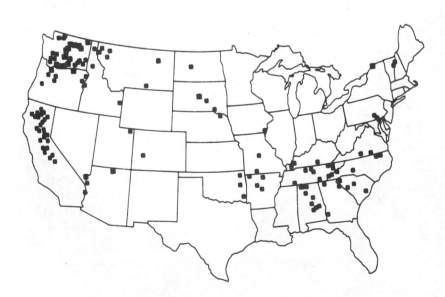

FIGURE 9-24 This map of the continental United States shows the location of many of the developed hydroelectric resources.

head height (the difference between the level of water inlet and water outlet). In actual practice, depending upon the geography of the land, the head can be as little as 20 feet for a low-head dam or 100 to 1,000 feet for a high-head dam. As prices for other fuels continue to increase, it becomes more economical to build low-head dams.

All hydroelectric dams used to produce commercial electrical energy must also have a reservoir to store the water. The amount of storage water needed is directly related to patterns of load peaks needed by electrical consumers. Figure 9-25 illustrates the monthly, daily, and hourly loads demanded by consumers for a power plant. During June, July, and August, the peaks were mainly caused by the use of electric air-conditioners. Refer to Figure 9-25a.

Figure 9-25b shows that daily electrical energy consumption increases to a high at midday and reaches a low at night. The hourly variations shown in Figure 9-25c suggest the times each day when additional loads are required. In light of these variations in demand, it is difficult to build an economically efficient power plant. Nearly 50 percent of the time, the power plant is running at less than full capacity.

To address this inequity, energy is stored for peak periods. Hydroelectric dams have the ability to meet varying demands easily. The reservoir serves as the storage medium. When demand reaches a peak, the gates are opened to allow more water to flow through the dam, thus producing more electricity. Figure 9-26 shows an aerial view of the Arizona A. Roosevelt Dam located near Phoenix. Note the large reservoir and the water being released on the side of the dam to control the level of the reservoir.

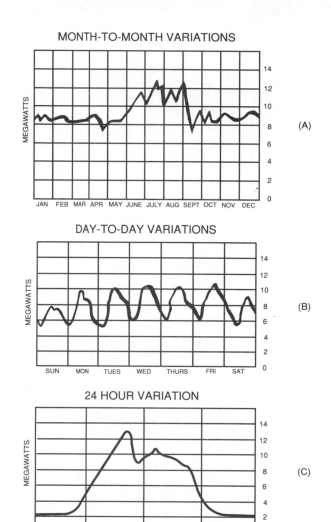

FIGURE 9-25 Hydroelectric power plants have the capability to meet variations in load demands. Load demands change on a month-to-month, day-to-day, and hour-to-hour basis.

A hydroelectric power plant is also more responsive than a steam-driven power plant. Water flow can be varied rapidly to meet the changing consumer load. Increases and decreases in steam pressure systems take more time to achieve.

FIGURE 9-26 Note the reservoir behind this hydroelectric power plant, located on the Salt River, near Phoenix, Arizona. (Courtesy of the American Petroleum Institute)

Pumped Storage

Several hydroelectric power plants, known as pumped storage plants, have been designed for load peaks. These are used in conjunction with fossil-fuel powered generators. A hydroelectric power plant is built on a large reservoir on top of a hill. During the night and at other times of low electrical demand, water is slowly pumped back into the reservoir and stored. When electrical demand is high, this additional water is then used to generate electricity.

Hydroelectric Power Plant Terminology

Figure 9-27 shows the major parts of a typical hydroelectric power plant. In operation, the water is sent through large pipes, known as *penstocks*, that direct the water to the turbine in the power plant. Water in the reservoir passes through a trash screen before going into a penstock in order to remove any debris.

The fast-moving water at the bottom of the penstock turns an impulse or reaction turbine. The total blade surface area of the reaction turbine is used for water pressure to turn the shaft.

However, the impulse turbine applies water pressure to one side or edge of a turbine blade. Because of this difference, reaction turbines are usually used in conjunction with low-head dams, and impulse turbines (needing high velocity water) are used with high-head dams.

The turning shaft of the turbine is used to rotate generators located below the dam, as shown in Figure 9-28. Whatever the type of turbine used, most hydroelectric power plants can obtain up to 92 percent efficiency when converting mechanical energy to electrical energy. This is considerably higher than the 30 to 35 percent efficiency of steam turbines used in fossil-fuel and nuclear power plants.

The high efficiency of hydroelectric conversion is its major advantage. One reason for the higher energy efficiency is a function of the density of water as opposed to steam. Water can produce more force on the turbine blades per volume of medium used. A second reason for the higher efficiency is that there are no thermal losses (from entropy) as there are with fossil-fuel generating power plants.

Smaller Hydroelectric Power Plants

Another method of extracting energy from moving water has centered on the use of smaller hydropower plants. These facilities have often been classified as micro, mini, or small. However, to date there appears to be little consensus on the definition of micro, mini, or small hydropower plants. For purposes of clarity, these plants have been defined in the following manner:

- Micro hydropower plants can produce electricity at levels below 100 kilowatts.
- Mini hydropower facilities produce electricity in a range from 100 to 1,000 kilowatts.
- Small hydropower plants produce in the range from 1 to 30 megawatts.

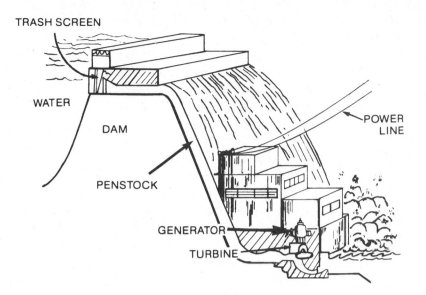

WATER

DAM

PENSTOCK

POWER
LINE

GENERATOR

TURBINE

FIGURE 9-27 A schematic of a typical hydroelectric power plant is illustrated here. Water flows through the penstock to the turbine and generator unit to produce electricity.

These hydropower plants have several advantages over larger hydroelectric dams. These facilities can be built at a lower cost, and they can be built on rivers with lower water flow. Also, these power plants have fewer environmental effects as compared to those with large dams.

Currently, China leads the world in smaller hydropower development. China has already developed some 58,000 smaller plants that provide in excess of 13,000 megawatts of capacity. Frequently, these plants provide power to rural communities that lack alternative sources to produce electricity. Other countries actively developing smaller hydropower plants include France, Italy, the United States, and Sweden. Each of these countries has already developed over 1,200 small-scale hydro plants, and each has plans for additions.

Costs Associated with Hydropower

The costs associated with developing hydropower are very site-specific. Meeting environ-

mental issues and the need to design the power plant to maximize its output vary from area to area. However, compared to other depletable and nondepletable energy sources, hydropower is among the least expensive of all the energy resources. Although the initial costs to develop and construct these facilities are not

FIGURE 9-28 These generators, rotated by hydroelectric turbines, are used to produce electrical energy. (Courtesy of Edison Sault Electric Company)

small, they have lower maintenance and operation costs. Taken together, the cost of electricity from hydroelectric plants ranges between 0.03 and 0.06 cents per kilowatt-hour. This makes these power plants attractive to meet the increasing need to supply electricity.

TIDAL POWER

Introduction to Tidal Power

Tidal power, also called ocean power, can be categorized as a type of indirect solar energy. Tidal power is created by the gravitational attraction of the sun and moon and the earth's rotation. These celestial bodies moving in relation to one another create tidal cycles all around the world. The cycles cause the surface of the oceans to be raised and lowered at certain times, depending upon a number of interacting cycles. For example, there is the half-day cycle, a result of the rotation of the earth within the gravitational pull of the moon. With the half-day cycle, there is a period of about 12.42 hours between successive high waters. Another cycle is called the 14-day cycle, a result of the gravitational fields of the sun and moon combining to provide the spring and neap tides (tides of the lowest range). Another cycle, called the half-year cycle, is affected by the inclination of the moon's orbit. This cycle is responsible for the 178 days between the highest tides in March and September.

Because the tides occur at predetermined intervals, the movement of the water can be used in the production of electricity. The energy available from tides is the kinetic energy of the water moving from high tide to low tide. Thus, the use of tidal power is similar to the use of energy created from the release of water in a hydroelectric dam.

In tidal power systems, a dam is constructed at the mouth of a tidal bay and is equipped with sluice and turbine gates. The sluice gates allow the water to enter the dam as the tide is rising. These are then closed at high tide. About six hours after high tide has passed, the water is allowed to flow through the turbine gates to produce electricity. This process is repeated throughout the day at intervals of about twelve hours. The energy generated depends upon the number of turbines installed and the amount of water that can be drained from the tidal basin.

History of Tidal Power

Tidal power was first used in the eleventh century to run mills constructed on the coasts of England, France, and Spain. Late, plants were also developed in the United States during the Colonial period. These plants operated with the use of a water mill and were used for centuries. However, these mills were gradually taken out of service during the Industrial Revolution with the increase of less expensive and more convenient fossil fuels. Today, there is a renewed interest in tidal power to meet the energy demands of the future.

Potential of Tidal Power

Before illustrating the potential of tidal power, note that this technology has already proven successful—the individual pieces of technology have been used in hydroelectric plants. Demonstration projects have been shown effective in the production of electrical power. Several countries have used tidal power including Canada, China, France, Japan, the United States, and the former Soviet Union. The last has a plant that can produce 400 kilowatts of electrical power. Another plant in France can produce over 500 million kilowatt-hours of electricity per year. Worldwide, several other large tidal energy projects are now under construction. An estimated 600,000 gigawatt-hours of energy could be produced if this resource is fully exploited.

Costs Associated with Tidal Power

The costs associated with constructing tidal facilities are extremely high. In the construction process large prefabricated units of concrete or steel must be towed to the dam site where they are placed into position on prepared foundations. However, these high capital costs are offset by low operation and maintenance costs over the lifetime of the facility. Remember that these plants have no fuel costs. Tidal power plants should be able to produce electricity at a cost between 0.04 and 0.06 cents per kilowatt-hour.

GEOTHERMAL ENERGY

Introduction to Geothermal Energy

Geothermal energy is considered a form of indirect solar energy, as previously identified. *Geothermal energy* is heat energy found below the crust of the earth. It is a result of the original formation of the planet some 5 billion years ago and the sun's gravitational forces that played a part in that formation. Sources of geothermal energy include the decay of radioactive elements, such as uranium and thorium, that have accumulated throughout the history of the earth. Other sources of this thermal energy include the heat generated by the compression of the earth's internal structures and the heat from the sinking of heavy metals as they move downward to form the core of our planet. Geothermal reservoirs are considered nondepletable sources of energy. For this reason, and the fact that geothermal energy cannot be classified into any of the other categories used in this text, geothermal energy is presented here.

Before reading a more detailed description of geothermal energy, note that the structure of the earth is layered. These layers consist of

1. the crust or the outer layer, which surrounds the earth;
2. the mantle or inner part, which consists of rocks, minerals and other solids, as well as liquids and gases; and,
3. the core, which is made of molten materials.

The temperature increases in areas closer to the center or core of the earth. The degree of temperature increase is termed the *normal thermal gradient*. The normal thermal gradient is estimated to be about 100° F for every mile one travels toward the core of the earth. To produce steam of approximately 400 to 500° F for a steam turbine/generator, wells must be drilled 3 to 5 miles deep. Figure 9-29 shows a geothermal rig used to drill such a well. However, most geothermal facilities in use today to power turbine generators are located in areas where faults and cracks occur in the earth's crust; therefore, the depth of the well can be much more shallow.

FIGURE 9-29 This rig is used to drill geothermal wells. (Courtesy of Unocal)

Geothermal Energy Development

The thermal energy near the core areas of the earth's structure is primarily generated by radioactive materials decaying over time. The crust of the earth is often very thin as a consequence of past geological cracks, shifting, or heaving. These geological characteristics cause the thermal energy under pressure to escape through the earth's crust and into the atmosphere. In the United States, the best example of geothermal energy is the geysers located in western regions, such as Yellowstone National Park in Wyoming. Other areas around the world that exhibit geothermal activity include Western Samoa, Fiji, Central Asia, Eastern Africa, and a large semicircular belt surrounding the Pacific Ocean known as the Ring of Fire.

History of Geothermal Energy

Geothermal energy has been used throughout human history. In ancient Rome these resources were used for public baths and spas. Such waters, which are rich in minerals, were regarded as therapeutic and were often used for medicinal purposes.

The use of geothermal energy remained largely for special bathing purposes until the early nineteenth century. It was during this period that the minerals contained within the heated water were used in the production of enamels for pottery. Later, geothermal energy was also used in industrial processing and the heating of homes by the steam produced. By 1904, electricity was produced in Italy by this resource.

Potential of Geothermal Energy Resources

The potential of geothermal energy resources is immense. The United States Department of Energy estimates that over 100 million quads are available in the earth's crust. Such a huge resource is many thousands times the world's annual energy consumption. Therefore, geothermal energy has been categorized as a nondepletable energy resource. However, despite its great quantity, much of this resource is either too diffuse or too deep to be practically and economically recovered. A more realistic estimate suggests that the current geothermal reserves may range from 400,000 to 148,000,000 megawatts.

The United States is the world leader in the use of geothermal energy. However, nearly all of the electricity commercially produced from this resource is done in northern California. This use of geothermal energy dates from about 1960. In 1974, output was 412 megawatts of electricity, which accounted for about 0.1 percent of the total United States' electrical power generation. Plans suggest that geothermal energy may eventually constitute about 2 percent of the total United States' electrical power after the year 2000.

Note that the overall cost of the electrical power generated by geothermal energy is less than that generated by fossil-fuel or nuclear power plants. Electricity produced by this method typically costs between 0.045 and 0.06 cents per kilowatt-hour. Figure 9-30 shows a typical geothermal energy power plant used to produce electricity.

FIGURE 9-30 This geothermal energy power plant supplies electrical energy to Manila in the Phillippines. (Courtesy of Unocal)

Location of Geothermal Energy Resources

Within the United States, many near-surface geothermal reservoirs exist in the western half of the nation—as shown in Figure 9-31. The eastern United States, along the Appalachian mountain range, also contains several small near-surface reservoirs.

With very few exceptions, a weakness or crack in the earth's crust (normally found near mountainous regions) will allow geothermal energy to escape. Geothermal energy is actually below the crust throughout the earth. It can be tapped if drilling technology is available and if such tapping is economically feasible and environmentally sound.

Types of Geothermal Reservoirs

Many types of reservoirs are considered part of the geothermal energy resource, as displayed in Figure 9-32. *Hydrothermal reservoirs* are those that have naturally circulating hot water or steam. Some hydrothermal reservoirs are dominated by steam while others contain a mixture of water and steam. *Dry steam hydrothermal reservoirs* are currently the most economically attractive. *Wet steam reservoirs* usually have lower temperatures and require various filters to extract the water, thus reducing turbine efficiency. However, wet steam reservoirs are about 20 times more common than dry steam wells.

Because wet steam reservoirs are of a lower temperature, they have been used with various low boiling-point refrigerants such as isobutane or freon™. In this system (called the binary cycle), the geothermal heat vaporizes the refrigerant at much lower temperatures. The vaporized refrigerant then is expanded to operate a turbine generator. The concept is basically the same as the ocean thermal energy conversion (OTEC) system presented later in this chapter.

Geopressured reservoirs are made of porous rock and sand containing water, usually at high pressures and temperatures. They are known to occur along the Texas and Louisiana gulf coasts. The energy in geopressured reservoirs is produced from hot water, hydraulic pressure, and dissolved natural gases. The hot water, when converted to steam, can be used in a turbine generator. The pressure can be used to turn a water turbine, much like that in a hydroelectric plant. The natural gas, if extracted, can be used as a fossil fuel. Although not currently exploited, geopressured reservoirs can and are expected to supply a meaningful amount of energy in the future.

Location of Near-Surface Geothermal Reservoirs

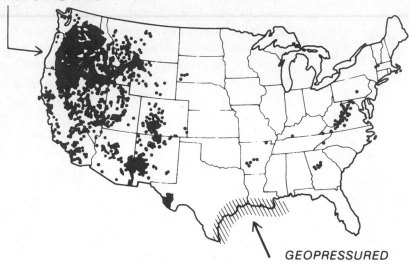

HYDROTHERMAL

GEOPRESSURED

FIGURE 9-31 This map shows the location of geothermal reservoirs within the continental United States.

Another type of geothermal reservoir is the *hot-dry rock*. The rock is heated by radioactive decay and is not in contact with water. Evidence suggests that hot-dry rock may be the largest available geothermal energy resource. Although no technology has been developed on a broad scale to extract this energy, several methods have been suggested. The most likely is that of hydraulic fracturing.

Hydraulic Fracturing

The process of *hydraulic fracturing* involves pumping cold water down into hot-dry rock (8,500 feet deep). The cold water will fracture the rock. Water is then pumped through the fractures, absorbing the thermal energy from the hot-dry rock reservoir. Finally, the hot water and steam are pumped back to the surface for use in a turbine generator. Figure 9-33 shows a cutaway of how this technology might operate.

Geothermal Energy in the Future—Energy Issue 9-8

Developers of an energy project would like to use the geothermal reserves found in Yellowstone National Park. These resources would be used for the production of electricity or for supplying small industries with heat. As a consumer and taxpayer, what concerns would you have about these plans? Develop a set of questions that you would like to have answered by the developers.

Types of Geothermal Reservoirs

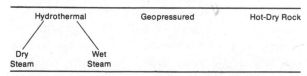

FIGURE 9-32 There are three types of geothermal reservoirs from which energy can be extracted: hydrothermal, geopressured, and hot-dry rock.

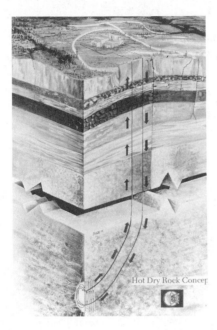

FIGURE 9-33 When water is pumped down into the hot-dry rock type of reservoir, it absorbs thermal energy which can be used for electric power production. (Courtesy of the Department of Energy)

ALTERNATIVE FUTURES

Ocean Thermal Energy Converstion (OTEC)

Ocean Thermal Energy Conversion (OTEC) is a promising energy resource for the future. In OTEC, power is produced by using the temperature difference that exists between warm water at the ocean's surface and colder, deeper waters. Since the oceans are warmed by the sun, OTEC is considered an indirect solar energy technology.

OTEC is defined as the extraction of the solar energy from oceans to drive an electricity-producing turbine generator. OTEC is not a new concept. In fact, it dates from the 1820s, when the French physicist, Sadi Carnot, demonstrated that mechanical energy can be derived from the flow of heat energy between a warmer and colder region.

History of OTEC

The first individual to propose the generation of electricity using the ocean's temperature differences was the French physicist Arsene d'Arsonval in 1881. However, it was not until 1930 that the first OTEC system was built and tested. Georges Claude, a student of Arsene d'Arsonval, built and tested a system at Matanzas Bay on the northwestern coast of Cuba. This system could generate in excess of 20 kilowatts of power; however, it could not be considered successful since it consumed more energy than it produced.

From the 1940s through the early 1970s, numerous experiments attempted to improve the operating efficiencies and design of OTEC systems. While design improvements were made, this technology was not adopted because of an abundance of inexpensive fossil fuels.

By the late 1970s, however, energy prices had risen to a level that encouraged additional experiments with OTEC technology worldwide. Countries such as France, Japan, Sweden, and the United States began actively to pursue OTEC research. One successful research project conducted by the United States involved the development of an OTEC plant for the state of Hawaii.

This Hawaiian plant was placed into operation on August 27, 1979. Known as MINI-OTEC, it was mounted on a barge and moored off Keahole Point, Hawaii. The plant was able to generate between 10 and 15 megawatts of net power (i.e., the power remaining after supplying its own energy requirements). Presently, OTEC research is continuing with hopes of becoming a significant energy resource in the future.

Potential of OTEC Plants

The oceans offer a tremendous collection system for solar radiation. It has been estimated that about one-fourth of the total solar energy reaching the earth's atmosphere is absorbed by the world's oceans. Since OTEC systems are only practical in regions where the temperature difference between the warmer and the colder sea water is 35° F to 50° F or more, the systems are limited to tropical waters. The most appropriate location is between the tropics of Cancer and Capricorn: a ribbon of land and (mostly) water around the equator. Some of the best locations for OTEC plants are in the Pacific Ocean where temperature differences greater than 70° F have been found.

Despite these location limitations, the potential for large-scale commercialization of OTEC is encouraging. The warm waters between the tropics of Cancer and Capricorn could provide up to 36 million square miles of potential OTEC development area—over two times the potential resource available in wind energy.

Obviously the energy available from OTEC resources is immense. The United States Department of Energy has estimated that some 170 trillion kilowatt-hours of electrical power could be generated annually from the Gulf Stream. This is more than 50 times the electrical generation used in the United States each year. In theory, the oceans could provide enough thermal energy to satisfy the world's energy requirements. Although the efficiency is extremely low, OTEC is still considered a future possibility because the resource is free (after the system is built), unlimited, and continually renewable. Efficiency is directly related to the temperature differential (delta T), between the warm and cold water. Normally, about a 35° F to 50° F differential is needed to make the operation economically feasible.

OTEC System Operation

The OTEC system in operation is rather simple. To date, several basic designs have been explored. These include the closed-cycle, the open-cycle, and hybrid OTEC systems. The OTEC plant may be operated on land, from a floating barge, or on a ship at sea. Although there is considerable debate over which is the best system, forecasts show that one or more of these systems will be in competitive operation with other generating facilities by the year 2005.

Closed-Cycle OTEC System

The closed-cycle OTEC system is illustrated in Figure 9-34. Warm water from the ocean's surface is pumped into the plant by large recirculating pumps. The amount depends on the output of the OTEC plant. A typical 1,000 megawatt plant will call for about 100,000,000 gallons of water to be pumped each minute. Both smaller and larger plants are planned.

The warm water is then fed past a series of heat exchangers to remove the thermal energy. After this energy has been transferred to a liquid refrigerant, the water is pumped out of the plant, away from the intake section.

When the liquid refrigerant reaches its boiling point, it becomes an expanding vapor or pressurized gas. (Most liquid refrigerants boil at very low temperatures, sometimes well below 0° F.) The boiling refrigerant, now a gas, and under about 150 psi pressure, is sent into a turbine generator. The turbine, turned mechanically by the expanding gases from the liquid refrigerant, turns a generator that can produce electricity.

The refrigerant is then sent through a series of condensers, where it is condensed into a liquid. To accomplish this, cold water (30° F to 35° F) is pumped from 3,000 to 5,000 feet

Closed-Cycle Ocean Thermal Energy Conversion Plant

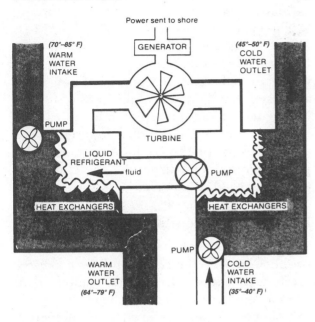

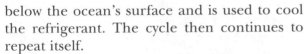

FIGURE 9-34 This diagram shows the basic operation of an OTEC power plant.

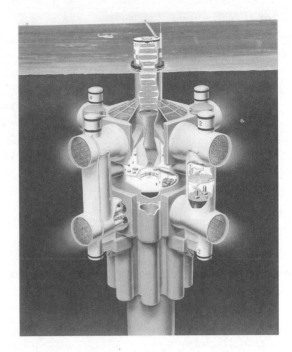

FIGURE 9-35 An artist's conception of the inside of an OTEC electric power plant is illustrated here. (Courtesy of Ocean Systems Marketing, Lockheed Corporation)

below the ocean's surface and is used to cool the refrigerant. The cycle then continues to repeat itself.

Figure 9-35 shows an artist's conception of such a power plant researched by Lockheed Corporation. Only a small equipment area and some living quarters are above the surface of the water; most of the power plant is below water. Buoyancy tanks are used for stability. Dimensions of a typical OTEC plant of the future will vary as technology advances, but estimates indicate that such a plant could be 390 feet in diameter and could extend 500 to 600 feet below the water's surface. The pipe that brings up the cold water would be 43 feet or greater in diameter and extend 3,000 to 5,000 feet below the station, depending upon the cold-water depth.

Open-Cycle OTEC System

The open-cycle system was the design originally conceived by Georges Claude in the 1920s. This system is very similar in design to the closed-cycle system with the exception that no liquid refrigerant is used. The open-cycle system exploits the fact that water boils below its ordinary boiling point when its pressure is reduced. In the open-cycle system, the warm surface water is converted to steam in a partial vacuum. The steam then passes through a turbine and is condensed by contact with cold seawater pumped from approximately 3,000 to 5,000 feet below the surface. Analogous to the closed-cycle system, the turbine will generate electricity for use by consumers.

Hybrid OTEC System

The hybrid OTEC system utilizes elements of both the open-cycle and closed-cycle systems to produce electricity and freshwater from the seawater. In this system the warm surface water is quickly evaporated and the resulting steam is condensed against ammonia, which then evaporates and is used to drive a turbine to produce electricity. A by-product of this system is the production of fresh water from the condensing action.

Overcoming Problems with OTEC Systems

The OTEC system operation is rather simple. However, the technology needed to accomplish the process is complex. For initial tests, several ships have been used to prove the concept instead of building a plant. Results so far show that the concept has good potential for producing energy at competitive prices in the long term. However, various problems must be overcome:

1. positioning and mooring of the plant in rough waters;
2. managing corrosion from saltwater on all parts;
3. unclogging of the pump inlet screens from algae and marine life;
4. improving heat exchangers;
5. developing long-distance underwater electrical transmission lines; and,
6. managing the varying or changing ocean temperature zones.

The Future of OTEC

The future of OTEC will be determined by the introduction of cost-effective technology and its competitiveness with other energy delivery methods. Since only several small demonstration plants have been built, it is difficult to assess what costs will be associated with the construction of a large plant. Estimates for capital costs range between $9,000 and $12,000 per installed kilowatt. Thus, an OTEC plant could conceivably be designed with operating costs of about 0.08 cents per kilowatt-hour. When capital and operating costs are taken together, consumer costs of 0.12 to 0.25 cents per kilowatt-hour could be realized. However, these costs could be lowered by creating markets for the freshwater or other by-products produced from the generation of electricity by OTEC systems.

By-products and Benefits of OTEC

OTEC systems offer several options for the production of electricity and by-products. Freshwater (up to 15 million gallons daily from a 100 megawatt plant), algae (to be sold for biomedical applications), and numerous mariculture operations (abalone, lobster, oysters, sea urchins, and seaweeds) are all by-products. Additionally, OTEC systems can provide buildings with air-conditioning and refrigeration by the systems that use the cold seawater. Finally, unlike other indirect solar energy technologies, OTEC systems can be operated 24 hours a day since the temperature difference between the warmer and colder ocean waters does not fluctuate significantly when the sun sets.

SOCIAL AND ENVIRONMENTAL COSTS OF USING INDIRECT ENERGY SOURCES

Social and Environmental Costs of Wind Energy

Although wind energy is one of the most benign energy choices available, it does have some social and environmental costs. One of the criticisms is aesthetics. For some individuals, wind farms are unpleasant visually, particularly when placed in residential or scenic areas. Wind power, additionally, produces noise levels that are objectionable to some. However, studies have shown that wind farms produce noise levels lower than those produced by busy street traffic. An environmental concern is that birds are killed by wind turbines. Although the number of birds killed per individual wind turbine site is not large, it is not uncommon to have 20 to 30 bird deaths at a wind farm during a single year. Research is currently under way to minimize this hazard. However, this threat will likely never be eliminated.

There are three additional social and environmental concerns associated with wind power: electromagnetic interference, land usage, and catastrophic failure. Wind turbines have, in the past, caused electromagnetic interference with televisions, radios, microwaves, and navigational systems. Since newer designs no longer use metal blades, these problems are expected to be minimized in the future. Regarding land usage, most wind farms need fewer than twenty acres per megawatt of power produced. This concern should be minimal since the land under the towers can still be productive for agriculture or other commercial pursuits. Finally, there is a concern of catastrophic failure (i.e., the blades shearing and falling to the ground). Although such an accident is considered a rare event and technology has developed to minimize the probability, the possibility of catastrophe is not negligible.

Social and Environmental Costs of Bioconversion

The use of bioconversion offers humanity a great potential to provide for its energy needs. It does not contribute to acid rain, nor does it increase the level of carbon dioxide in the atmosphere. If grown on a large scale, biomass fuels can reduce carbon dioxide in the atmosphere. Carbon monoxide emissions are also lower with alcohol fuels made from biomass. Also, the use of energy crops and waste products can provide employment opportunities for individuals in industrialized or developing nations, as well as providing a mechanism for the disposal of municipal solid wastes. However, there are several social and environmental costs of developing large-scale biomass projects.

If improperly managed, this resource is nonrenewable and can be considered a depletable resource. The history of depletion of forest areas is a good example. If biomass developers do not replant trees to assure a continual supply, species that rely on these forests could be harmed, a disruption to the ecosystem could occur, and the resource could be depleted.

Furthermore, increases in biomass production from energy crops compete with food production. This is a serious concern for both developing and industrialized countries. Politics, economics, and overpopulation already contribute to the difficulty of feeding the world's population. An increased production of energy crops could decrease the amount of food produced worldwide, with drastic results.

Social and Environmental Costs of Hydropower

One might think that hydroelectric energy is free from environmental problems. There is no waste, no radioactivity, and no pollution. Nonetheless, hydroelectric plants have several major environmental problems to solve. Generally, however, the social and environmental costs of hydropower are more prevalent for large-dam projects. Small-scale hydro projects have the least social and environmental effects.

Since it takes large amounts of land to build a hydroelectric power plant and dam, the natural and scenic beauty of rivers and streams is often destroyed. Towns and their inhabitants, also, often must be relocated. For example, the Three Gorges Dam in China will require the displacement of hundreds of towns and 1.4 to 10 million people. This resettlement can have long-lasting negative consequences because, in rural areas, it may involve a dismantling of people's lives. Along with this displacement of humans are also the problems associated with species loss and deforestation. Finally, the quality of water from dams may be quite poor.

The damming of rivers has been known to alter the water quality in several negative ways. Although moving streams tend to aerate and cleanse themselves, dammed water can become stagnant. This warmer water that has been oxygen depleted is harmful to plant and fish species and can lead to a concentration of minerals, such as salt, in the water.

Dams also damage or destroy fish spawning areas. Fish are unable to migrate upstream to spawn, and hatchlings have difficulty moving downstream to the sea. Although fish ladders can be incorporated into new dam projects, these add an additional 50 percent to the cost of a project. Even with these devices, fish will still die when they are drawn into the turbines of the hydroelectric plant.

Downstream rivers are also affected by hydroelectric dams. When small amounts of water are released during nonpeak hours, downstream water temperatures may increase enough to kill certain water species.

Hydroelectric dams often provide flood control downstream. However, during low-water periods downstream, some land areas no longer get the nutrients that they contained before the hydroelectric dam was built, thus altering land usage.

Finally, note that dams are subject to siltation. In this process, the waterways that feed into the reservoir carry silt to them as well. Silt fills in the reservoir behind the dam, causing a decrease in operating efficiency and eventual closure of the dam. In many cases, this is a negative consequence of poor land management techniques upstream from the dam. An example of this can be seen in the Sanmenxia Dam on the Yellow River in China. Within just four years of the dam's completion, the reservoir had become almost completely filled by silt. Furthermore, without dams, floods and high water often deposit rich nutrients to soils. Dams stop this natural process. These consequences may not seem significant in comparison with nuclear radioactivity and waste. Nevertheless, they are causes for concern that must be addressed.

Social and Environmental Costs of Tidal Power

Because the ocean is a constant source of energy from the gravitational forces of the sun, moon, and earth's rotation, tidal power can be classified as a nondepletable energy source. Like many other nondepletable or renewable energy sources, tidal power avoids the pollution problems associated with the burning of fossil fuels. This resource also does not use up any natural resources in the production of electricity. However, tidal power does have several environmental and social costs.

There are dangers to fish and animal species dependent upon the natural ecosystem. By constructing tidal power plants there are concerns that the tidal dynamics in the area will be altered. Although the severity of the damage is dependent upon the geography of the region, the effects on the local ecology will depend upon how nutrients are distributed throughout the area. Changing tidal patterns may have a negative result on fish, shellfish, and birds in the region.

Other concerns that need to be addressed before the introduction of tidal power occurs on a large scale include how these plants will disrupt navigation systems. Also, there are some who have expressed a displeasure over the potential dispoiling of scenic areas with tidal power plants.

Social and Environmental Costs of Geothermal Energy

Economic and environmental problems also plague the progress of geothermal energy. As presented earlier, geothermal energy can be found anywhere below the crust of the earth. However, much of the resource cannot be extracted because of the prohibitive cost of tapping the deeper reservoirs. This drawback, in addition to the problem of localization, causes some energy experts to shy away from promoting geothermal energy.

One environmental dilemma with geothermal energy is that of air pollution. Geothermal fluids contain dissolved gases, which include carbon dioxide and hydrogen sulfide, that can make their way into the atmosphere. While the release of carbon dioxide is not expected to cause significant harm (due to the scale of these plants), hydrogen sulfide has an objectionable odor (a rotten-egg smell). In addition, hydrogen sulfide is corrosive and has noxious properties that can cause serious health effects. With concentrations at 667 parts per million (ppm), hydrogen sulfide gas can cause death from respiratory paralysis. However, it is expected that most geothermal plants would emit concentrations in the order of 1 ppm with the installation of abatement systems to treat the gases before discharge into the atmosphere.

Another problem associated with the use of geothermal energy is the possible contamination of waters at the plant site. The water and steam extracted from a geothermal reservoir contain minerals such as arsenic, mercury, or boron. The discharge of this water and steam to the surface or to streams and rivers could cause significant environmental harm. Current operational practices reinject this water into the reservoir to recharge the aquifer. With this practice, there is a danger of contaminating the surrounding groundwaters near the geothermal plant.

Land subsidence and the possibility of increased seismic activity are also concerns to large-scale development of geothermal power. Land subsidence occurs when large quantities of a fluid are withdrawn from the earth without reinjection. Increased seismic activity is a possibility that results from reinjection of water into the reservoir. While this reinjection may help to reduce or eliminate land subsidence, it can result in added lubrication along fault lines. The concern expressed by geologists and energy experts is that this added lubrication could cause the faults to shift. Land shift could cause damage to the plant structures and result in pipeline ruptures, geothermal well blowouts, and earthquakes.

Lastly, the development of geothermal power plants is likely to occur at or near scenic areas. Although it is possible to minimize the visual impact of the power plant, the drilling rigs, steam pipes, and substations can be unsightly. This is of particular concern in areas that are unspoiled or have been valued for their natural beauty.

Despite the environmental and economic problems, geothermal energy continues to be viewed as part of the future of United States' energy resources. A modest annual goal of 7,500 to 20,000 megawatts of electricity is the geothermal contribution estimated to be available by the early twenty-first century.

Social and Environmental Costs of OTEC

The introduction and large-scale development of OTEC systems could have several environmental effects. As with any other unproven technology, problems are difficult to assess until small demonstration plants are instituted and operated for several years. Despite the unproven nature of OTEC systems, the following are a sample of several concerns to be addressed. First, the large use of hot and cold water might alter global weather patterns. Second, carbon dioxide, contained in deep ocean water, could be released during the operation of open-cycle plants. (Carbon dioxide has been determined to be a contributor to global warming.) Third, there is concern over the release of either chemicals or wastewater back into the ocean. Fourth, in closed-cycle or hybrid systems there is concern that the working fluid might escape into the sea if the system is floating, or into the ground in a land-based system. Additional safety precautions must be built into system designs to counteract this potential problem.

Two more environmental and social concerns are the effects on marine life and the use of chlorine to clean the heat exchangers. OTEC systems would release both warm and cold water of a temperature different than the surrounding water. Although the release would be regulated to maintain a minimal effect on the environment, the water would still not be the same temperature as the surrounding water. The concern expressed by experts is that this temperature change may affect surrounding organic species. Additionally, the changes in salinity to the water from the release of freshwater might affect the local ecosystem. A final concern is that the heat exchangers are periodically cleaned by chlorination. If chlorine is released into the environment, it would be detrimental to marine life. Despite all of these concerns, the International Energy Agency has stated that, based on studies already conducted, the negative effects of OTEC systems would be limited.

POINT/COUNTER POINT 9-1

TOPIC:

Fossil-Fuel Dependency vs Indirect Solar Energy Resources

Theme: During the past 20 years, many renewable and nondepletable energy resources have been tested for their economic and technical feasibility. The majority of these resources have been able to be converted into useful energy and are clearly more abundant than nonrenewable resources. Despite their abundance, they have been incorporated only minimally in most industrialized societies. In a discussion or classroom debate, attempt to answer the following questions:

1. Which of the renewable and nondepletable energy resources have potential for large-scale commercial use in the next 5 to 10 years? Which of the renewable and nondepletable energy resources have potential for large-scale commercial use in 20 years?

2. What social, technical, and environmental factors control the speed of development of any renewable or nondepletable energy resource? How can consumers alter this process of development?

3. What effect does the price of existing nonrenewable resources have on the implementation of renewable or nondepletable energy resources?

4. How can the implementation of renewable or nondepletable energy resources help offset the economic difference between industrialized and nonindustrialized countries?

5. What environmental concerns need to be addressed in the development of renewable and nondepletable energy resources?

6. What social strategies must be implemented to convince the public of the need to develop and adopt renewable and nondepletable energy resources?

7. In the late 1970s and early 1980s there was more funding by government and industry allocated to develop renewable and nondepletable energy resources. Identify the factors that caused this research funding to be reduced.

8. Identify or develop an action plan and the steps necessary that would encourage the rapid development and implementation of renewable and nondepletable energy resources. After identifying or developing this action plan, debate its merits and shortfalls.

9. If you were to develop a plan for the implementation of renewable or nondepletable energy resources for the next 20 years, which specific resource(s) would be used and why? Identify possible social changes that may also have to occur to allow a given society to be energy independent. In attempting to answer these questions consider the environmental, economic, political, social, and technical factors necessary for large-scale implementation.

SUMMARY & REVIEW 〜〜〜〜〜〜〜〜〜〜

Summary/Review Statements

1. Indirect solar energy is energy from the sun that must be changed into another form before it can be utilized.

2. Wind is produced because of sunlight. As the sunlight falls on different areas of the earth, the atmosphere is unevenly heated. Since warm air weighs less than cool air and rises as the temperature increases, air currents are created.

3. Of the solar energy that reaches the earth from the sun, only 2 percent is converted into wind. This small percentage contains more energy than humanity uses in one year.

4. In theory, the wind generated in the Great Plains could meet all of the United States' electricity requirements several times over.

5. Winds can be classified as prevalent or energy winds. Energy winds range in speed from 10 to 25 miles per hour and blow at this rate about two to three days per week. Prevalent winds are slower winds averaging 5 to 15 miles per hour and blow three to five days each week.

6. Wind turbines can be constructed with either vertical or horizontal shafts and in upwind or downwind configurations.

7. Each type of wind machine has a specific power coefficient, which is a measure of the energy that can be extracted from the wind by the turbine.

8. Theoretically, only 59.3 percent of the energy in wind can be extracted.

9. Wind energy farms are clusters of wind turbine generators used to produce electrical energy.

10. Bioconversion is the process of producing usable energy from organic matter, such as plant and organic wastes.

11. Bioconversion takes chemical waste and organic materials and converts them into thermal energy.

12. Bioconversion processes may be divided into four major types, depending upon the source of organic materials: urban and industrial wastes, agricultural and forestry wastes, land and freshwater energy farming, and ocean farming.

13. Energy experts estimate that early in the twenty-first century, 10 to 50 percent of the world's energy requirements could be met through bioconversion processes.

14. Waste can be decayed by either aerobic or anerobic processes.

15. Ocean Thermal Energy Conversion (OTEC) plants use the thermal energy near the surface of the ocean to operate the plant. Cold water from the bottom of the ocean helps to cool and condense the gases used to drive the turbine generator.

16. To date, several basic OTEC systems have been investigated: the closed-cycle, the open-cycle, and the hybrid.

17. One of the oldest methods used to extract energy from the earth's resources is known as hydro energy. Hydropower is really a form of kinetic energy, using the gravitational forces of falling water to turn water turbines to generate electricity.

18. Today, hydroelectric power is the largest nondepletable energy source in use. It supplies almost 27 percent of the world's energy needs.

19. Tidal power, also called ocean power, can be categorized as a type of indirect solar energy. It is created by the gravitational attraction of the sun and moon and the earth's rotation.

20. Geothermal energy is the heat energy located in the core of the earth. It is used to turn moisture into steam, which is then extracted and used to operate a turbine generator system.
21. There are three types of geothermal reservoirs: hydrothermal, geopressured, and hot-dry rock.
22. Geothermal reservoirs are considered nondepletable sources of energy.

Discussion Questions

1. How feasible is the use of wind energy on both a small and large scale?
2. What are some of the environmental problems associated with the use of wind energy?
3. Why are major public utilities building smaller-sized hydroelectric dams for producing electricity? What are the factors that lead to decisions in favor of building smaller dams?
4. List at least three advantages and three disadvantages of using geothermal energy.

IV

Introduction to

Energy Utilization

To this point in this textbook, all energy resources have been defined and categorized. Section IV constitutes the final process: considering how these energy resources are converted for use, stored, and conserved. These stages are collectively called energy utilization. This final section consists of one chapter divided into three parts that deal with energy utilization: energy conversion, energy storage, and conservation.

To help introduce this chapter, energy utilization has been shown graphically in Figure S-1, the Energy Flow Diagram. Stage One deals with energy resources. Stage Two identifies both conversion and storage of energy. Stage Three involves energy conservation.

This diagram can be used to identify how energy flows from resources to application. For example, after the energy resource has been developed or produced, energy can either be sent to storage or be converted to another form of energy. An example of energy in storage is that of coal stored at the plant site or water held by a hydroelectric dam. A gasoline engine is an example of energy converted to another form. The gasoline engine converts petroleum energy resources (chemical energy) into mechanical energy to move an object.

Note that there are arrows between storage and conversion. This is to illustrate that after the energy has been stored, it may be converted to another form before utilization. Energy can also be sent to storage after it is converted. For example, coal energy resources can be converted to electricity (conversion), then sent to a battery for storage.

Energy Flow Diagram

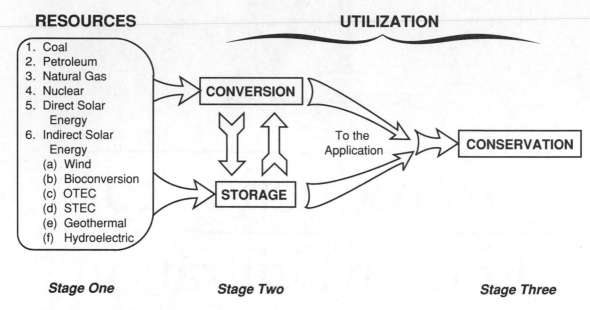

FIGURE S-1 Energy utilization is shown graphically in this Energy Flow Diagram. Note the three stages of using or conserving energy.

Conservation of energy (Stage Three) is necessary whether the energy resource is converted or stored. The purpose of energy conservation is to be aware of how energy is degraded during both the storage and conversion of energy. Energy conservation is concerned with utilizing the energy in the most efficient manner based upon social values.

chapter 10

Energy Conversion, Storage, and Conservation

PURPOSES

Goals of Chapter Ten

The overall goal of this chapter is to examine how energy resources are converted for use, stored, and conserved. This chapter provides an overview of various conversion technologies: chemical to thermal conversion, mechanical conversion, and electrical conversion. Storage techniques reviewed include chemical, thermal, and mechanical systems. The final part of this chapter presents information on conserving energy sources. In that part, specific information is presented on methods of reducing energy consumption and changing values for greater energy awareness. At the completion of this chapter, you will be able to

1. Define the concept of energy conversion in relation to efficiency.
2. Compare and contrast the various forms of energy conversion.
3. Examine the technologies used to convert chemical energy into mechanical energy.
4. Apply the concepts of electrical converters.
5. Evaluate the major types of energy storage.
6. Determine how human values affect the success of energy conservation.
7. Analyze the methods in which energy can be conserved within the various energy sectors of society.

Terms to Know

By studying this chapter, the following terms can be identified, defined, and used:

Air-fuel ratio
Alternating current
Brush
Chimney effect
Coils
Collector rings
Comfort level
Electrolysis
Electrolyte
Energy management

Energy storage
Heat gain
Heat loss
Induction
Neutral walls
Personal energy consumption
Polyphase
Regenerative system
Rotor
Static energy
Stator
Universal motor
Waste management

INTRODUCTION

The process of using energy from resource to application follows distinct paths. One path is conversion. After extraction of the resource, it is necessary to convert the energy to some usable form. For example, the automobile needs mechanical energy to produce motion of the vehicle. Since mechanical energy is not easily transported, the chemical energy in petroleum is converted to mechanical energy using an internal combustion engine.

A second path that may occur is to place the energy in storage. This is often necessary until it is needed for application. An example of storage occurs in hydroelectric power systems that are used to meet high energy demands. These systems are able to store the potential energy of water behind a dam until the water is needed to flow through a turbine at a peak load period.

A component of either path is conservation, which may be applied to an energy source at various times. For example, resources may not be used, thus saving them for future generations. In another method of conservation, energy sources may be used, but in such a manner as to maximize the efficiency of the conversion processes. This has the effect of helping to extend the reserves for future use. To a large extent, energy conservation is concerned with social values.

THE CONCEPT OF ENERGY CONVERSION

Need for Energy Conversion

The method in which a society utilizes energy becomes easy to understand when one considers the concept of energy conversion. As illustrated in chapter 1, there are six forms of energy. Energy conversion technology is necessary because the energy resources available to a society are often not in the form needed for application. For example, a boat needs mechanical energy to produce motion of the vessel. Therefore, the chemical energy in petroleum is converted to mechanical energy using a gasoline engine.

A great deal of technology associated with energy converters has been developed. Often the type of fuel that was available during the initial stages of development of a conversion technology determine new technical advancements in the area. For example, when piston engines were developed, gasoline was available. In the commercial and residential sectors, electrical and thermal energy forms are used. Again, because of the energy resources available at the initial stages of development, energy converters were designed to convert coal, oil, and natural gas into electrical and thermal energy. The industrial sector requires electrical, thermal, and chemical energy, depending upon the type of industry and its products. The energy resources available during the beginning of the industrial revolution dictated the type of converters that were developed and used. Many energy problems experienced today stem from these past developments. Fuels that were once readily available are dwindling today, and their cost is rapidly increasing.

Energy Conversion Examples

Changing energy from one form to another can involve a number of conversion processes. For example, the energy required to light an incandescent bulb can be used to illustrate the importance of the conversion processes. Initially the sun's radiant energy is converted (via photosynthesis) by plants into chemical energy. Eventually (over millenia) decay changes the plants into fossil fuels such as coal, oil, and natural gas. The stored chemical energy found in fossil fuels can be converted into heat (thermal) energy in a boiler or combustion chamber. The heat is then converted to mechanical energy, which may subsequently be converted into electrical energy by a generator. In contrast, the technology associated with a solar cell has reduced the number of conversion processes. A solar cell converts the sun's energy (radiant) directly into electrical energy that can be used for illumination (radiant energy), heat (thermal energy), or to operate appliances (mechanical energy).

There are many types of converters used today. Some of the more common ones are listed on the left side of Figure 10-1. Along the left side of the grid, the net efficiency is shown. Along the top and forming the remainder of the grid, the type of conversion (e.g., chemical to thermal, thermal to mechanical, and so on) is shown for each converter and marked at its level of net efficiency. Consider the electrical generator as a converter. It can obtain up to 99 percent efficiency when converting mechani-

cal energy to electrical energy. At the bottom of the chart, the incandescent lamp (as a converter of electrical to radiant energy) is as low as 5 percent efficient.

The steam turbine on the other hand, which utilizes coal, natural gas, or nuclear energy, is capable of converting thermal energy to mechanical energy at an efficiency of about 45

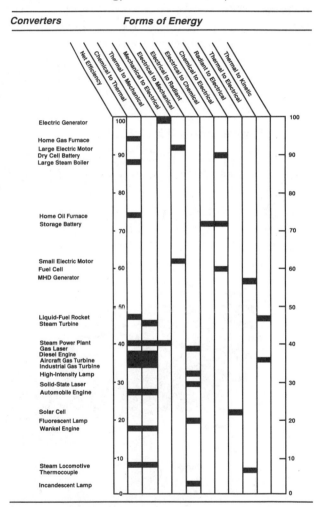

FIGURE 10-1 This chart shows the many ways in which energy can be converted. The applications are on the left and the type of conversion is at the top. Efficiencies are indicated by the placement of bars on the grid.

to 46 percent. Figure 10-2 shows a cutaway of a steam turbine generator system used to make this conversion.

Although numerous converters are shown in Figure 10-1, the principle of converting the available resources, at specific efficiencies, remains the same. Note that these efficiencies may increase as technology improves. For example, solar cell efficiency is projected to double in the next ten years as new materials and processes are developed. Only ten years ago, solar cells were only about 5 to 10 percent efficient; now they are closer to 25 percent efficient.

Energy Conversion—The Problem

Conversion of energy becomes a critical problem in societies that are reliant upon chemical energy, such as fossil fuels. Because the effects of entropy limit efficiency, as numerous conversions occur efficiency is lowered. Only converters with the highest efficiency should be built. This constitutes the real challenge for scientists, technologists, and researchers who must work within economic, political, environmental, and social constraints.

Converting the Best Form of Energy—Energy Issue 10-1

Energy must always be converted from one form to another to be useful in society. When considering the concept of entropy, what would be the best initial form of energy to use in each sector of society?

Charts and Terms

Performance Charts

In chapter 3, several terms were defined related to the power output of converters. Terms such as torque, horsepower, force, and work all play an important role in the study of the performance of energy converters. Performance

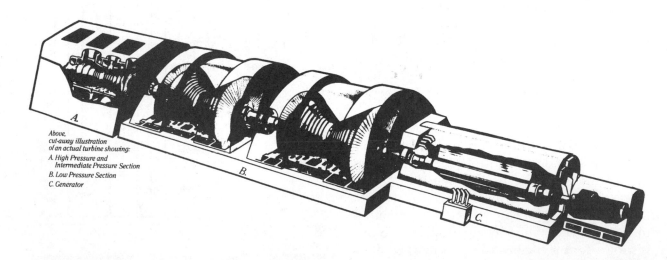

Above,
cut-away illustration
of an actual turbine showing:
A. High Pressure and
   Intermediate Pressure Section
B. Low Pressure Section
C. Generator

FIGURE 10-2 This coal-fired steam turbine generator system converts coal (chemical energy) into thermal energy, then into mechanical energy used to generate (convert mechanical energy into) electrical energy. (Courtesy of Cincinnati Gas and Electric Company and Lawler-Ballard Advertising)

charts are used in the power converter industry to display these output characteristics for each individual converter. In fact, almost every converter, from gasoline engines to high speed turbines, utilizes performance charts. From these charts, optimum speeds (rpm) can be selected for the highest fuel efficiency at specific power outputs. This data allows engineers to select the most efficient converter for the best application.

Many types of performance charts are utilized in the converter industry, but all have certain features in common. Generally, the charts are much like the one illustrated in Figure 10-3. On the horizontal axis, the engine speed or rpm is shown. The left vertical axis shows the power output in either watts or horsepower (1 horsepower is equal to 746 watts). The right vertical axis illustrates torque output measured in Newtons or in foot-pounds. Fuel consumption (in pounds per brake horsepower per hour) is often graphed on the performance chart, usually below the horsepower and torque measurements.

Types of Horsepower

Numerous types of horsepower are commonly used in the converter industry. Some of the more important horsepower terms used in the industry include

1. Frictional horsepower—horsepower lost internally in a converter or engine due to friction and resistance from moving parts.
2. Maximum theoretical horsepower—horsepower available without considering frictional losses within the converter.
3. Brake Rated Horsepower—horsepower available considering the internal frictional losses. This is also referred to as brake horsepower.
4. Continuous horsepower—25 percent less than the brake rated horsepower, used for long-time continuous duty operation.
5. Intermittent horsepower—the range between brake rated horsepower and continuous horsepower.
6. Shaft horsepower—horsepower available at the external shaft of the converter.

Performance Chart

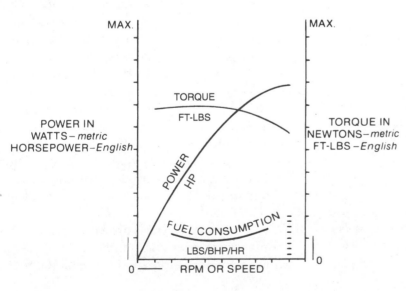

FIGURE 10-3 An engine performance chart consists of three curves: torque, horsepower, and fuel consumption.

7. Road horsepower—horsepower available at the drive wheels of a vehicle utilizing a gasoline or diesel converter.

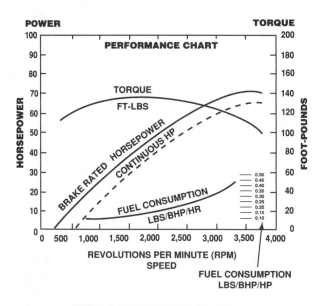

POWER · TORQUE

STANDARD ATMOSPHERIC CONDITIONS:
ALTITUDE, 500 FEET
TEMPERATURE, 88°F

FIGURE 10-4 This performance chart of a small gasoline engine shows brake horsepower, continuous horsepower, torque, and fuel consumption throughout the rpm range.

Reading a Performance Chart

The performance chart shown in Figure 10-4 illustrates the performance of a four cylinder gasoline engine. Horsepower is shown on the right vertical axis, and torque is shown on the left vertical axis. Rpm is shown on the horizontal axis.

Several converter characteristics can be identified from this performance chart:

1. As engine rpm increases, horsepower will also increase until a maximum is reached. The formula for calculating horsepower at any one torque rating is as follows:

$$HP = \frac{T \times RPM}{5252}$$

Where:
HP = Horsepower
T = Torque
RPM = Revolutions per minute of the engine

2. Torque tends to remain relatively constant throughout the rpm range, but it does peak at a specific rpm. Air intake contributes to this relationship. As the engine speed changes, it may be more or less difficult to get the air into the engine efficiently. The ease with which air moves into and out of the engine is defined as volumetric efficiency. The highest volumetric efficiency is near the peak of the torque curve. Things that can affect the volumetric efficiency include: intake valve opening, exhaust restrictions, intake restrictions, shape of the intake and exhaust manifolds, and so on.

3. All performance charts are based upon a converter operated at standard specifications, which include an altitude of 500 feet

above sea level and an air temperature of 88° F. Engine characteristics vary with altitude and air temperature. For example, if the engine were run at a higher altitude, the air would be less dense per equal volume. This would cause a change in the air-fuel ratio (richer fuel). The *air-fuel ratio* is defined as the ratio of air to fuel (in weight) entering a combustion chamber. Optimal combustion efficiency is a function of the air-fuel ratio. Theoretically, it takes 14.7 parts of air to burn one part of fuel. The result of a change in the air-fuel ratio is a change in torque and horsepower rating. Also, if the air temperature were to change, the air density would change— thus causing a change in the air-fuel ratio again.

4. The optimum speed for producing maximum torque on the engine illustrated in Figure 10-4 is about 1400 rpm. The best speed for maximum horsepower would be about 2,800 rpm.

5. When the engine is cut back by 25 percent to operate on continuous horsepower ratings (see dotted line), the torque characteristics are reduced.

Calculating Data from Performance Charts—Math Interface 10-2

The data found on a performance chart can be used to calculate cost per year for companies that use many engines. Solve these performance chart problems to gain additional understanding of *horsepower, torque, rpm,* and *Btu's per BHP/hr.*

1. When making a dynamometer test on a multiple cylinder engine, a load is applied so that the torque reading is 90 foot-pounds at 4,500 rpm. Under these conditions, how much horsepower is produced? How many watts are produced?

Horsepower produced is equal to

_____.

Watts produced are equal to

_____.

2. A large trucking firm needs to purchase 120 new vehicles. Each vehicle will be operated for 10 hours per day, 6 days per week, over a period of 1 year. The average horsepower during operation is 50 BHP at 2,000 rpm. Two engines are to be compared. Under the stated conditions, engine A has a fuel consumption of 0.2 pounds per brake horsepower per hour. Engine B has a fuel consumption of 0.23 pounds per brake horsepower per hour. How much money will be saved per year if engine A is used instead of engine B? Note that fuel cost is $1.15 per gallon and that there are 8 pounds of fuel in a gallon. (Carry all calculations to 3 points beyond the decimal.)

a. How many hours per year must *each* engine run?

Hours each engine is run per year are equal to _____.

b. How much fuel in pounds per hour is required for *each* engine under the stated conditions?

Engine A fuel pounds per hour equal _____.

Engine B fuel pounds per hour equal _____.

c. How many gallons are required for *each* engine per hour?

Engine A gallons per hour equal

_____ .

Engine B gallons per hour equal

_____ .

d. How many gallons are required for *each* engine per year?

Engine A gallons per year equal

_____ .

Engine B gallons per year equal

_____ .

e. How much will it cost the trucking firm to operate *each* engine per year at $1.15 per gallon fuel cost?

Engine A fuel cost per year equals

_____ .

Engine B fuel cost per year equals

_____ .

f. How much in savings will be realized *per vehicle* each year if engine A is used rather than engine B?

Savings per vehicle each year would equal _____ .

How much will be saved per year in total _____ .

CHEMICAL TO THERMAL TO MECHANICAL CONVERTERS

Internal Combustion Engines

A common type of conversion used in technical systems is that of a hydrocarbon fuel (chemical energy) converted into thermal energy, then further converted into mechanical energy to produce motion. The gasoline engine and diesel engine are designed to accomplish these conversion processes. They use chemical energy (petroleum) as the input. Their output is in the form of mechanical energy and can be measured in torque and horsepower. A brief description of the gasoline and diesel engine converters are illustrated in the following sections.

Gasoline Engines

Gasoline Engine Converter

The gasoline engine, also called a spark ignition or Otto cycle engine, has been in existence for more than 90 years. It is the most common converter used to change chemical energy into thermal and, then, to mechanical energy. The gasoline engine is an intermittent combustion converter. This means that the combustion is not continuous, but occurs intermittently.

Gasoline engines are used for prime movers in the transportation sector, including use in automobiles, boats, and certain small aircraft. Other applications include recreational vehicles, garden equipment, agricultural machinery, and almost any application requiring torque.

In principle, gasoline (the chemical fuel) is brought into the engine along with air. The carburetor or a fuel injection system mixes the two at a 14.7 to 1 air-fuel ratio. A piston compresses the mixture, which is then ignited by a spark plug. The expanding gas pushes downward on the piston. This push causes the downward movement of the piston to turn a crankshaft. The crankshaft then converts this downward movement to rotary motion, or torque.

Four Stroke Cycle Engine Design

There are several types of gasoline engine designs. One is called the four stroke cycle design. Figure 10-5 shows the operation of a typical engine of this type. There are four strokes identified throughout its operation: intake, compression, power, and exhaust. The intake stroke is shown in Figure 10-5(a). As the piston is moved downward, the intake valve is opened to allow a mixture of air and fuel to enter the combustion chamber. With the valves closed, the piston moves upward to produce the compression stroke, as shown in Figure 10-5(b). During this stroke, the volume of the air and fuel is compressed—in this case to about 1/9 its original size. This engine is said to have a compression ratio of 9 to 1. When the piston is near the top of the cylinder, the spark plug ignites the compressed air-fuel mixture. It is at this point the chemical energy is converted to thermal energy.

After ignition by a spark plug, the power stroke occurs—as shown in Figure 10-5(c). The power stroke is the result of combustion that converts the thermal energy into mechanical energy. The piston is, thus, pushed downward. The crankshaft turns, which moves the piston upward on the exhaust stroke as shown in Figure 10-5(d). This upward motion forces the exhaust gases out through the exhaust valves. Most automobile engines today operate on this four cycle principle.

The four stroke cycle engine design only yields a power stroke every other revolution. This means that if the engine is turning at

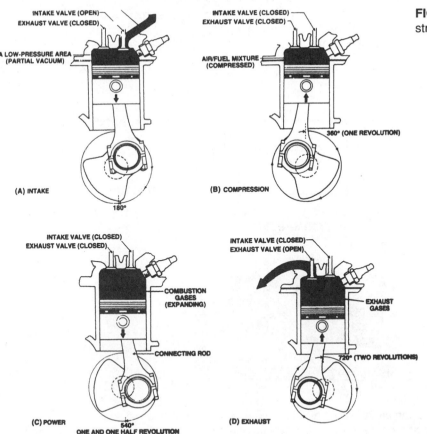

FIGURE 10-5 A schematic of a four stroke cycle engine design is shown.

5,000 rpm, each piston is producing 2,500 power pulses per minute. This causes the overall response (quickness of rpm change) of the engine to be somewhat slower than other types of engines.

Gasoline Engine Limitations

Gasoline engines, although refined technically, have certain inherent limitations. Because of the physical design of the engine and gasoline used as a fuel, there will always be a certain amount of incomplete combustion. This unburned fuel produces significant amounts of pollution. Therefore, pollution-control devices were designed and installed on engines. Although pollution has been reduced, these controls sometimes cause greater gasoline consumption.

To overcome the pollution and poor fuel consumption, engines are controlled by computers both to increase their efficiency and reduce pollutants. Today, the air-fuel ratio and the ignition timing are controlled by computers. Ignition timing is controlled and must be changed because as the engine goes faster the spark plug ignition must occur sooner to lessen pollution and increase engine efficiency. Figure 10-6 shows various inputs needed to control both air-fuel ratios and ignition timing. For example, the onboard engine computer is able to sense various inputs including

1. oxygen in the exhaust (via an oxygen sensor);
2. vehicle altitude;
3. coolant temperature;
4. outside temperature;
5. throttle position;
6. rpm of the engine; and,
7. load on the engine.

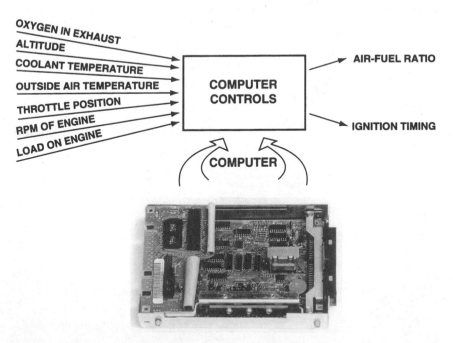

FIGURE 10-6 To maintain a proper air-fuel ratio and to ignite the fuel mixture at the right time, computers are used on gasoline engines.

Each of these inputs, if changed, will have a direct effect on the air-fuel ratio and the ignition timing. Based upon these inputs, the computer constantly adjusts the air-fuel ratio and ignition timing several times per second. In addition, in many automobiles, the computer controls the valve opening and timing for greater efficiency.

Internal temperatures also affect the efficiency of gasoline engines. These internal temperatures can be increased only to a certain point because of the design of the parts and their materials. It would be advantageous to operate an engine at a much higher temperature since thermal energy is needed to make mechanical energy. However, to keep the engine from being overheated and damaged, about 33 percent of the thermal energy in the combustion process is removed through a cooling system. Another 33 percent of the thermal energy in the combustion chamber is removed by the exhaust. This gives the engine a relatively low efficiency rating.

Another problem with piston type engines is the motion involved. The piston must change directions twice per stroke, once at the top of the stroke and once at the bottom of the stroke. This is not mechanically efficient. It would be more advantageous if the motion were continuous rather than reciprocating. Taking heat and motion into account, the overall efficiency of the gasoline engine is only about 22 to 30 percent.

Diesel Engines

The diesel engine shown in Figure 10-7 is known as a compression ignition engine. As with the gasoline engine discussed in the last section, it is also an intermittent combustion converter. The diesel engine uses a slower burning fuel than gasoline. This makes the diesel engine more

FIGURE 10-7 The efficiency of a diesel engine is slightly higher than that of a gasoline engine. (Courtesy of Cummins Engine Company, Inc.)

efficient. Experts note that the efficiency of a diesel engine is near 32 to 38 percent.

Several recent design factors have helped to improve diesel engine efficiency, enhancing its attractiveness in many markets. Such factors include higher compression ratios, higher efficiency, high-pressure fuel injection, and higher energy fuel.

Gasoline Versus Diesel Engines

Figure 10-8 shows a comparison between the diesel and the gasoline engine. These include:

1. Gasoline contains less energy per gallon than diesel fuel. Typically, diesel fuel has about 130,000 Btu's per gallon; gasoline has about 100,000 Btu's per gallon. Using diesel fuel means that more energy can be extracted from an equivalent volume of fuel.
2. The intake stroke in a gasoline engine involves both air and fuel; the diesel engine intakes only air.

3. The compression ratio of a diesel engine is much higher than a gasoline one, in the range of 20:1 or 25:1. This high compression ratio can be used in the diesel engine because no fuel is present in the intake mixture. Thus, the fuel cannot preignite—a typical problem with the gasoline engine. The high compression ratio causes higher pressure within the combustion chamber. This means that better mixing of the air and fuel will occur.

4. In the diesel engine, the air and fuel are mixed immediately after the injection of fuel by high-pressure injectors. In the gasoline engine, the air and fuel are mixed by low-pressure fuel injectors (or in some applications by a carburetor) before going into the cylinder.

5. The combustion occurs from the heat of compression in a diesel engine. (Ignition comes from a spark plug on the gasoline engine.) When a volume of air is compressed it heats up. Because of the high compression ratio, the heat is enough to ignite the fuel in the diesel engine. The power stroke produces more pressure on the top of the piston of the diesel engine because of the high compression ratio and the type of fuel used.

6. The exhaust temperatures in a diesel engine are cooler than those in a gasoline engine. This is because the diesel engine burns the fuel more slowly and efficiently. On a gasoline engine, as the exhaust stroke occurs, many partially burned hydrocarbons are exhausted—causing the increased temperature. Unburned or partially burned fuel produces more carbon monoxide emission in the gasoline engine. The diesel engine, on the other hand, produces very little carbon monoxide; however, it produces a great deal more nitrogen oxide as a pollutant because of the higher internal combustion temperatures.

7. Considering all factors, the diesel engine is about 10 percent more efficient than the gasoline engine.

Although only two types of chemical to thermal to mechanical converters were presented in this section, note that other converters are used to meet specific performance (horsepower, torque, rpm) and/or efficiency needs. Other chemical to thermal to mechanical converters include the two stroke cycle engine,

		GASOLINE	DIESEL
	Fuel BTU'S	100,000 BTU/Gallon	130,000 BTU/Gallon
1.	Intake	Air-Fuel	Air
2.	Compression	8–10 to 1 130 psi 545°F	16–25 to 1 400–600 psi 1000°F
3.	Air-Fuel Mixing Point	Carburetor or injector before valves	Near top dead center in combustion by injection
4.	Combustion	Spark Ignition	Compression Ignition
5.	Power	464 psi	1200 psi
6.	Exhaust	1300°–1800°F CO = 3%	700°–900°F CO = .5%
7.	Efficiency	22–28%	32–38%

NOTE. These are average figures and may change with engine size and type, fuel, and operating characteristics. Because of the differences as listed above, the diesel engine is about 10 percent more efficient than the gasoline engine.

FIGURE 10-8 This chart shows a comparison between a gasoline and a diesel engine.

rotary engine converter, coal-fired boiler, gas turbine converter, and rocket engine converter. More detailed information on these converters and their applications can be found in power technology textbooks. Figure 10-9 shows photos of some of the different types of converter technology.

MECHANICAL TO ELECTRICAL CONVERSION

Often it is necessary to convert mechanical energy into electrical energy. For example, the gasoline engine in an automobile needs to convert some of the mechanical energy of the engine to electrical energy. This energy can then be used for operating the radio, lights, and other electrical applications.

One of the most common methods of converting mechanical energy to electrical energy is to use generators. A generator converts mechanical energy in the form of torque into electrical energy. In fact, large-scale electrical generating power plants utilize generators in this manner.

Boilers are used to convert coal (chemical energy) into thermal energy and finally into steam that turns a turbine (mechanical energy). This technician is inspecting the pipes used to carry the steam in a boiler to the turbine. (Courtesy of Wisconsin Electric Power Company) From Wisconsin Electric Power Company.) (b)

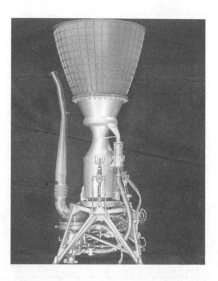

An Agena rocket engine is shown. (Courtesy of Bell Aerospace Textron) (c)

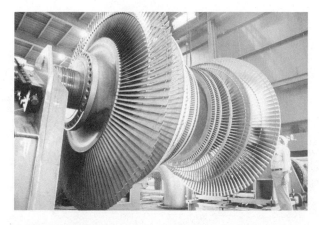

FIGURE 10-9 This turbine is used in large-scale production of electricity. (Courtesy of Kentucky Utilities) (a)

Generators

Generators are the most common method used to produce electrical energy. Because of the number of applications utilizing electricity, many types of generators are currently in use. Generators work on basic electromagnetic (electrical and magnetic) principles. Electromagnetic principles state that when a wire conductor is passed or cuts through an electromagnetic field, a voltage is induced in the wire.

Figure 10-10 shows how voltage is produced in a generator. In the top drawing, there are four wire positions shown (a, b, c, and d). Think of the wire as rotating inside of a magnetic field. The magnetic lines of force flow from the north to south pole. When the wire is at position *a*, there are no lines of force being cut because the movement of the wire and the lines of force are in the same direction. This is represented on the lower graph as point *a* or 0 degrees rotation. As the conductor rotates to point *b* on its axis, or 90 degrees rotation, maximum lines of force are now being cut downward. This produces a positive voltage, shown on the graph at point *b*. As the conductor rotates to point *c*, or 180 degrees rotation, again, there are no lines of force being cut and no voltage is produced. This is shown on the graph at point *c*. As the conductor rotates toward point *d*, or 270 degrees rotation, maximum lines of force are again being cut, only in a reverse direction. This produces a negative voltage as shown at point *d* on the lower graph. Because of this overall design, the voltage is alternating or reversing direction every revolution. The rotating conductor finally returns to point *a*, where no lines of force are cut. At this

Generator Design

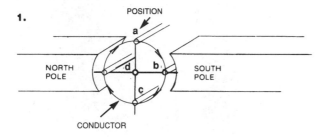

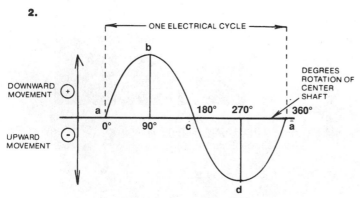

FIGURE 10-10 As a conductor rotates 360 degrees on its axis, through a magnetic field, it produces one electrical cycle, also called alternating current (AC).

point the cycle will repeat itself. When the AC voltage is tapped off the rotating wire, the voltage is changed to DC.

Although many conductors and stronger magnetic poles are used, this is the basic operation of any generator used to produce alternating current. The electricity is then used for residential, commercial, industrial, or transportation applications.

Automobile Alternator

The automobile alternator uses many of the electromagnetic principles of the generator. The alternator uses a rotating set of poles, however, rather than stationary ones. The rotating sets of poles are manufactured by wrapping wire around a soft iron core, as shown in Figure 10-11(a). Two iron segments with interlacing fingers are placed on each end of the electromagnet, producing seven N and seven S poles, Figure 10-11 (b). An electric current is fed into the rotating poles by the use of two slip ring brushes shown in Figure 10-11 (c). A stationary conductor, also called the stator, is placed around the rotating magnetic field, Figure 10-11 (d). To produce electricity in the alternator, electrical energy is sent to the rotor (rotating magnetic field). As the seven N and seven S poles rotate, their magnetic fields pass the stationary conductor or stator. As the conductors cut the magnetic field, a voltage is induced into the conductors. Note that the alternator produces AC voltages, which are then converted to DC for use in the automobile.

Alternator

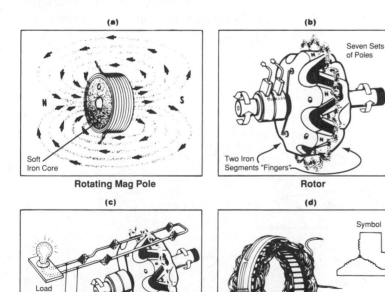

Rotating Mag Pole

Rotor

Stator

Three Phase Stator Assembly

FIGURE 10-11 The alternator used in automobiles changes mechanical energy into electrical energy by rotating a magnetic field with seven sets of poles. (Courtesy of Delco Remy)

ELECTRICAL TO MECHANICAL CONVERSION

Motors

Often it is necessary to convert electrical energy into mechanical energy for turning shafts, engine starters, or other mechanical needs. Electric motors are used for this conversion process. Electric motors consume about 70 to 80 percent of all electric power generated in society. They are, also, the chief source of mechanical energy in most industrial and manufacturing plants.

Motor Operation

Motors operate by using three electrical and magnetic principles: voltage, current, and an opposing magnetic field. Figure 10-12 illustrates basic motor operation. In this diagram, both a North and South pole are shown; a simple conductor or wire has been placed in between the poles, in the magnetic field.

When an electrical current passes through the wire placed within a magnetic field, another magnetic field is developed and flows in a clockwise circle around that wire. This is illustrated graphically by the circle shown around the wire. Look closely at the bottom of the wire. Notice that now there are two magnetic fields working against each other (see the direction of the arrows). One is from the poles, and one is from the circular magnetic field. Also, on top of the wire, two magnetic fields are aiding each other (again, see the direction of the arrows).

The difference between the opposing and aiding magnetic fields causes the wire to move upward. Since the wire in the middle of the magnetic field is attached to a rotating shaft, the wire will rotate. If the direction of electricity within the wire reverses or if the stationary North and South poles reverse, the movement of the wire and the direction of the motor reverse. Motors, therefore, are able to change electrical energy into mechanical energy (measured in torque).

Motor Operation

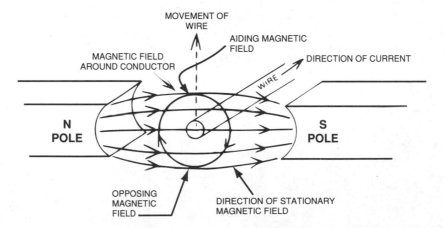

FIGURE 10-12 When current is passed through the center conductor, a magnetic field surrounds the conductor. This field in conjunction with the North and South magnetic fields opposes and aids each other so as to make the wire move.

Motor Terms

There are many types of motors in use. The following terminology is used with motors:

Alternating current (AC)—the regularly reversing flow of an electrical charge through a conductor. Most electric power delivered to homes, offices, and factories by utilities in the United States is AC.

Brush—a conductor used to maintain electrical contact between stationary and moving parts in a motor.

Coils—one or more turns of wire wound as a component in an electrical circuit.

Collector rings (slip rings)—metal rings that conduct current via stationary touching brushes into or out of a rotating member of a motor.

Induction—the production of an electric current in a conductor by a magnetic field in close proximity to it.

Polyphase—current in more than one phase, usually three. Single phase current is simply a single alternating current. Three phase (a type of polyphase) current has three alternating currents. Each of the three goes through its own cycle, reversing at a different time from the other two.

Rotor—the rotating part of most AC motors.

Stator—the stationary parts of a magnetic circuit with associated windings.

Universal Motor—a motor that can be operated on either AC or DC.

CHEMICAL TO ELECTRICAL CONVERSION

Fuel Cells

Often it is necessary to convert chemical energy directly into electrical energy. Although batteries are able to do this conversion effectively, they are a type of storage technology.

Fuel cells convert chemical energy directly into electrical energy. They are used commercially as well as in the space program. Known for over 100 years, the fuel cell has recently been revived because of increasing societal energy needs. Given the energy issues facing society, the fuel cell may play an important role in the conversion of energy.

Some of the advantages of fuels cells include

1. higher efficiency than batteries;
2. low maintenance requirements;
3. variable sizing;
4. ease of adding components;
5. environmental safety; and,
6. the ability to augment existing technology.

Not many energy converters today have all of these advantages.

Some of the fuel cell's diversified applications include

1. space programs (flight of remote or human assisted rockets);
2. communications (microwave relay, radio, television, and telephone transmissions);
3. aids to navigation (buoys and lighting); and,
4. standby power stations.

In the past, various experimental fuel cell projects included applications to provide electric power to apartment houses, commercial establishments, and small industries. From this research it was found that fuel cells could be adapted to the residential and commercial sectors.

Because they can efficiently produce electrical energy on both small and large scales, fuel cell systems are also candidates for on-site power installations. For example, 80 percent of commercial and residential buildings have a maximum power rating of under 200 kilowatts. On-site fuel cells capable of producing this type of power requirement could save 25 to 30 percent of the fuel required to transmit electricity to such buildings.

Fuel Cell Operation

The fuel cell uses an electrochemical process. The chemical energy that bonds atoms of hydrogen (in a hydrocarbon source) and oxygen (in air) is converted directly into electrical energy. In Figure 10-13 a simplified illustration of a fuel cell is shown. Hydrogen, carbon dioxide, and oxygen are inputs. A single fuel cell consists of an anode and a cathode plate with a chemical mixture between the two. Individual cells such as this generate 0.5 to 1 volt and must be stacked and connected to boost voltage and capability.

In operation, fuel (hydrogen and carbon dioxide) from a tank or processor enters the cell. It flows down along the anode plate, losing electrons and giving the plate a negative charge. When oxygen as a fuel is supplied to the cathode, oxygen molecules pick up electrons on the cathode plate, leaving it positively charged. When an external load is connected to the two electrodes, excess electrons on the anode discharge through the load to the cathode. During this process, positive hydrogen ions at the anode transfer through an electrolyte solution and combine with oxygen ions from the cathode to produce water. The *electrolyte* solution is a substance that dissociates into ions, thereby becoming electrically conducting. (An ion is an atom or atomic group that is electrically charged as a result of having gained or lost electrons.)

There are several by-products from a fuel cell. These include water and heat, and carbon dioxide from the fuel input. There is generally enough heat for the water to be changed to steam during the process.

The electrolyte, composed of a phosphoric acid solution, circulates through the porous matrix separating the electrodes (which are

Fuel Cell

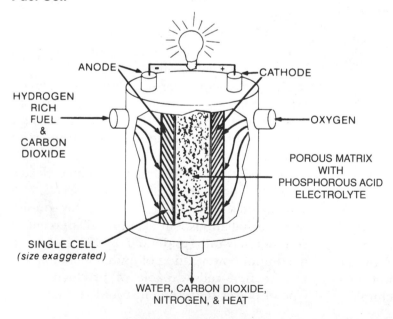

ANODE — CATHODE

HYDROGEN RICH FUEL & CARBON DIOXIDE

OXYGEN

POROUS MATRIX WITH PHOSPHOROUS ACID ELECTROLYTE

SINGLE CELL
(size exaggerated)

WATER, CARBON DIOXIDE, NITROGEN, & HEAT

FIGURE 10-13 A fuel cell uses hydrogen, carbon dioxide, and oxygen as the input fuels. In addition to electrical current, carbon dioxide, nitrogen, and thermal energy are output.

also porous). The electrode plates are actually paper-thin carbon sheets containing a metal catalyst, such as platinum. These aid in the ionization of the hydrogen fuel and oxygen.

Fuel Cell System

The fuel cell generator system has three main parts. These are shown in Figure 10-14. They are the fuel processor, the fuel cell, and the power inverter—each of which is a necessary component of the fuel cell system.

Hydrocarbon fuels and steam recycled from the fuel cell operation are initially fed into the fuel processor. The fuel processor separates the hydrogen from the hydrocarbons for use as a fuel. Then, the mixture is fed into the fuel cell where the cells are connected to each other. As electricity is produced, waste water from the stack can immediately be sent back to the fuel processor.

The DC (direct current) power from the fuel cell must then be converted to AC (alternating current) power for utility application. The power inverter, which accomplishes this conversion, operates at nearly 90 percent efficiency.

These three parts—processor, cell, inverter—can produce electricity from fuel to utility grid at an overall efficiency of near 38 to 40 percent. This percentage is certainly comparable to the efficiency of large power-plant converters commonly used today. Researchers estimate that fuel cell efficiency may reach 50 to 55 percent in the early twenty-first century. Figure 10-15 shows an on-site fuel cell unit.

> **Fuel Cells in the Future—Energy Issue 10-2**
>
> The use of fuel cells has great potential for use in the future. What economic, environmental, and social factors seem to control the future use of fuel cells? When do you think fuel cells will become commercially available as an electrical energy converter? What factors would have to change to get fuel cells into commercial application?

INTRODUCTION TO ENERGY STORAGE

Energy Storage Defined

In this section, *energy storage* is defined as the placing of energy into a state so that it can be held for future use with minimal energy dissipated. Often such stored energy is referred to as *static energy*. Although energy storage is important, it is nearly impossible to accomplish this without having some of the energy degraded. For example, batteries will dissipate the energy stored within them over a period of time.

Examples of Energy Storage

There are numerous examples to illustrate the forms in which energy can be stored. Figure

Fuel Cell System

FIGURE 10-14 The major components of a fuel cell generating system are shown.

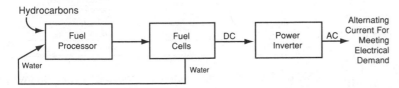

FIGURE 10-15 This fuel cell unit is capable of producing 40 kilowatts for on-site power application. (Courtesy of Power Systems Division, United Technologies Corporation)

The energy used to produce a nuclear fission reaction is stored within the atom. Energy from organic waste or bioconversion, after its conversion processes, is stored as methane and/or alcohol. Geothermal energy is stored in the ground or the surrounding water within the earth.

Solar thermal energy converters (STEC) store thermal energy, which is hot water, air, oil, or rock held in a large tank. The thermal energy is converted into electricity, which can be stored in batteries.

The solar energy used in ocean thermal energy converters (OTEC) is stored in the ocean water. The electrical energy produced by OTEC can be stored in batteries. No matter which form of energy is used, storage can occur. Many storage forms, both natural and manufactured, exist. In fact, by following the

10-16 lists different forms of energy and their storage methods. For example, solar energy is absorbed into plants through photosynthesis. It is stored (after plant decay) as chemical energy such as coal, petroleum, and natural gas. Plants not decayed still store energy in the forms of wood and grain. Of course, it takes millions of years for coal, oil, and natural gas to develop. But the chemical energy stored in wood and grain take much shorter periods of time.

Solar energy can be converted into thermal energy (by thermal solar collectors) and stored in rock, water, or eutectic salts. Solar energy can also be converted directly into electricity through photovoltaic cells, which can then be stored in batteries as a chemical form of energy or used to produce hydrogen.

A hydroelectric dam holds the potential energy held or stored within a reservoir. Wind energy is initially converted to electricity and then stored in batteries or used in a utility grid.

Forms of Energy Storage

Resource or Form of Energy	Storage Method Examples
1. Solar energy through photosynthesis	Coal Petroleum Natural gas Wood Grain
2. Solar-thermal energy	Rock Water Eutectic salts
3. Solar photocells	Battery Hydrogen
4. Hydroelectric energy	Water reservoir
5. Wind energy	Battery
6. Nuclear energy	Atoms
7. Bioconversion	Methane or alcohol
8. Geothermal energy	Hot compressed water and air—steam
9. STEC	Hot compressed water and air—steam Oil and rock
10. OTEC	Water (ocean)
11. Electricity	Battery

FIGURE 10-16 Various energy resources, shown on the left, can be stored in different forms, as shown on the right.

flow of energy through its conversion process, one may find that several storage periods may exist. The resources of energy, their present form, and how they are to be converted all play an important role in the number of storage stages.

Energy Stored Per Weight

Many of the developments in storage technology center around the storage per weight of energy. The goal of any storage medium is to store as much energy in as small a space as possible. For example, when storing thermal energy used in solar space heating, rock storage was first utilized. Now, eutectic salts are more promising because they are able to store more heat per unit of weight than rock.

Forms of Energy Storage

In examining the methods in which energy storage occurs, three categories emerge:

1. Chemical Storage
2. Thermal Storage
3. Mechanical Storage

Chemical storage includes such technologies as batteries, hydrogen, atoms, coal, petroleum, and natural gas. Thermal or heat storage includes rock, water, and eutectic salts. Mechanical storage includes water reservoirs, compressed steam, air storage, and flywheels. Since thermal storage was previously presented in conjunction with solar energy resources in chapter 8 of this text, it will not be reiterated here.

CHEMICAL ENERGY STORAGE TECHNOLOGY

Battery Storage Technology

One type of chemical storage is via the battery. This device, developed around 1800, is a method of storing chemical energy to be converted later directly into electrical energy. Many types of batteries are used today in an assortment of applications. A battery is made of one or more cells. A cell is the part of a battery that stores chemical energy for later use as electricity. Batteries are generally made up of a number of cells. These are connected either in series, to increase the voltage, or in parallel, to increase the current capacity while keeping the voltage the same.

A cell consists of three components: two dissimilar metals and an electrolyte solution. The electrolyte is used as a medium to allow ions to pass electrons through the cell.

Battery Types

Batteries can be identified by the type of cells used: primary and secondary cells. Primary cells are those that are nonrechargeable. They are used until the chemical reactions that store the energy have been exhausted. Secondary cells are those that can be discharged and recharged repeatedly by reversing the normal electrical current flow.

Examples of primary cells are those most often used in flashlights, calculators, smoke alarms, radios, and the like. Various metals and electrolytes are used to produce the needed voltage (e.g., 9 volt) developed from these primary cells. Examples of secondary battery cells are those used in automobiles, trucks, snowmobiles, all terrain vehicles, and motorcycles. Note also that some secondary battery cells have been developed for numerous small applications, such as those listed above.

Determining Battery Voltage

Voltages that are developed in the cell within a battery are determined by the energy level of each of the metals in the battery. This is indicated by the amount of dissimilarity between the metals as shown on the periodic table of elements. For example, when zinc and lead are used, approximately 0.5 volts will be developed in each cell. When zinc and silver are used, about 1.2 volts can be developed.

Lead-Acid Battery

Primary cells are used in many smaller applications; secondary cells are used in larger power output applications. The most familiar example of a secondary cell battery application is the automotive lead-acid battery. It consists of two dissimilar metals, lead and lead peroxide, and an acid as the electrolyte. These dissimilar metals are able to produce about 2 volts. Each battery usually consists of three or six cells in series, which produce 6 or 12 volts (respectively).

The positive plate of each cell is made of lead peroxide (PbO_2), and the negative plate is made of sponge lead (Pb). Antimony is added to strengthen the lead plates. This battery is referred to as the lead-antimony battery. Today's maintenance-free lead-acid batteries have calcium added rather than antimony. A trace of tin is also added. These new batteries will not produce as much hydrogen gas during charging and discharging, which helps to make the batteries safer as hydrogen is explosive.

Several plates are connected in parallel to increase the amperage capacity of the cell, but to keep the voltage per cell the same. The reason to increase the amperage capacity of a battery is to lengthen the time it takes to become discharged. The cells are placed in series within the casing of the battery to produce, commonly, 12 volts per battery. Series means to connect the cells so that a current passes from one to the next without branching to other cells.

A typical cell is shown in Figure 10-17. The negative and positive plates are grouped together with a separator placed between them to eliminate a short circuit. The electrolyte that the cell is placed in is generally made of a mixture of sulfuric acid (H_2SO_4) and water (H_2O), in a ratio of 40 percent H_2SO_4 to 60 percent H_2O.

Hydrogen Storage Technology

Energy in a chemical form can also be stored in hydrogen gas. Hydrogen, the lightest of all the basic chemical elements, is an essential component of water and organic compounds.

Lead-Acid Battery Cell Construction

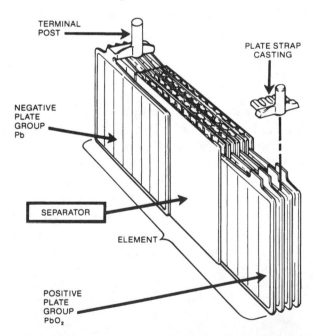

FIGURE 10-17 A typical lead-acid battery cell is made of a series of positive and negative plates separated by an insulator. (Courtesy of Delco Remy)

Hydrogen is also rare in its natural form and must, therefore, be produced or manufactured.

At ambient temperatures, hydrogen is a colorless, odorless, tasteless, and transparent gas that weighs only 7 percent as much as air. It also will burn readily if ignited. The only products of combustion, when burned with oxygen, are water and a slight amount of nitrogen oxide (considerably less than most hydrocarbon fuels).

When burned in a gaseous state, the Btu content of hydrogen per cubic foot is about two thirds less than that of natural gas. However, its energy content per pound (as a liquid) is almost three times that of gasoline. Figure 10-18 shows a comparison of the Btu's per pound for several known energy resources and liquid hydrogen.

Comparison of Btu/Pound of Various Energy Sources (Average Figures)

Gasoline	17,000
Jet Fuel	19,000
Liquid Propane	22,000
Liquid Hydrogen	50,000
Hydrides	8,000

FIGURE 10-18 Each fuel has a different energy content in liquid form (Btu's per pound). Note that liquid hydrogen has approximately three times as much energy as gasoline.

When hydrogen gas is cooled to below its boiling point, about −423° F, it changes to a liquid. In this state, it takes up approximately 1/700 as much space. This provides a valuable type of fuel for space programs, which rely on high-energy and low-weight fuels.

Production of Hydrogen

Several million tons of hydrogen are produced annually within the United States. Hydrogen is primarily used in petroleum refining and in making ammonia and methyl alcohol (two major industrial chemicals). Most of the hydrogen is produced by reacting natural gas or certain types of oil with steam at high temperatures. Hydrogen can also be produced by a process called *electrolysis*, which is the decomposition of a substance by means of an electric current (as in the production of hydrogen and oxygen from water). This is the same basic process that takes place when charging or storing electricity in a lead-acid battery.

Electrolysis means that electrical energy has been passed through water. This process causes the separation of oxygen and hydrogen from the water molecules. The result is that hydrogen and oxygen can be manufactured by using electrolysis.

Water (H_2O) is made of two hydrogen molecules (H_2) and one oxygen molecule (O). When an electric current is passed through water, the molecules separate, producing oxygen and hydrogen gases. The hydrogen gas can be collected and stored in specially manufactured containers.

Hydrogen as an Energy Resource

Hydrogen can be used as a fuel by burning it, or it can be used to power a fuel cell. The use of hydrogen for storage and then conversion to electricity by fuel cells shows greater efficiency than by burning it. This is because, when using hydrogen as a fuel to burn, the power plant is only about 33 percent efficient. However, a fuel cell system is about 38 to 40 percent efficient.

Storing Hydrogen

Hydrogen is typically stored in a gaseous state and has very low energy content per cubic foot. However, when changed into a liquid, it takes up less space and has a higher energy content per cubic foot. Thus, it is necessary to liquefy it. Hydrogen gas can be liquefied by reducing its temperature to below its boiling point. To do this, extremely low temperatures (–423° F) are needed. Although this method is currently used by some industries and NASA's space programs, it is expensive for commercial use. Also, the energy needed to keep the hydrogen refrigerated to this temperature reduces its overall efficiency. The main advantage of this type of storage is that large amounts of hydrogen can be stored in a relatively small space.

A second method considered for hydrogen gas storage is to use large natural underground caverns such as depleted oil and gas fields. Large quantities of hydrogen gas could be stored in a gaseous form under high pressure. While natural gas is currently stored by this means, hydrogen gas storage in underground caverns still needs to be researched in terms of safety, feasibility, and site location.

A third storage method considered is to store hydrogen at very high pressures in metal tanks. Although more expensive than using natural gas caverns, tanks offer portability.

A fourth method considered for storing hydrogen gas is to use hydride compounds, such as magnesium-nickel, magnesium-copper, and iron-titanium. These compounds are capable of absorbing hydrogen gases. Once absorbed, new chemical hydride compounds are formed. When hydrides are heated, they release hydrogen gas. This method of storing hydrogen eliminates large volumes of gas handling, refrigeration systems, and pressurized vessels.

Figure 10-19 illustrates the inside of part of a 900-pound hydride storage vessel holding as much as 105 gallons of hydrogen in pressurized tubes. The tubes are surrounded by iron-titanium hydride crystals to absorb and store the hydrogen gas.

In operation, hydrogen gas is pumped into the vessel through porous metal tubes and into the hydride crystals to be absorbed. Heat released in this process is removed by cold water flowing through another set of tubes. These tubes will later carry hot water to heat and decompose the hydride, releasing hydrogen for use.

Although energy experts note that eventually hydrogen will be a fuel and storage medium, a great deal of research must still be accomplished in this area. This research includes energy systems utilizing hydrogen storage, improved storage methods, and transportation mechanisms.

Hydrogen Storage Using Hydride Compounds

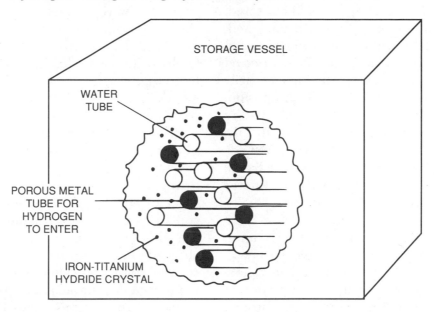

FIGURE 10-19 Hydrogen can be stored using hydride compounds.

Calculating Energy in Hydrogen—Math Interface 10-3

Hydrogen has 333 Btu/cubic foot in a gaseous form and 107,000,000 Btu/ton in liquid form. Natural gas has 1030 Btu/cubic foot in a gaseous form and 47,000,000 Btu/ton in the liquid form. What percentage less in energy is gaseous hydrogen than natural gas in the gaseous form? What percentage more in energy is liquid hydrogen than natural gas in liquid form?

Percent less energy in hydrogen in a gaseous form equals _____.

Percent more energy in hydrogen in a liquid form equals _____.

MECHANICAL ENERGY STORAGE TECHNOLOGY

Flywheel Storage

The flywheel is one method used to store mechanical energy. Flywheels are large, spinning, heavy wheels that rotate on bearings that have low friction. Mechanical energy is stored within a flywheel as its rotation speed increases. Once the flywheel turns, mechanical loads can be placed upon the wheel. The stored mechanical energy is then released to perform work.

The operation of a small vehicle with a flywheel would act in the following way. While the vehicle is stopped or parked, electrical energy from an electrical generating power plant or a public utility would turn a motor. This electrical energy would be delivered to the vehicle from overhead wires (much like those used in trolley cars). As the speed of the flywheel increases, mechanical energy would be stored. With the electrical energy input disconnected, the flywheel would continue to turn because of inertia. A generator could then be connected to the spinning flywheel, producing electricity to turn an electric motor used for propulsion. Storage batteries could also be used to store excess electrical energy generated by the flywheel as the vehicle is stopped. This flywheel-usage concept is known as a *regenerative system*.

The United States Department of Energy has conducted research on such vehicle flywheel systems. Automobiles someday may be designed to operate from public utility systems. These flywheel-type vehicles would be adaptable to short, intercity, start and stop operations. It is estimated by the Department of Energy that if only 15 percent of the automobiles in the United States were powered by flywheel propulsion systems, the oil savings could reach 200 to 500 million barrels of oil per year. Figure 10-20 illustrates an example of a flywheel-propulsion vehicle.

In this vehicle illustrated, the motor and generator were designed as an integral unit able to operate as either a motor or a generator. An AC propulsion motor was also used to drive the rear wheels. The vehicle was 33 feet long, weighed about 24,200 pounds and was designed for 35 seated and 35 standing passengers.

In terms of operating characteristics of flywheels, the size is not as important as the speed at which the wheel rotates. For example, if one flywheel is rotating at twice the speed of another of equal size and shape, the faster one contains almost four times the energy stored in the slower wheel. Also, flywheels can be charged (increased in speed) and discharged (loaded down) quickly with almost an infinite number of cycles available.

Larger applications using flywheel storage are also considered. Mechanical energy from the wind could be collected and stored using flywheels. The advantage would be in overcoming intermittent wind characteristics. Power plants could also spin flywheels during

Flywheel Vehicle Application

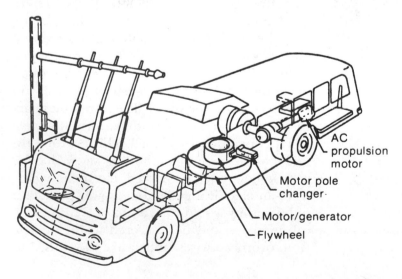

FIGURE 10-20 This vehicle, developed years ago by a Swiss company, is powered by a flywheel that is periodically increased in speed to store mechanical energy used for propulsion.

low energy demands. The mechanical energy would be released to generate electricity during high demand periods. As research continues, eventually these and other applications of flywheel storage may be implemented.

> **Flywheel Storage—Energy Issue 10-4**
>
> *What economic, environmental, social, and technical factors must be overcome to cause manufacturers to build flywheel-type vehicles to replace urban automobiles? Describe a technical system that would work in the United States.*

Compressed-Air Storage

Compressed air, although used in industry for transporting mechanical power, can also serve as a mechanical energy storage medium. For example, a standard air compressor uses electrical energy to operate a motor that pumps air into a tank. Electrical energy is thus converted to pressure (compressed air) and mechanically stored until it is needed to perform work in some application. Because of this capability, compressed air is under consideration for large-scale energy storage.

At present, compressed-air storage is used on small, individual applications. If used on a large scale for commercial energy storage, natural reservoirs must be located for storage. Since air may be practically compressed to as high as 600 pounds per square inch, the size of storage caverns would only need to be a small fraction of the equivalent volume of water used for a pumped storage system.

Research is currently in progress on the use of compressed-air storage systems to be used in conjunction with a power plant. In these systems, air would be compressed during low-energy demand periods. Later, during high-energy demand periods, the compressed air would be used to operate a turbine generator without the frictional losses of a typical compressor. The result could be an efficiency boost in turbine operation by some 300 percent. Current research indicates that these systems can be built using existing technology. However, proper site selection, storage reservoirs, and cost considerations must be researched further.

ENERGY CONSERVATION

Energy Conservation Awareness

Energy conservation has been encouraged for many years. The study of energy conservation began immediately after the oil embargo of 1973 and has continued for more than 20 years. Jimmy Carter, President of the United States at that time, stated the problem of energy conservation clearly when he noted

1. It is cheaper in the short run to save oil than to produce it.
2. Production of oil and gas will keep pace with demand only if the government reduces demand.
3. Increased production will, in the end, inevitably overwhelm the recoverable reserves.

In the late 1970s, the federal government initiated a national comprehensive energy plan, directed at the flow of energy from resources to utilization. The plan emphasized energy conservation. More specifically, the plan had seven energy goals:

1. Reduction of the annual growth rate in energy demand to less than 2 percent.
2. Reduction of gasoline consumption by 10 percent.

3. A reduction in foreign oil imports to 6 million barrels per day.
4. Establishment of petroleum reserves of 1 billion barrels, about a ten-month supply—at the time of the plan's writing.
5. An increase in coal production from about 400 million tons annually to more than 1 billion tons per year.
6. Insulation of 90 percent of homes and all new buildings.
7. Use of solar energy in more than 2.5 million homes.

Despite the energy goals established by former President Jimmy Carter, the United States has not become energy independent, and most consumers have generally not embraced the concepts of energy conservation. Regarding the foregoing list of goals, note that

1. The annual growth rate in energy demand continues to increase between 2 and 3 percent per year.
2. Although today's automobiles get better gasoline mileage compared to those in service during the oil embargo of 1973, many more cars are offsetting overall gasoline conservation.
3. The United States has failed to curb its dependence on foreign petroleum.
4. While petroleum reserves have been accomplished with the establishment of a Strategic Petroleum Reserve (review chapter 5), this reserve will only last some 100 days should another oil shortage occur.
5. The increased production and use of coal has not been realized due to production limitations and environmental concerns.
6. Although many buildings have been insulated since the oil embargo, the majority of all dwellings have relied upon construction methods designed prior to 1987. These

methods are inferior to the energy efficient designs possible today.
7. The large-scale use of solar energy (both active and passive systems) has not been realized. This is due in large part to the low prices charged for fossil fuels, a lack of tax credits to researchers and consumers, and a failure to educate the populace about the importance of energy conservation and protecting the environment.

It should be apparent that energy conservation programs are a significant part of any comprehensive energy plan if a reduction and conservation of fossil fuels is to be realized. This reduction and conservation is even more important when one realizes the consequences of not reducing the demand for energy.

Consequences of Burning Fossil Fuels and Using Energy

Scientists studying the effects of energy use note that, as a result of the burning of fossil fuels, the environment and various species are suffering from the cumulative effects of air pollution. Scientists now report that in the next 20 years we may see a decline in the earth's capacity to support life. There could be a steady loss of croplands, fisheries, forests, and plant and animal species as the degradation of the earth's water and atmosphere continues. In fact, some scientists estimate that as many as 20 percent of all animal and plant species on earth could be lost by the early twenty-first century.

Methods for Conserving Energy and Protecting the Environment

Generally, there are two accepted methods used to conserve energy. The first is by improving various technologies to make them more efficient. For example, an automatic setback thermostat in a home saves energy automatically by setting the thermostat back at various times through the day or night when heat is not needed. In this case, technology has been designed to save energy.

Secondly, and more importantly, people can be educated to use energy in a different fashion. Then, based upon their knowledge of how energy is used, they can take steps to conserve energy. Thus, people should always be thinking about how they are using energy and the impact of this use. Through education, an individual can understand the need to live more in harmony with other life forms. This education could provide people with a basic understanding of ecological systems and their interaction with human-made communities. Once this education had occurred, individuals might choose to use this knowledge to make changes in their own lifestyles and in their purchasing of goods and services.

Using the foregoing example of a thermostat, the conservation-educated person who wishes to live more in harmony with other life forms may purchase a setback thermostat. They know that they are conserving energy by keeping their homes cooler in the winter and warmer in the summer, or they may decide not to purchase this device and instead make a conscious effort to turn the thermostat down in the winter and up in the summer—thus achieving the same effect.

Human Values and Energy Conservation

With only 5 percent of the world's population, the United States consumes about 28 percent of the world's energy. The United States utilizes the most energy per capita in the world and is the likely choice, therefore, to lead the way to a sound energy conservation program.

This challenge will not be easily accomplished.

As the United States has developed, energy has been produced to meet increased demand. This energy is often taken for granted. As technology developed, the style of living generally increased in terms of material wealth. With this material wealth came an increase in the rate of personal energy consumption.

The term *personal energy consumption* indicates the amount of energy utilized per person. The United States has one of the highest personal energy consumption rates in the world. In fact, it is estimated that an average four-member family spends between $1,500 and $3,500 for energy each year. Much more is spent indirectly because other goods and services are purchased that require energy for production and delivery.

United States consumers have rarely been asked to use less of a commodity or to cut back in usage—the exceptions being during World War II and during the 1970s energy crisis. In fact, most citizens in the United States view energy consumption as a right of living in an industrialized society. However, one part of any energy conservation program is to use less. The degree to which consumers conserve energy depends directly upon their attitudes and values concerning energy and the environment.

The process of conserving energy by the public must be continuous to be effective. Assistance comes from various energy conservation programs, economic incentives, and increased energy awareness through educational programs. Obviously, energy conservation cannot be accomplished over a short period of time. It may take years before energy conservation is not thought of as a passing fad and people conserve energy naturally.

Changing values is not only necessary for consumers but also for business, government, and industrial leaders if energy conservation is to work. When energy conservation is a high priority, programs on national, state, and local levels can be incorporated. For example, immediately following the first oil shortage in 1973, several midwestern states were granted over $24 million for loans and grants by the federal government to improve insulation in some 300,000 homes.

Besides insulation, other methods of energy conservation have also been studied and implemented. For example, infrared photography (called thermography) can now be used to detect heat losses in homes and buildings. Also, new energy-saving building codes have been implemented. New Mexico, for example, has improved both its thermal and lighting efficiencies, as well as changing almost all of the state vehicles to compact cars.

In California and other states, appliance efficiencies have been raised continually. In addition, minimum operating efficiencies have been established for refrigerators, freezers, room air-conditioners, water heaters, washers, dryers, and ranges. All of these efforts are due primarily to the changing values about energy and conservation in society.

> **Values and Energy Conservation— Energy Issue 10-5**
>
> Human values seem to be the cornerstone for extensive energy conservation efforts. What are the best methods to change values about energy conservation?

Defining Energy Conservation

As energy is converted from one form to another, it eventually changes to a form that is no longer usable to society. For example, in the transportation sector, when an automobile engine burns gasoline, about 28 percent of the fuel is converted to useful work while the remaining 72 percent is converted to thermal energy. This thermal energy is no longer useful to the propulsion of the automobile. Thus, it can be said that the thermal energy is so random that it can no longer be used technically or economically.

Energy conservation can be accomplished through many approaches. However, one basic guideline is economics. For example, in the early 1900s most houses were not insulated. Fuel was so inexpensive at that time that it cost the homeowner more to insulate than it did to use the energy. By 1950 the price of energy had increased to the extent that it was economically feasible to use enough insulation in a home to produce an R value of 5 to 10 in the ceiling. (The R value represents the resistance to heat flow. This value is presented in greater detail later in this chapter.) Because of the price of existing fuels today, it is economically advantageous to insulate the ceiling to an R value of 32 to 60.

Economics, human values, and technological advancements have had a direct effect on energy conservation. It has developed in the following directions:

1. Saving thermal energy that normally would become random.
2. Utilizing energy to its best advantage.
3. Using less energy per person.
4. Improving efficiencies of converter technology.
5. Operating technology at its best efficiency characteristics.

CONSERVATION IN THE RESIDENTIAL/COMMERCIAL SECTORS

The residential and commercial sectors in the United States use approximately 18 percent of the energy consumed in that society. This energy is used in houses, apartments, schools, office buildings, and so on. The amount of energy wasted depends upon the condition of the building, user lifestyles, personal energy habits, and the condition of the equipment and appliances that utilize energy.

To determine what energy conservation measures can cause a significant and immediate effect, it is important to determine first where the energy is being used. In many buildings, both electrical energy and natural gas (or fuel oil) are used as the primary energy resources.

Figure 10-21 shows various electrical appliances and energy conversion devices. Percentage of total residential energy use and kilowatt-hours per year are given for each. For example, about 55 percent of the energy used in an all-electric building is used to heat the structure. These statistics represent only average energy usage and can vary with time of day, time of year, geographical location, family size, lifestyles, appliance efficiencies, and type of fuel. Not all appliances are listed. A more complete breakdown of energy usage in a typical home is shown in Appendix B.

The major energy consumption devices in a building are the water heater, electric resistance heating, and central air-conditioning. After reviewing how energy is used, it would be advantageous to look closely at those devices that use higher amounts of energy. In this way, when energy is conserved, a more noticeable decrease in energy consumption can be immediately realized.

Heat Loss and Heat Gain

Heat loss is defined as the amount of thermal energy passing out of a building. *Heat gain* is defined as the amount of thermal energy coming into a building. Heat gain can be from a heat source such as a furnace or wood burning stove, or from solar gain. Heat losses occur through windows, doors, cracks in the foundation and walls, and so on. Generally, the less heat loss occurring, the less heat gain necessary to keep a constant temperature within the dwelling. Several variables affect the amount of heat loss:

1. The wind outside of the building.
2. Temperature differential between the outside and the inside of the building (also referred to as delta T [i.e., ΔT]).

Average Residential Electrical Energy Usage

	Kilowatt-Hours/Year	Percentage of Total
Water Heater	4,811	12.8%
Refrigerator/Freezer automatic defrost 16.18 cubic feet	1,795	4.7%
Freezer (15–21 cubic feet) manual defrost, chest type	1,320	3.5%
Central Air-Conditioning	4,800	12.8%
Electric Resistance Heat	21,000	55.7%
Home Entertainment		
Radio	86	0.2%
Color T.V.	320	0.9%
Comfort Conditioning		
Dehumidifier	377	1.0%
Humidifier	163	.4%
Laundry		
Automatic Washing Machine	103	.2%
Clothes Dryer	993	2.6%
Food Preparation		
Range, self-cleaning oven	1,205	3.2%
Dishwasher	263	.9%
Other Appliances	200	.5%
Miscellaneous	100	.2%
TOTAL	37,636	over 99.6%

Source. Adapted from Potomac Electric Power Company

Note. Average kilowatt-hours/year consumption of an all-electric home.

FIGURE 10-21 Average percentages of electrical usage by appliances in an all-electric house within the United States are presented. (Courtesy of Potomac Electric Power Company)

3. The type of materials conducting the heat out of the building.
4. The length of time over which heat losses occur.

Heating a structure consumes the greatest amount of energy because of the large amount of heat loss through the structure itself. Once it is known how and where heat losses occur in a building, various strategies can be established to help conserve energy. Figure 10-22 shows typical heat losses throughout a building. The most common occur through the ceiling, walls, floors, windows, chimney, and vents. The optimum building would have what is called neutral walls.

A *neutral wall* is one in which little or no heat or thermal energy could escape. With today's technology and economics, it is not possible to obtain a neutral wall. However, many structures are called super insulated. These buildings typically use 12-inch thick walls and add more insulation to reduce heat losses. Although not a neutral wall, the added insulation keeps more energy inside the structure, rather than letting it escape into the atmosphere.

Heat Loss by Infiltration

Heat losses in a building can come from either winter or summer infiltration. Winter infiltration refers to cold air seeping in and warm air passing out of the structure. Summer infiltration occurs when warm air from outside seeps into the building or cooler air-conditioned air seeps out. Infiltration can occur through leaks and cracks around windows, cracks in the walls, the fireplace opening, the chimney and other vents, and through the ceiling. Figure 10-23 shows examples of common places in which air can seep into and out of a structure.

The ideal object of reducing infiltration is to seal the building so that energy is conserved. Typically, an old and loose-fitting structure may undergo approximately three air changes per hour due to infiltration. An average building may undergo two air changes per hour. A relatively tight building typically has about one air change per hour. Minimum infiltration should be no less than one-half air change per hour.

Many super insulated structures have reduced their infiltration to as low as one-half air change per hour. A building that is tighter than this may be dangerous to the occupants

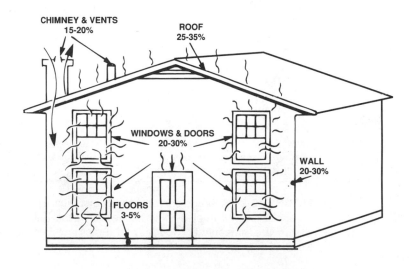

FIGURE 10-22 A typical house has various locations in which thermal energy is lost. These losses are expressed, here, in average percentages.

CHIMNEY & VENTS 15-20%

ROOF 25-35%

WINDOWS & DOORS 20-30%

WALL 20-30%

FLOORS 3-5%

Air Leakage in a House

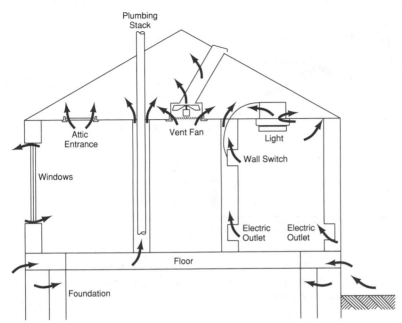

Plumbing
Stack

Attic
Entrance

Vent Fan

Light

Wall Switch

Windows

Electric
Outlet

Electric
Outlet

Floor

Foundation

FIGURE 10-23 Infiltration occurs because of the many cracks, holes, vents, and so on, in a house. Arrows show various infiltration spots in a typical house.

because of a lack of oxygen and a possible buildup of toxic gases. These toxic gases can be produced from the building material contained within the structure. In addition, it has been found in certain geographical locations that buildings that are too tight have the possibility of high amounts of radon. Radon is a naturally occurring gas that has been found to be a cause of cancer.

Causes of Infiltration   Infiltration is the result of several factors:

1. The wind.
2. Opening of outside doors.
3. The building acting as a chimney.
4. Heating systems that take their combustion air from inside the structure.

Figure 10-24 shows that as the wind blows across the building, high air pressure causes cold air to infiltrate into the structure on the left side. This causes warm air from the build-

ing to leak out on the right side, aided by a low pressure.

The effect of opening doors is apparent. However, the chimney effect may not be as obvious. The ***chimney effect*** is caused when the air inside a building is warmer that the air outside of it. Warm air rises and leaks from the cracks at the upper levels, while drawing in cold air through cracks in the lower levels.

The amount of heat loss from the chimney effect is related to the temperature difference between the inside and outside of the building and the height of the building. A two-story house having a 68° F inside temperature and a 32° F outside temperature will produce a chimney leakage equivalent to a 10 mph wind blowing against the outside of the building. Figure 10-25 illustrates the chimney effect. As heat rises to the upper portion of the building and escapes through cracks, cool air from below the structure infiltrates inward. The net effect is an increase in infiltration.

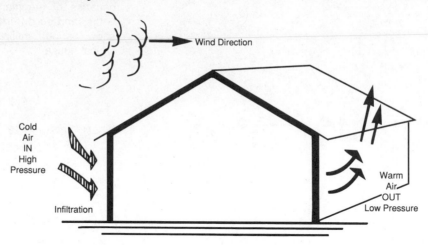

Infiltration By Wind

Wind Direction

Cold
Air
IN
High
Pressure

Warm
Air
OUT
Low Pressure

Infiltration

FIGURE 10-24 Infiltration by wind is one of the major causes of heat loss in the winter. Conversely, heat gain in the summer is caused by air of the opposite temperatures shown being forced by the wind.

Cold Air for Combustion   Infiltration is also caused from heating systems using inside air for combustion. For example, Figure 10-26 shows that as an older type furnace is operated, it uses inside air for combustion. As this inside air is used, a low pressure is created within the building, which increases infiltration.

To offset this type of infiltration, heating systems—furnaces, fireplaces, and wood-burning stoves—should use outside air for combustion. The net effect is to reduce infiltration and increase the building efficiency. Figure 10-27 shows an efficient fireplace. As combustion occurs, only cold outside air is used, rather than the heated air from inside the building. Using this system, less infiltration occurs and a substantial energy savings is realized.

Determining Infiltration   There are several mathematical formulas that energy auditors use to determine how much heat loss occurs

through infiltration. However, one can also visually inspect a building for certain characteristics. Things to examine during a visual building inspection include the following:

1. Check the doors and windows for cracks, poor caulking, broken frames, and the like.
2. Check the exterior of the building. Look for evidence of cracked caulking or the lack of caulking around doors, windows, sills, and other outside corners and joints.
3. Check doors for air leakage near the threshold; look for cracks and broken seals.
4. Check for drafts at basement windows.
5. Check electrical outlets on outside walls and feel for drafts near the corners and baseboards.
6. Check each window for fit to the sill and casings.

Not all infiltration can be stopped. However, many homes were built in such a manner that a small investment in caulking, weather strip-

Chimney Effect

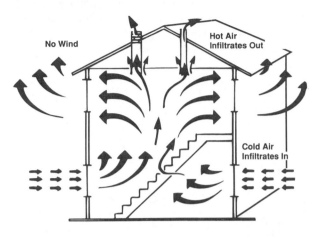

FIGURE 10-25 The chimney effect occurs when cold air infiltrates the lower section of the building, drawn in by warm air rising (because of convection) and escaping out of the upper part of the building through cracks.

Infiltration Caused By Inside Combustion Air

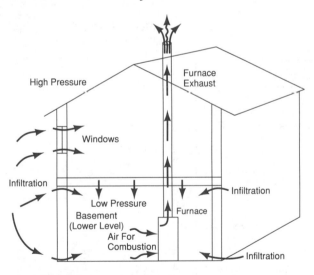

FIGURE 10-26 When any combustion process consumes air within a building, additional infiltration occurs. As oxygen is used in combustion, a low pressure building interior causes the infiltration.

ping, and insulation will reduce infiltration. Most often, the return on investment can be realized in less than one heating season.

Figure 10-28 shows building conditions that cause one, two, or three air changes per hour. By reviewing these conditions, one can determine the condition of a building and the average amount of air infiltration.

Infiltration and Conservation—Energy Issue 10-6

One of the easiest methods to conserving energy is to reduce infiltration by caulking, weather stripping, and so on. Identify reasons why owners of buildings do not perform these maintenance tasks. How could the federal or state government help people to make their buildings more energy efficient?

Heat Losses Through Conduction

Heat losses also occur through the process of conduction. Note that heat always flows from a warmer to a cooler temperature. This concept was previously presented in chapter 3 of this text.

Heat loss through conduction in a building refers to thermal energy being conducted through outside windows, ceilings, floors, roofs, walls, and so on. For example, when there is a temperature differential across a wall, a certain amount of heat loss by conduction will occur. Certain variables affect how much heat loss will occur through conduction. Some of the important variables include:

1. The temperature differential across the wall or material.
2. The type of material in the wall.
3. The thickness of material in the wall.
4. The length of time during which heat losses occur.

To help minimize heating or cooling losses

Combustion Air From Outside

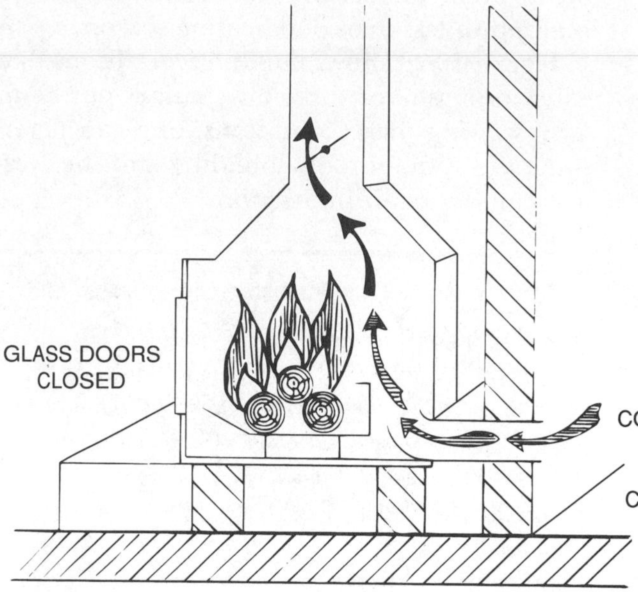

GLASS DOORS
CLOSED

COLD OUTSIDE
AIR
FOR
COMBUSTION

FIGURE 10-27 When outside air is used for combustion, the fireplace will not deplete warm room air or lower the interior air pressure of the house.

Conditions Producing Infiltration

FIGURE 10-28 The condition of any building should be visually checked for infiltration to estimate the average air changes per hour.

Building Component	One Air Change Per Hour	Two Air Changes Per Hour	Three Air Changes Per Hour
Building with cellar OR	Tight, no cracks, caulked sills, sealed cellar windows, no grade entrance leaks	Some foundation cracks, no weather-stripping on cellar windows, grade entrance not tight	Stone Foundation considerable leakage area, poor seal around grade entrance
Building with crawl space or on posts	Plywood floor, no trap door leaks, no leaks around water, sewer and electrical openings	Tongue-and-groove board floor, reasonable fit on trap doors, around pipes, etc.	Board floor, loose fit around pipes, etc.
Windows	Storm windows with good fit	No storm windows, good fit on regular windows	No storm windows loose fit on regular windows
Doors	Good fit on storm doors	Loose storm doors, poor fit on inside door	No storm doors, loose fit on inside door
Walls	Caulked windows and doors, building paper used under siding	Caulking in poor repair, building needs paint	No indication of building paper, evident cracks around door and window frames

due to conduction, architects and contractors select and use building materials and techniques appropriate for a given geographic area. In addition, these individuals will select building materials based on their measure of heat resistance (their R value) and their measure of heat conductivity (their U values).

R Values

The R value of a material represents the resistance to heat flow in terms of the thickness of the material used. Thus, R values measure the ability to stop heat flow. In addition, R values commonly are used to show the resistance to heat flow through a structure. That means that when various materials are used in a wall or ceiling, the R values are added together to get a total R value of a wall. Product manufacturers provide R value information. Higher R values mean greater resistance to heat flow; lower R values mean less resistance to heat flow. R values are typically expressed in Btu's per hour, per square foot, or per degrees Fahrenheit across the material.

U Values

The U value of a material represents how much heat transfers through any single material or an entire assembly of materials. For example, a wall may have insulation, sheet rock, paint, and so on. The U value shows how much heat transfers through the total wall. U values are often expressed as Btu's per hour, per square foot, or per degrees Fahrenheit across the entire assembly of materials.

R Values Related to Location and Economics

To construct a building with the proper R values and materials, two items must be considered: the geographical area and the economic balance between insulation and energy costs.

In terms of the geographical area, buildings in the northern states require insulation for the winter heating season. However, structures in southern states require insulation to reduce solar heat gain when running air-conditioners.

Economics also influence the R value of materials used in a building. Again, there is a balance between the cost of increasing the R value and the cost of energy. Today, many experts believe that the price of energy is high enough so that insulating beyond the minimum recommended levels usually will pay off in the long run.

Figure 10-29 shows the minimum R values to be used for heating and cooling insulation throughout the United States. In this figure, the country has been divided into various geographic zones. As an example, if one lived in Minneapolis, Minnesota, the minimum R value for the ceiling would be R 38; the wall, R 19; and the floor, R22.

These recommended R value figures are based upon a home heated with natural gas. The R value will be different if heated by other means, such as oil or electric resistance heating, because these forms of energy generally cost more than natural gas.

Types of Insulation

Many types of insulation are on the market today, each having a different R value. The type depends upon where the insulation is to be used, the existing insulation in the building, and the price of each type. Figure 10-30 lists several types of insulation, the R value per inch, and the necessary inches needed for different R values.

Loose fill is one of the most common types of insulation. It can be poured or blown into unfinished attic floors, or blown under finished attic floors. Blown insulation usually

Insulation R Values for Various Geographic Locations

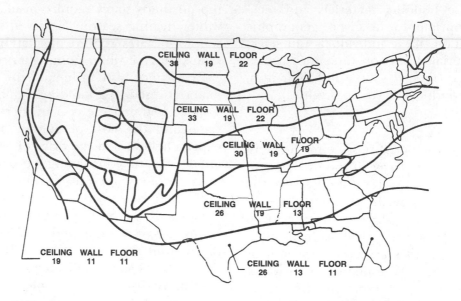

FIGURE 10-29 Different zones are identified across the United States to determine the R values of insulation needed in winter months. (Courtesy of FMC Corporation)

requires a machine to blow the insulation into the correct location. These machines can be rented, usually at the place where the insulation is purchased. Cellulose, a form of loose fill, has greater insulation value than fiberglass or mineral wool. Cellulose is used when there is limited space, such as under the floors or in walls. It is also a recycled material and is more economical than fiberglass.

Vermiculite, another form of loose-fill insulation, tends to be more expensive than the others; but, it can be placed into smaller areas. Its R value is lower than all other forms of insulation.

Batt and blanket are popular for insulating between standard ceiling joists or rafters, especially when the space is relatively free of obstruction. Batts can be purchased in precut sections or in long rolls. Batts and blankets also come with or without a vapor barrier backing, which is used to keep moisture from condens-

ing inside of the insulation.

Rigid board insulation—made from polystyrene, urethane, and fiberglass—is not used normally in attic floors. However, rigid board insulation is placed on the inside and outside surfaces of basement walls or crawl spaces and, in some cases, underneath the siding for increased wall insulation.

Rigid board insulation generally has a higher insulation value per equal thickness than other types of insulation. It also can be purchased to fit various widths and is sold in thicknesses up to 4 inches. Most rigid board insulation comes in 2' × 4' or 4' × 8' sizes. The materials used for rigid board insulation also act as a vapor barrier.

Foam insulation made from urea formaldehyde can be used in attics, forced into walls without insulation, and pressurized into floors if spacing permits. However, it tends to be

Typical R Values of Different Forms of Insulation Sold Within the United States

	R/Inch	R11	R19	R22	R34	R38	R49
				Inches Needed For			
Loose Fill							
Blown-Machine							
Fiberglass	R2.25	5	8.5	10	15.5	17	22
Mineral Wool	R3.125	3.5	6	7	11	12.5	16
Cellulose	R3.7	3	5.5	6	9.5	10.5	13.5
Loose Fill							
Poured-Hand							
Cellulose	R3.7	3	5.5	6	9.5	10.5	13.5
Mineral Wool	R3.125	3.5	6	7	11	12.5	16
Fiberglass	R2.25	5	8.5	10	15.5	17	22
Vermiculite	R2.1	5.5	9	10.5	16.5	18	23.5
Batts or Blankets							
Fiberglass	R3.14	3.5	6	7	11	12.5	16
Mineral Wool	R3.14	3.5	6	7	11	12.5	16
Rigid Board							
Polystyrene beadboard (Styrofoam)	R3.6	3	5.5	6.5	9.5	10.5	14
Extruded polystyrene	R4–5.41	3–2	5–3.5	5.5–4	8.5–6.5	9.5–7	12.5–9
Urethane	R6.2	2	3	3.5	5.5	6.5	8
Fiberglass	R4.0	3	5	5.5	8.5	9.5	12.5
Foam							
Ureaformaldehyde	R4.8	2.5	4	4.5	7	8	10.5

FIGURE 10-30 Many types of insulation are on the market today. Each has a different R value per inch.

more expensive than other types of insulation. Foam is usually fire resistant, and it also has a higher R value per equal thickness than more conventional materials. On the negative side, foam type insulation can produce dangerous gases that could accumulate in tight buildings. Thus, when using foam insulation, always work with the retail store for correct installation procedures.

Vapor Barriers

Although all buildings have varying degrees of moisture in them, it is important to keep the moisture at the correct level. Vapor barriers are used for this purpose. When installing insulation, it is important to consider the location of vapor barriers. In the winter, moisture has a tendency to move from the inside of the building to the exterior surface. Vapor barriers are installed to reduce the flow of moisture through the insulation so that condensation of the moisture will not occur in the insulating material. If this were to happen, the material could become saturated, losing some of its R value. Retention of this water in the wall would also cause degradation of the structure.

To protect against condensation, vapor barriers should always be installed on the warm side of the building wall. Thus, the moisture will be stopped before it reaches the insulation. If possible, vent the cold side of the insulation material to outside air to remove any moisture that does escape through the insulation.

Many buildings today use a thin film material that is wrapped around the external part of a structure, usually under the outside siding. Its purpose is to allow moisture to flow out of the building, yet keep air from infiltrating into the building.

Attic Ventilation

As an important part of energy conservation, there should be adequate ventilation in the attic to allow summer thermal energy and moisture to escape. In warm weather, attic ventilation lessens attic heat build-up, which can decrease air-conditioning effectiveness. As a general rule, the ratio between vent area and roof area is about 1 to 150. This means that for each 150 square feet of roof area to be vented, at least 1 square foot of ventilator area should be used.

Calculating Attic Ventilation—Math Interface 10-4

One side of a roof on a structure measures 12 feet by 40 feet. The other side of the roof measures 20 feet by 40 feet. How many square feet of attic ventilation should be used on this building?

Square feet of attic ventilation equal

_____.

In moderate climates, power ventilators can serve as a partial substitute for air-conditioning in the summer because frequent changes of attic air will help keep a house cooler. Even in hot climates, the fans are very useful as a substitute for air-conditioning. A power fan should be humidity and temperature controlled so that it will run whenever the humidity or temperature is high. Although the power fan consumes electrical energy, it contributes to energy savings, especially when used with air-conditioning. A typical power fan, drawing

about the same amperage as a 75-watt light bulb, can reduce air-conditioning costs 10 to 30 percent—assuming that the attic is isolated from the rest of the house and does not exhaust cooled air.

Figure 10-31 shows the importance of a power ventilator in a building during hot days. During the day when the temperature outside is 95° F, the temperature inside the attic is near 150° F without a power ventilator. When the temperature is reduced with a power ventilator, less conduction of heat goes into the house, keeping it cooler. This reduces the need for air-conditioning and conserves energy.

Power Ventilators

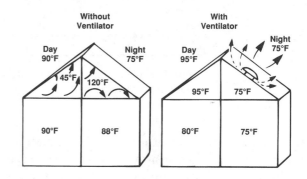

FIGURE 10-31 Power ventilators can help reduce the air temperature in the attic (and thus in the house) enough to save 10 to 30 percent on air-conditioning costs.

A second method of attic ventilation is the use of turbine ventilators, which take advantage of the wind to exhaust air and humidity from the attic. A turbine ventilator is a device having a series of blades on a rotary shaft mounted in a tube on the roof. As wind moves across the fan blades, the warm air is drawn out of the attic area.

A third type of attic ventilation system is the vented ridge and vented soffet. Figure 10-32 shows how this system works. As the attic area is heated by the sun, the warm air rises to the center of the roof by convection. The hot air is then vented out the top of the roof. Cooler air is then drawn into the attic through the soffit screens. Light breezes increase the air flow. This system provides a uniform cooling effect in the attic area of a building, again reducing energy costs for air-conditioning.

Temperature Control

Approximately 55 percent of all energy consumed in an all-electric structure is used for heating the building. The next largest consuming device is the air-conditioner. Even if the building uses oil, natural gas, or other heat sources, the energy needed for heating and cooling still makes up the majority of the energy used.

One method for conserving energy is to control the air temperature in the structure. For example, setting the thermostat back an aver-

age of seven degrees Fahrenheit during the winter will save energy. The savings range between 3 and 10 percent of the monthly fuel cost. The exact amount of savings will depend upon the outside temperature, heat losses through conduction and infiltration, and the lifestyle of the occupants. Some savings estimates are even greater. For example, in Chicago, Illinois, it is estimated that the monthly savings for heating costs range between 2 and 3 percent for every degree the thermostat is set below 68° F.

Turning the thermostat up seven degrees during the summer when the house is air-conditioned can also save 1 to 3 percent of the monthly fuel costs. Again, the exact savings will depend upon the amount of solar heat gain, outside temperature, and the lifestyle of the occupants.

Humidity Control and Comfort Level

Comfort level is defined as the temperature at which the occupants in the building feel comfortable. The most common comfort level in a building is between 68° to 70° F. Often this comfort level may change. For example, if the moisture in a structure decreases, the comfort-level temperature may increase. Moisture in a building helps to transfer the thermal energy through the air. If the moisture level is inadequate, the occupant may feel cool at 68° F. Thus, the comfort level temperature would need to be increased. Adequate humidity and a room temperature of 68° F has the same comfort level as a drier 72° F. Thus, with adequate humidity, the temperature in a building can be reduced while still maintaining comfort and warmth.

One method to lower a comfort-level temperature is to add moisture to the air. Humidity in the home has a tendency to drop during the

Vented Ridge Ventilation

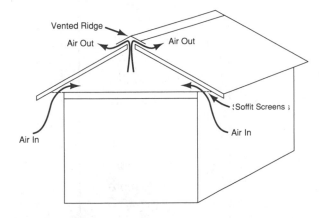

FIGURE 10-32 The ridge ventilation system has screens on the soffits and a vented ridge on the roof. As warm air rises, it flows out of the vented ridge, while cooler air comes through the soffet screens.

winter months because cool outside air does not retain moisture well. In addition, the furnace inside the structure removes moisture even further. It is, therefore, often wise to invest in and use a humidifier during the winter months. Figure 10-33 illustrates a conceptual relationship among relative humidity, comfort-level temperature, and additional fuel costs. As the relative humidity increases, the comfort-level temperature decreases, and fuel costs are reduced.

For example, if there is 0 relative humidity in a building, the temperature must be set at 76° F for a person to feel comfortable. The result is that about 25 percent more fuel will be needed for heat than that necessary at 40 percent humidity. When the relative humidity is increased to approximately 30 percent, the thermostat need only be set at 70° F for a person to feel the same amount of comfort.

Although higher humidity levels enable the temperature inside a structure to be lower, too much moisture can also cause problems. Excessive moisture comes from a number of causes, including

1. lack of infiltration (super insulated building);
2. the use of household appliances such as dishwashers, dryers, and humidifiers;
3. the use of showers and tubs and from cooking vapors; and,
4. excessive numbers of plants.

High humidity levels cause the moisture to migrate through the walls and ceilings and into the insulation. The moisture often will then

A Conceptual Relationship Between Temperature and Relative Humidity Cost

Inside Relative Humidity Percentage	Comfort-Level Temperature	Additional Fuel Costs
0	(76°F)	25.0% or more
5	(75°F)	21.9% more
10	(74°F)	18.7% more
15	(73°F)	15.6% more
20	(72°F)	12.5% more
25	(71°F)	9.4% more
30	(70°F)	6.2% more
35	(69°F)	3.1% more
40	(68°F)	0%

FIGURE 10-33 As relative humidity increases in the building, the comfort-level temperature decreases. Thus, the thermostat can be lowered to save energy and reduce fuel costs.

freeze and reduce insulation efficiency. Besides soaking insulation and making it less efficient, excess moisture can stain and crumble ceilings, blister exterior paint, and rot structural members. Freezing and thawing moisture can also damage roof shingles as well.

To eliminate these problems, control of the amount of humidity in a building is extremely important. Figure 10-34 suggests relative interior humidity levels for various outside temperatures, assuming a 68° F indoor temperature. If condensation problems in the form of fog or frost are evident on the inside of windows at the recommended levels, it may be necessary to cut back on the humidifier setting.

Energy Conservation Tips

There are many simple conservation tips that can have a significant effect on the energy used in the residential/commercial sector of society. Although each suggestion by itself may not seem important, the more that are used, the greater the energy savings.

In the area of lighting, the following suggestions will help to save energy:

1. Strategically locate lights where they can be of best use, such as in stairwells or on desks.

2. Dust all light bulbs and their dust covers so that more radiant energy can be released.
3. Install fluorescent light where practical. These devices are four times more efficient than incandescent bulbs. Natural light is the most efficient.
4. Switch lights off when not in use. It takes less energy to turn them on and off than to leave them on.
5. Use large-watt bulbs in lamps. A few large-watt bulbs are more efficient than using several smaller ones.
6. Consider using light colored paint and furniture. Lighter colors reflect up to 85 pecent of the light; whereas, dark paint reflects only 15 percent.
7. Use motion detectors for control of lights. These work well for outdoor lighting and can be effective for security measures.

In the area of heating and cooling, consider these energy conservation suggestions:

1. Turn down the thermostat and use more blankets at night.
2. If possible, in moderate climates, open windows and use fans in the summer instead of air-conditioners.
3. In winter, close curtains at night to act as an insulator, but during the day leave

Suggested Humidity Levels at 68° F Indoor Temperature

Outside Air Temperature	Recommended Relative Humidity
−20°F or below	not more than 15%
−20°F to −10°F	not more than 20%
−10°F to 0°F	not more than 25%
0°F to 20°F	not more than 30%
10°F to 20°F	not more than 35%

FIGURE 10-34 Suggested humidity levels within a building according to various outside temperatures.

them open to benefit from the sun's rays (if the sun is shining).

4. Close doors and vents in unoccupied rooms. Heat and cool only occupied rooms. This is called zone heating.

5. Replace or clean filters on heating and cooling equipment at least once a month. This will allow better air flow through the system and produce less of a load on the motors and fans.

6. Consider passive solar energy by landscaping. Plant trees that shade the house in the summer and lose their leaves in the winter to allow the sun to shine through.

7. Install awnings, louvers, or reflective screens on the outside of windows during the summer.

8. Use darkening shades, curtains, shutters, or sun-control film on inside of windows in summer to eliminate heat gain.

9. When buying heating or cooling appliances, heat pumps, and air-conditioners, always check the energy efficiency of the appliance. The higher the efficiency, the greater the energy savings.

10. Insulate or carpet floors above unheated crawl spaces. Any heating of crawl spaces is wasted energy, if not used for storage.

11. Insulate heating ducts or pipes whenever possible, especially in unheated areas.

12. Install double-pane or triple-pane windows. The R value increases proportionately.

13. Consider off-peak heating systems when replacing older space heating. Off-peak heating systems are available through many power companies. They are designed to use electricity for heating during off-peak times, usually at night. For example, some heating systems use off-peak electricity to heat a small thermal storage unit at night. Then during the day (when electricity is more costly), the stored heat is distributed throughout the house. Electrical cost during off-peak times may be between 50 and 75 percent less than during peak periods.

When doing laundry, these suggestions will help to conserve energy:

1. Wash and dry only full loads. It takes the same amount of energy to wash a full load as it does a half load.

2. Cold-water washing will clean clothes sufficiently in many cases. This will save the energy needed to heat water.

3. If possible, dry clothes outside.

When conserving hot water, these suggestions will be helpful:

1. Insulate water pipes. When purchasing a home, make sure the water heater and furnace are centrally located to reduce pipe length.

2. Adjust the hot-water thermostat to 120° F on water heaters or 140° F for systems that use dishwashers or serve larger families. For every 10° F above 140° F, the hot-water bill increases 3 percent on average.

3. Insulate the water heater, especially if it is located in a nonheated area.

4. Take quick showers instead of baths.

5. Use cold water for all dish or other rinsing.

6. Repair leaky faucets. Ninety drops per minute could cause a loss of 1,000 gallons of water per year.

7. Larger families should shower one member after the other to effectively utilize hot water in the pipes.

8. Install a flow restrictor in the showers to reduce the flow of water. Install aerators in sink faucets.

9. Flush out the hot water heater once every year to reduce the sediment buildup that lowers heating efficiency.
10. Make sure that natural gas pilot lights burn with a blue flame for maximum efficiency. If the flames are yellowish, this means that an adjustment is needed. Pilot lights and gas can also be turned off during long periods of absence.
11. When replacing the water heater, consider using off-peak water heating.

When using kitchen appliances and during cooking, follow these conservation tips:

1. Plan menus so that the oven is filled every time it is used.
2. Keep the heat reflectors clean on the range. They will reflect more energy, thereby improving efficiency.
3. Preheat the oven only when necessary. When cooking, use a timer to avoid overcooking.
4. Use cold water to operate the garbage disposal. It not only solidifies grease but also saves hot water.
5. Fill refrigerators and freezers to capacity without blocking air circulation. A full freezer is more efficient than a partially empty one.
6. Wash only full loads in the dishwasher and consider not using the dry cycle.
7. With any appliances, open the doors only when necessary.
8. Plan what is being put into and taken out of the refrigerator before opening the door.
9. Cool warm foods to room temperature before putting them in the refrigerator or freezer.
10. Use the manual defrost rather than using frost free refrigerators and freezers. It is more efficient when maintained properly.

With any appliances, open the doors only when necessary.

11. Defrost freezer when ice is ¼–½ inches thick.
12. Check gaskets on refrigerators and freezers. If a paper can slip through the door when closed, replace the gasket.
13. Keep the grill and evaporator coils clean on all freezers and refrigerators.
14. When cooking, fit the pot to the right size burner or heating element.
15. Thaw foods before cooking—if appropriate.
16. Use tight fitting lids on cooking utensils.
17. Self-cleaning ovens are better insulated and, therefore, more efficient than regular ovens. Use the self-cleaning cycle only when necessary because it consumes large amounts of electricity.
18. Use small appliances—including toasters, electric fry pans, and small portable ovens—in lieu of the oven.

19. Consider the purchase of a microwave oven. It can reduce energy usage by some 75 percent when compared to traditional ovens.

CONSERVATION IN THE INDUSTRIAL SECTOR

The industrial sector is the second largest user of energy within our society. This sector currently consumes about 27 percent of the energy used in the United States. Thus, conservation within the industrial sector has become very important. Generally, the conservation strategies in this sector fall into two major categories:

1. reducing energy waste; and,
2. improving the efficiency of machines and processes.

Energy Management

Reducing energy waste can be accomplished in much the same way as described in the residential sector—including lowering thermostats, reducing infiltration, and incorporating better insulation. The process of saving energy in the industrial sector is often referred to as *energy management* or *waste management*.

On a national level, many energy management programs have been started to help employees conserve energy. Primarily, waste management looks at how energy that is wasted can be captured and reused. In this process, efficiency improvements are more involved and complex than in the residential and commercial sectors. In the industrial sector, energy management programs often include redesigning technology and changing the processes of manufacturing. For example, one method of change deals with reducing the frictional power losses in the machines. Another aspect involves the use of smaller and more efficient electric motors designed to replace old, inefficient ones. In one process, a tanning company originally used eight electric motors to turn eight tanning drums. These drums were eventually replaced with one large tanning drum, and turned by one high-efficiency electric motor. The energy consumption in kilowatts was significantly reduced, and better tanning processes improved the leather. In addition, there was more room for new processes.

Conservation Methods and Technology

Within the industrial sector, approximately 39 percent of all energy is used for commercial heating, typically to obtain steam. In addition, 26 percent of the energy is used as direct heating. Thus, 65 percent of the energy used is thermal energy. This strongly implies the need for more insulation, more heat recovery, and more efficient heating technology.

Energy conservation technology in the industrial sector can be subdivided into five categories:

1. Heat confinement
2. Scheduling
3. Heat recovery
4. Proper equipment
5. Maintenance

Heat Confinement

Heat confinement can take on many forms depending upon how the energy is used. When energy is to be saved through heat confinement, the following suggestions should be considered:

1. Use insulation. Increasing energy costs over the years have made insulation cost effective.

2. Reduce unnecessary ventilation where safety is not an overriding factor.
3. Cut air infiltration by installing weather stripping and caulking to seal cracks around windows and doors of industrial buildings.
4. Reduce air-conditioning in unoccupied space.

When the energy is used for industrial processes, heat confinement can be obtained by several methods:

1. Insulate steam pipes and pipes that carry hot liquids or air.
2. Direct or confine flames used for heating to the area where it is needed.
3. Optimize flame geometry (shape) to help accomplish conservation.

Scheduling

The scheduling of various processes can become an important part of energy conservation within the industrial sector. The following suggestions offer ways that energy can be conserved for space heating by scheduling techniques:

1. Reduce heating or cooling during times when the space is unoccupied. Time clock controls are very effective in reducing costs.
2. Reduce or eliminate lighting wherever and whenever possible. Time clock controls, motion detectors, or even photocell switches could be considered for exterior and decorative lighting. Most indoor lighting close to the perimeter of the building can be turned off during the daylight hours.
3. Avoid short runs on thermal processing equipment and shut down or idle equipment during production interruptions.
4. Schedule proper loads so that equipment can perform at optimal levels of efficiency. Survey the plant for improperly loaded electric motors, possibly also optimizing their capacity as part of the review.
5. Use time clock controls on heating devices, and electric motors where practical, and use automatic combustion systems controls where needed.
6. Use electricity more efficiently by diverting to off-peak periods. When heavy loads exist, voltage may drop slightly, causing a significant decrease in equipment efficiency.

Heat and Waste Material Recovery

Many industries could recapture more wasted thermal energy. Often thermal energy is put into the air, various materials, and water. The following suggestions will help in the area of waste heat recovery:

1. Waste heat from burner stacks and industrial processes can, in many cases, be recovered and used. One company heats an entire warehouse with heat from the waste gas of the boiler burner, which previously had been vented directly to the environment. Another company utilizes waste heat in a greenhouse. An electric-generating plant design, built by Northern States Power, uses water heated from the main power plant to heat a greenhouse in Becker, Minnesota. The waste heat enables vegetables and flowers to be grown all year. Several nurseries are currently leasing tracts of land to grow their products year round using such greenhouses.
2. Some industrial processes are also able to use the high temperatures of waste heat to operate smaller turbine generators. In this case, the electrical energy generated by the waste heat can help reduce the electrical load on public utilities.
3. Recycled waste materials, such as sawdust or other materials, can also be used as a

supplemental fuel. One company was able to run a 15-kilowatt boiler by using the fumes from the varnish and paint kettles. Air pollution from the plant was also decreased.

4. Waste fuel and waste heat can be used for space heating.
5. Exhaust recycling (recuperation) for combustion equipment can be installed to preheat incoming combustion air.

Proper Equipment

Often energy is used excessively because the wrong equipment is employed. The following suggestions will help determine the proper equipment to conserve energy:

1. Check with equipment suppliers and qualified engineers to determine the extent to which the heating, ventilating, and air-conditioning equipment is optimally suited for the application. In some cases, equipment may be oversized and, occasionally, one boiler in a multi-boiler system can be eliminated by the efficient use of the others.
2. Consider the use of infrared heaters in high-ceiling areas and warehouses. Infrared heating provides comfort and prevents condensation on materials and parts stored. It requires less fuel than other kinds of heating because it heats solid objects in the radiation path without heating the air in the room. Savings of more than 20 percent can be realized when compared to conventional heating.
3. Use fluorescent lighting where practical.
4. Convert to higher voltage electrical systems where practical.
5. Use load shedding devices to shut off automatically less-crucial machinery when voltage drops. These devices both help to save energy and maintain power to crucial machines.

Maintenance

One area that is easily overlooked is that of proper maintenance. Cleanliness, equipment condition, load factors, leaks, and the like, all play an important role when conserving energy. Several suggestions can help to eliminate unnecessary energy usage through improved maintenance:

1. Emphasize boiler maintenance. Cleanliness of the heating surfaces and maintenance of the proper air-fuel ratio at the burner are two important variables that affect boiler efficiency. Boiler suppliers are a good source of information on proper maintenance of this equipment.
2. Monitor exhaust stack temperatures and the concentration of carbon dioxide (CO_2), oxygen (O_2), and carbon monoxide (CO) in the flue gas from any combustion process. A stack temperature that is excessively high means wasted thermal energy. The presence of carbon monoxide in the flue gas indicates incomplete combustion. Oxygen should be kept below 2 percent, and carbon dioxide should be in the range of about 8 to 12 percent depending upon the fuel used. Also, any flames in the flue represent unnecessary thermal energy losses.
3. Check and maximize the power factor in the plant. Power factor is a measure of the number of motors used in the plant. As more motors are put on the electrical line, efficiency drops because voltage and current go out of phase. A power factor of .8 is generally the lowest acceptable level.
4. Make sure the plant is equipped with enough properly functioning capacitors or fluorescent lighting for correcting the power factor.
5. Emphasize boiler efficiency and maximize it through good maintenance.

6. Check steam systems regularly for leaks and poor insulation, and provide steam traps where needed.
7. Make a special effort to keep all heat transfer surfaces clean.
8. Continuously watch for hot spots on furnaces for signs of deterioration in the lining; reline or insulate accordingly.
9. Be alert for opportunities to conserve fuel-related products used in other ways. One company discovered that about 90 percent of its hydraulic oil outlay was lost through leakage.

Numerous suggestions have been presented to illustrate the techniques in which energy can be conserved within the industrial sector. Many others will be suggested and implemented in the future, thereby helping to conserve energy. The critical component, as with the residential sector, is the attitudes and values of the employers and workers. This is especially true with those who make major decisions about energy conservation.

CONSERVATION IN THE ENERGY SECTOR

The Cogeneration Process

As defined in chapter 2, the energy sector represents the industries and utilities that produce electrical energy. One method of improving efficiency in the energy sector is to apply cogeneration units to existing power plants. Cogeneration is a process in which a power plant is able to produce electricity as well as thermal energy (hot water or steam) from a single fuel source. Figure 10-35 illustrates the process of cogeneration. In this particular illustration natural gas is the input, and the output is both electricity and hot water for space heat-

ing or other uses. However, other fuels—such as coal, wood, industrial or municipal solid waste—may be used in a cogeneration unit. Note that cogeneration units are suited for specific applications and that the cogeneration user must have a need for thermal energy as well as electricity. This means that the systems are designed to meet a specific energy need (providing both electricity and thermal energy) and that the systems are designed for maximum efficiency.

A typical, small cogeneration system consists of an internal combustion engine operating on natural gas (or other fuel). A generator attached to the engine is used to produce the necessary electrical power for the load. Heat exchangers are used to extract the heat from

Cogeneration

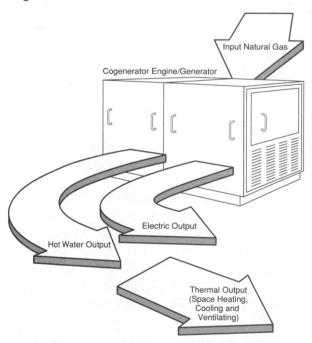

Input Natural Gas

Cogenerator Engine/Generator

Hot Water Output

Electric Output

Thermal Output (Space Heating, Cooling and Ventilating)

FIGURE 10-35 Cogeneration is a process in which natural gas (or other fuel) is converted to electricity and also thermal energy (hot water) for use in industrial processes.

the engine exhaust and cooling system. A microprocessor system is used to control all functions. This is necessary because the load is variable.

Cogeneration System Design

Cogeneration systems range in size from large megawatt systems to small systems of 20 to 100 kilowatts of electrical power. Figure 10-36 illustrates a comparison of a typical power generation system and a cogeneration system. In a standard power generation system (shown in Figure 10-36 a), for every 100 Btu's put into the system, about 48 percent is rejected heat, 20 percent goes into boiler losses, and 2 percent is degraded to miscellaneous sources. Only about 30 percent of the input is actually converted to usable electrical power.

Cogeneration improves the efficiency sig-nificantly. As shown in Figure 10-36 (b), approximately 84 percent of the input energy is converted to usable power. Only 14 percent of the energy is lost to engine heat and 2 percent is degraded to miscellaneous sources. Also, since cogeneration units are often located near points of use, transmission line losses are held to a minimum.

CONSERVATION IN THE TRANSPORTATION SECTOR

The transportation sector uses approximately 23 percent of the energy in the United States. The majority of this energy comes from petroleum. It is used for airplanes, trucks, buses, and (most importantly) the automobile.

Internal Combustion Engine

In each of the transportation applications, the primary consumption device is the internal

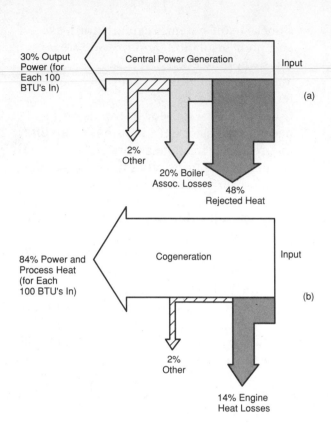

FIGURE 10-36 When comparing a typical central power generation station (a) with a cogeneration station (b), the latter is much more efficient.

combustion engine. Gasoline, diesel, gas turbine, and jet engines are the most common types of energy converters used in this sector.

The internal combustion piston type engine is not very efficient. Only 26 to 30 percent of the chemical energy input is eventually converted to mechanical energy or work. The rest is radiated or exhausted into the atmosphere. Although thermodynamic principles will not allow extremely high efficiencies, there is still room for significant energy-conservation improvement.

Improving Fuel Mileage

Several methods are currently used to help improve fuel mileage in the transportation sector, especially within the automotive market. There are more than 100 million registered vehicles in the United States that average between 15 and 30 miles per gallon of gasoline. Improvements are continually implemented to improve fuel economy—such as weight reduction, making the vehicles smaller, improving fuel injection systems, enhancing computer combustion controls, and building smaller engines.

Gasoline fuel consumption can be improved approximately one mile per gallon for each 400 pounds reduction in weight. The result is smaller vehicles, shorter wheel bases, and lighter construction materials. With lighter vehicles, the automotive manufacturers can reach the fuel consumption guidelines mandated by the federal government. The Corporate Average Fuel Economy (CAFE) requirement of 27.5 miles per gallon during the 1990s was easily met by automobile manufacturers.

> ### Improving Fuel Mileage—Energy Issue 10-7
> Are automotive manufacturers doing everything they can to increase gasoline mileage on new cars? What factors affect the amount and speed of funds allocated by automotive manufacturers to improve gasoline mileage?

An attempt at conserving energy in the transportation sector was to lower the speed limit. The national speed limit was lowered to 55 miles per hour back in the mid-1970s. This effort caused gasoline consumption to drop approximately 4 percent, and highway deaths decreased 17 percent. However, because of the additional supplies of oil worldwide and driving distances being long in this very large country, speed limits have since been raised to 65 miles per hour on many interstate highways.

Another approach to reducing gasoline consumption is by car pooling. Note that the average passenger load is 1.2 people per commuter car. This is not an efficient use of petroleum. People can be educated to the importance of car pooling.

A fourth approach to conserving fuel is to emphasize the use of mass transportation. Although considered the least expensive means of travel, mass transit has still not become usable or popular in many parts of the United States. Granted, many large metropolitan areas have developed people movers, subways, and the like. However, most people still want the luxury of their own personalized transportation system, their car. Taking this luxury away cuts deep into the lifestyles to which most have become accustomed. Furthermore, mass transit is, at present, not suitable to most rural areas—which constitute a significant portion of this country.

Another approach to conserving fuel is to increase the price of fuels used in the transportation sector. It is clear from the data collected that the automobile is not used only as a means of transportation to and from work. Vacations and social and recreational pursuits have increased the daily vehicle miles per household to over 45. Experts note that there is a correlation between miles driven and fuel costs. As fuel costs rise, consumers tend to drive less and to purchase more energy-efficient vehicles.

The transportation sector also uses energy to move products and materials. For example, trucks move products and goods around the country and within cities. One major effort to help increase fuel economy is by using air foils designed to reduce wind resistance of the vehicle (the coefficient of drag). The coefficient of drag is a measure of the amount of air moved by a vehicle as it cuts through the air. Wind resistance absorbs a significant part of the available power in vehicles, particularly at higher speeds.

There have been and continue to be improvements made in the design of engines for automobiles. Some improvements include the use of computers for controlling:

1. Electronic ignition and timing.
2. Electronic carburetors.
3. Electronic fuel injection.
4. Air flow into the engine.
5. Exhaust temperatures.

In addition, the use of other engine designs—such as the diesel, two-cycle, Stirling, and gas turbine are constantly evaluated by automotive manufacturers.

Energy Conservation Tips

In addition to technical changes that are necessary for energy conservation, many suggestions can also help each individual to conserve energy. These include:

1. Accelerate smoothly when driving in the city to improve efficiency.
2. Drive at a steady pace if possible, avoiding stop-and-go traffic.
3. Minimize braking.
4. Do not let the engine idle for more than one minute. It takes less gasoline to restart the engine than it takes to let it idle. Also, when cold starting, put the engine under a gentle load as soon as practical. The engine will heat up quickly and use less gasoline.
5. Keep the engine air filter clean. An air-starved engine makes a rich air-fuel mixture, thus wasting gasoline.
6. Check the tire pressure regularly. Under-inflated tires increase frictional power loss-

es, thereby increasing gasoline consumption.
7. Remove unnecessary weight from the car.

As the field of energy conservation is studied, human values and attitudes are changed more and more by energy awareness. Even though the consuming public has been in the habit of using energy freely, the habit can just as easily be that of conserving energy.

ALTERNATIVE FUTURES

New Battery Technology

Many types of batteries are under investigation, especially with more attention being given to battery-powered vehicles. As such vehicles come into use in the future, battery pack replacement will be as commonplace as filling the gas tank is today. Estimates indicate that battery-powered vehicles could account for 20 percent of new car sales and 10 percent of all vehicles on the road by the early part of the twenty-first century. Note that electric vehicles need batteries with a minimum of 20 watt-hours per pound or 45 watt-hours per kilogram in order to operate with acceptable performance for highway driving.

Most battery manufacturers are trying to improve three characteristics of batteries:

1. Watt-hours per pound (or watt-hours per kilogram).
2. The number of cell life cycles.
3. The cost of the batteries.

Of these three characteristics, watt-hours per pound seems to be the most important factor. Watt-hours per pound is a measure of battery cell density or specifically, the amount of energy stored per weight and volume. Figure 10-37 compares six different batteries that are currently under research and testing

Battery Comparison

	Lead-Antimony	Calcium-Lead	Zinc-Chloride	Sodium-Sulfur	Lithium-Sulfur	Molten Sodium
Cell Density (watt-hours/lb)	(3–4)	(6–8)	(10–15)	(15–18)	(15–18)	(45)
Operating Temperature	Near ambient	Near ambient	Near ambient	(572°–662°F)	(716°–842°F)	(482°–626°F)
Cell Life Cycles	500	600	600	400	500	2000–3000

FIGURE 10-37 Batteries in the future will have increased cell density and more cell life cycles. However, higher battery temperatures will need to be controlled for safe use in vehicles.

for use in the automotive and other similar markets.

This figure shows comparisons between the different batteries tested and their density, operating temperature, and cell lifecycles. The high temperatures shown on the sodium-sulfur, lithium-sulfur, and molten-sodium batteries cause severe operating conditions. Sealing and corrosion problems must be solved before being commercially successful. Research does indicate that these high-temperature problems can be solved. However, the cost of these batteries will be significantly higher.

One futuristic type of battery that is under research is the molten-sodium battery. Within this battery, the electrodes are molten sodium (Na) and sulfur (S). The electrolyte is a solid ceramic material that allows atoms to pass through it. The cell densities are very high and the cell life cycle has been increased substantially.

The lithium-sulfur battery cell shown in Figure 10-38 can also produce a large amount of energy per unit weight and size. These cells could be placed in a stacked arrangement on an electric vehicle. They could also be used for energy storage by an electric utility company. For off-peak energy storage, large banks of batteries in an electric utility system would be charged during periods of low demand. Then, the batteries could be discharged, and converted to AC during periods of high demand. The effect of this type of storage would be the leveling of a load on the power-generating facilities, thus keeping the power plant at higher efficiency. Of course, there would be some energy lost when the electricity is converted back to AC for commercial use.

Numerous other possibilities exist for energy to be stored in batteries. As battery manufacturers continue to improve cell densities and cycles, batteries will be used more for industrial and commercial applications that require large energy outputs. Such improvements could have a significant effect on the storage and efficiency of energy systems.

Lithium-Sulfur Battery

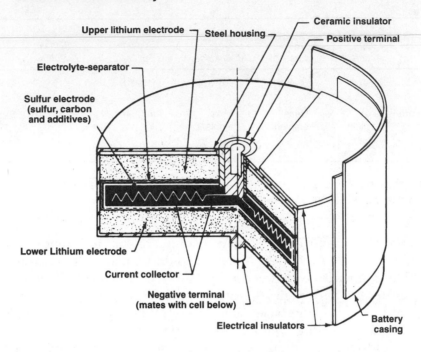

FIGURE 10-38 A cutaway of a lithium-sulfur battery shows the terminology of many internal parts.

POINT/COUNTERPOINT 10-1

TOPIC:

Efficiency and the Process of Energy Conversion

Theme: In the past few years there have been many new technologies developed to convert energy from one form to another. In the transportation sector, most of the conversion technologies utilize gasoline or diesel fuel and are relatively inefficient. In a discussion or classroom debate, attempt to answer the following questions:

1. Why have industries continued to research and develop further gasoline and diesel engines in light of their efficiency losses and environmental effects?

2. What economic factors cause these types of engines to be further developed?

3. Why are there so many applications that require mechanical energy, and why have gasoline and diesel engines been applied to these applications?

4. Could commercially attractive converters that produce mechanical energy replace the gasoline or diesel engine? If so, what are their costs and are they environmentally attractive?

5. What influence does the federal government have on the continued development of gasoline and diesel converters?

6. What influence do oil companies have on the continued development of gasoline and diesel converters?

7. What are examples of new types of converters to produce mechanical energy, and can these be incorporated into society at a reasonable cost to the consumer and to industry?

8. If consumers in the United States were to change to another more efficient type of mechanical converter than the gasoline or diesel engine, what additional technologies and social changes would be needed?

POINT/COUNTERPOINT 10-2

TOPIC:

Battery-Powered Automobiles

Theme: In the past few years there has been an interest in developing battery-powered vehicles to be used for intercity driving and short- to medium-length trips. Some researchers believe that eventually gasoline engines will be replaced by battery-powered vehicles. Others feel that as long as there is gasoline available, the gasoline engine should be used since it is less expensive. In a classroom discussion or debate, attempt to answer the following questions:

1. What would be the main reason for changing from gasoline automobiles to battery-powered vehicles?
2. Would a battery-powered transportation economy have less pollution?
3. Several years ago, many industries were resisting research on battery-powered vehicles. Today, it seems that more emphasis is being placed on this new development. Why is this so?
4. Would it cost less in the long run to operate automobiles on battery power than on gasoline?
5. What effects would be felt in employment and jobs if the federal government mandated the use of battery-powered vehicles (at least 30 percent of all new vehicles off the assembly line) within a 10-year period? What would happen to all of the service sector jobs associated with gasoline engines?
6. Are battery-powered vehicles capable of most driving demands placed on vehicles today?
7. What technologies would have to be improved to make battery-powered vehicles more efficient?
8. Should battery-powered vehicle technology be held back until the battery manufacturers improve designs?
9. Should battery-powered vehicles be designed in conjunction with solar cells for daytime charging of the batteries?

POINT/COUNTERPOINT 10-3

TOPIC:

Federal Incentives to Encourage Energy Conservation

Theme: In the past few years energy conservation has not been very popular with the public. The supply and demand of energy could be controlled easily if society were to conserve. Recently, there has been very little incentive for people to conserve. Should the federal government establish an incentive program—through tax breaks, and so on—to encourage more energy conservation? Or, should energy conservation be a personal goal for each individual, without pressure from government? In a classroom discussion or debate, attempt to answer the following questions:

1. Why is it so important to conserve energy in all sectors of society?
2. Why haven't industries designed homes to conserve even more energy?
3. People still go faster than the speed limit. Most people know that more energy is used at faster speeds, thus reducing the effects of energy conservation. How can the speed limit be better enforced? Or, should it?
4. If the federal government were to provide tax incentives, how would the programs be regulated, controlled, and monitored? Who would pay for this service?
5. Are there other ways in which people learn to save energy? Who would provide the necessary funds for these methods?
6. Why haven't the large, medium, and small industries taken steps to encourage more energy conservation in their manufacturing processes, in product use, and so on?
7. In your opinion, what is the most effective method used to get the general public to conserve energy?
8. What relationship exists between increased energy conservation efforts and economic factors?
9. What is the effect of a society not placing energy conservation before economic costs?

SUMMARY & REVIEW

Summary/Review Statements

1. Energy conversion technology is necessary because the energy resources available to a society are usually not in the correct form needed by the application.
2. Changing energy from one form to another can involve a number of conversion processes.
3. The most common type of conversion used in technological systems is that of a hydrocarbon fuel converted into thermal energy and then into mechanical energy to produce motion.
4. Most automobile engines operate on the four stroke cycle engine design.
5. Often it is necessary to convert mechanical energy into electrical energy. One of the most common methods of converting mechanical energy into electrical energy is to use generators.
6. Electromagnetic principles state that a wire conductor passed through an electromagnetic field induces a voltage into the wire.
7. Electrical energy may be converted to mechanical energy through the use of electric motors.
8. Fuel cells are able to convert chemical energy into electrical energy.
9. Energy storage is defined as the placing of energy into a particular state so that it can be held for future use with minimal energy loss. This state of energy may also be referred to as static energy.
10. Energy storage can be classified into three categories: chemical storage, thermal storage, and mechanical storage.
11. Hydrogen gas is an example of energy stored in a chemical form.
12. Electrolysis is a process that separates the oxygen and hydrogen from water molecules.
13. Flywheels may be used to store mechanical energy for later use.
14. The United States has not become energy independent, and consumers have not wholeheartedly embraced the concepts of energy conservation.
15. Two accepted methods used to conserve energy are improving technologies to make them more efficient and increasing education to develop a set of social values related to energy use.
16. The United States utilizes the most energy per capita in the world.
17. The term *personal energy consumption* is the amount of energy utilized per person.
18. One method of conserving energy is to capture random thermal energy and use this heat in other energy systems.
19. The residential and commercial sectors in the United States use approximately 18 percent of the total energy consumed in that society.
20. The major energy consumption devices in a building are the water heater, electric resistance heating, and central air-conditioning.
21. Heat loss is defined as the amount of thermal energy passing out of a building. Heat gain is defined as the amount of thermal energy coming into a building.
22. A neutral wall is one through which little or no thermal energy can transfer.
23. Heat losses in a building can come from either infiltration or conduction.
24. Heat always flows from a warmer to a cooler temperature.
25. The R value of material represents the resistance to heat flow in terms of the thickness of the material used.

26. The U value of a material represents how much heat transfers through any single material or an entire assembly of materials.

27. Comfort level is defined as the temperature at which the occupants in the building feel comfortable. The most commonly accepted comfort level in a building is 68° F.

28. The process of saving energy in the industrial sector is often referred to as energy management or waste management.

29. The industrial sector consumes about 27 percent of the energy used in the United States.

30. One method of improving efficiency in the energy sector is to apply cogeneration units to existing power plants.

31. Cogeneration is a process in which a power plant is able to produce electricity as well as thermal energy from a single fuel source.

32. The transportation sector uses approximately 23 percent of the energy in the United States. The majority of this energy comes from petroleum.

33. The internal combustion piston type engine is not very efficient. Only 26 to 30 percent of the input chemical energy is eventually converted to mechanical energy or work.

Discussion Questions

1. Identify the most important trends in the technical development of energy conversion systems. What are the primary goals of scientists and technologists in this area of research?

2. How does the study of conservation of resources fit into a study of the general principles of energy technology? What is the purpose of energy conversion, and why is its study so significant?

3. In what applications would fuel cells have a practical use?

4. What is the purpose of developing storage technology? What problems would occur without appropriate energy storage?

5. After developing a futuristic proposal for a hydrogen-based economy, attempt to answer the following questions: What applications could not be met by the use of this fuel? What social and environmental concerns need to be addressed before adopting this proposal?

6. In the United States, what human values would need to change in order to adopt a long-term conservation attitude with regard to energy use?

7. What methods can be used to change consumer values to encourage conserving energy? What are the short- and long-term consequences of not moving in this direction?

8. What are five significant lifestyle changes that you can adopt to conserve energy?

Appendix A—English and Metric Conversions

TEMPERATURE CONVERSIONS

$F° = (C° \times 1.8 + 32)$

$C° = \dfrac{F° - 32}{1.8}$

SPEED CONVERSIONS

3. Multiply By	1. To Convert From							2. Into
	Miles per HR	Feet per SEC	Miles per SEC	Knots	KM per HR	Meters per MIN	CM per SEC	
	1	.68182	3600	1.1516	.6214	.03728	.02237	= Miles per HR
	1.467	1	5280	1.6889	.9113	.05468	.03281	= Feet per SEC
	.000278	.000189	1	.00032	.000173	1.04×10^{-5}	6.21×10^{-6}	= Miles per SEC
	.8684	.5921	3126	1	.5396	.03238	.01943	= Knots
	1.6093	1.0973	5793.6	1.8532	1	0.06	0.036	= KM per HR
	26.82	18.29	96540	30.88	16.67	1	0.6	= Meters per MIN
	44.70	30.48	1.61×10^{6}	51.48	27.78	1.667	1	= CM per SEC

LENGTH CONVERSIONS

3. Multiply By	1. To Convert From											2. Into
	Inch	Feet	Yard	Rod	Mile (statute)	Mile (naut.)	Milli-meter	Centi-meter	Meter	Kilo-meter	Light Year	
	1	12	36	198	63360	72962.4	.03937	.3937	39.37	39370	3.73×10^{17}	= Inch
	.08333	1	3	16.5	5280	6080.2	.00328	.03281	3.2808	3280.8	3.10×10^{16}	= Feet
	.02778	.3333	1	5.5	1760	2026.7	.00109	.01094	1.0936	1093.6	1.03×10^{16}	= Yard
	.00505	.06061	.18182	1	320	368.49	.000199	.00199	.19884	198.84	1.88×10^{15}	= Rod
	1.58×10^{-5}	.000189	.000568	.003125	1	1.1516	6.21×10^{-7}	6.21×10^{-6}	.00062	.62137	5.88×10^{12}	= Mile (statute)
	1.37×10^{-5}	.000164	.000493	.00271	.86839	1	5.40×10^{-7}	5.40×10^{-6}	.00054	.53959	5.10×10^{12}	= Mile (naut.)
	25.400	304.80	914.402	5029.2	1.61×10^{4}	1.85×10^{4}	1	10	1000	1×10^{4}	9.46×10^{18}	= Millimeter
	2.540	30.480	91.4402	502.92	160935	185235	0.1	1	100	1×10^{5}	9.46×10^{17}	= Centimeter
	.02540	.30480	.914402	5.0292	1609.35	1853.25	0.001	.01	1	1000	9.46×10^{15}	= Meter
	2.54×10^{-5}	.000305	.000914	.00503	1.60935	1.85325	1×10^{-4}	1×10^{-5}	.001	1	9.46×10^{12}	= Kilometer
	2.69×10^{-18}	3.23×10^{-17}	9.69×10^{-17}	5.33×10^{-16}	1.70×10^{-13}	1.96×10^{-13}	1.06×10^{-19}	1.06×10^{-18}	1.06×10^{-14}	1.06×10^{-13}	1	= Light Year

AREA CONVERSIONS

1. To Convert From

3. Multiply By

SQ Inch	SQ Feet	SQ Yard	SQ Rod	SQ Mile	SQ Centimeter	SQ Meter	SQ Kilometer	Acre	2. Into
1	144	1296	39204	4.01×10^9	.1550	1550	1.55×10^9	6.27×10^6	= Sq Inch
.00694	1	9	272.25	2.79×10^7	.00108	10.764	1.08×10^7	43560	= Sq Feet
.00077	.11111	1	30.25	3.10×10^6	.00012	1.1960	1.20×10^6	4940	= Sq Yard
2.55×10^{-6}	.00367	.03306	1	102400	3.95×10^{-4}	.03954	39536.7	160	= Sq Rod
2.49×10^{-19}	3.59×10^{-6}	3.23×10^{-7}	9.77×10^{-6}	1	3.86×10^{-11}	3.86×10^{-7}	.3861	.00156	= Sq Mile
6.452	929.03	8361.3	252930	2.59×10^{19}	1	10,000	1×10^{10}	4.05×10^7	= Sq Centimeter
.000645	.0929	.83613	25.293	2.59×10^4	.0001	1	1×10^6	4046.9	= Sq Meter
6.45×10^{-10}	9.29×10^{-8}	8.36×10^{-7}	2.53×10^{-5}	2.590	1×10^{-10}	1×10^{-6}	1	.00405	= Sq Kilometer
1.59×10^{-7}	2.30×10^{-8}	.000207	.00625	640	2.47×10^{-6}	2.47×10^{-4}	247.1	1	= Acre

VOLUME CONVERSIONS

1. To Convert From

3. Multiply By

CUBIC INCH	CUBIC FEET	CUBIC YARD	CUBIC CM	CUBIC METER	2. Into
1	1728	46656	.06102	61023	= Cubic Inch
.00058	1	27	3.53×10^{-5}	35.314	= Cubic Feet
2.14×10^{-5}	.03704	1	1.31×10^{-6}	1.3079	= Cubic Yard
16.387	28317	764559	1	1×10^6	= Cubic Cm
1.64×10^{-5}	.028317	.764559	1×10^{-6}	1	= Cubic Meter

CAPACITY CONVERSIONS (LIQUID)

1. To Convert From

3. Multiply By

Ounce	Gill	Pint	Quart	Gal. (U.S.)	Gal. (Imp.)	Milliliter	Liter	Cubic Inch	Cubic Feet	Tablespoon	Cup	2. Into
1	4	16	32	128	153.73	.03381	33.815	.55411	957.51	.05	8	= Ounce
0.250	1	4	8	32	38.432	.00845	8.454	.138528	239.38	0.125	2	= Gill
0.0625	0.250	1	2	8	9.608	.00211	2.113	.03463	59.844	0.0313	0.5	= Pint
.03125	0.125	0.5	1	4	4.804	.00106	1.057	.01732	29.922	0.016	0.250	= Quart
.007812	.03125	0.125	0.250	1	1.201	.000264	.2642	.00433	7.481	.0034	0.0625	= Gallon (U.S.)
.006503	.02602	.1041	.2082	.8327	1	.00022	.220	.00360	6.229	.0028	0.052	= Gallon (Imp.)
29.573	118.29	473.17	946.33	3785.3	4546	1	1000	16.387	28316	14.786	237	= Milliliter
.02957	.11829	.47317	.94633	3.785	4.546	.001	1	.01639	28.316	0.015	.237	= Liter
1.80469	7.21875	28.875	57.75	231	277.9	.06103	61.025	1	1728	0.902	14.432	= Cubic Inch
.001044	.00417	.01671	.03342	.13368	.16054	3.54×10^{-5}	.0353	.000579	1	.000522	.00835	= Cubic Feet
2	8	32	64	256	307.46	.06760	67.64	1.108	1915	1	16	= Tablespoon
0.125	0.5	2	4	16	19.20	.00424	4.232	0.0693	119.68	.0625	1	= Cup

CAPACITY CONVERSIONS (DRY)

3. Multiply By	1. To Convert From								2. Into
	Pint	Quart	Peck	Bushel	Liter	Cubic Feet	Cord	Barrel	
	1	2	16	64	1.816	51.43	6582.9	210	= Pint
	0.5	1	8	32	.9081	25.714	3291.4	105	= Quart
	.0625	0.125	1	4	.11351	3.214	411.43	13.13	= Peck
	.01562	.03125	0.25	1	.02838	.80356	102.86	3.281	= Bushel
	.5506	1.1012	8.810	35.24	1	28.316	3625	115.624	= Liter
	.01944	.0389	.3111	1.2445	.03532	1	128	4.0833	= Cubic Feet
	.000154	.0003	.00243	.00973	.000276	.00781	1	.03186	= Cord
	.00476	.0095	.0762	.3048	.00856	.2446	31.32	1	= Barrel

WEIGHT (OVER 1 LB) CONVERSIONS

3. Multiply By	1. To Convert From							2. Into
	Pound	Kilogram	Short Ton	Long Ton	Metric Ton	(Short) 100 Wt	Slug	
	1	2.2046	2000	2240	2204.6	100	32.174	= Pound
	.4536	1	907.18	1016	1000	45.36	14.594	= Kilogram
	.0005	.00110	1	1.12	1.102	.05	0.016	= Short Ton
	.000446	.00098	.8929	1	.9842	.0446	.01428	= Long Ton
	.000454	.001	.90718	1.016	1	.0454	.01459	= Metric Ton
	.01	.0220	20	22.40	22.05	1	.32175	= Short 100 Wt.
	.03108	.06852	62.143	69.619	68.481	3.108	1	= Slug

WEIGHT (UNDER 1 LB) CONVERSIONS

3. Multiply By	1. To Convert From							2. Into
	Milligram	Grain	Gram	Dram (avdp.)	Dram (apoth.)	Ounce (avdp.)	Pound	
	1	64.80	1000	1771.8	3887.9	28350	453592	= Milligram
	.01543	1	15.432	27.34	60	437.5	7000	= Grain
	.001	.0648	1	1.772	3.888	28.35	453.6	= Gram
	.00056	.0366	.5644	1	2.1943	16	256	= Dram (avdp.)
	.00026	.0167	.2572	.4557	1	7.292	116.67	= Dram (apoth.)
	3.53×10^{-5}	.00229	.0353	.0625	.1371	1	16	= Ounce (avdp.)
	2.20×10^{-6}	.000143	.0022	.00391	.00857	.0625	1	= Pound

Appendix B—Major Electrical Uses and Demand in Your House

	Kilowatt-hour/Year		Kilowatt-hour/Year
Water Heater	4,800	Laundry	
Refrigerator/Freezer		Automatic Washing Machine	103
Manual Defrost (10–15 cubic feet)	700	Iron	60
Automatic Defrost (16–18 cubic feet)	1,800	Food Preparation	
Freezer (15–21 cubic feet)		Range	
Chest Type, Manual Defrost	1,320	With Oven	1,175
Upright Type, Automatic Defrost	1,985	With Self-cleaning Oven	1,205
Clothes Dryer	993	Broiler	85
Room Air-Conditioner (typical)	688	Coffee Maker	106
Central Air-Conditioning	4,800	Deep Fryer	83
Electric Resistance Heating	21,000	Dishwasher	363
Heat Pump	14,000	Microwave Oven	190
Home Entertainment		Roaster	60
Radio	86	Trash Compactor	50
Radio/Record Player/Tape/CD	109	Waffle Iron	20
Television		Miscellaneous	
Color (solid state)	320	Clock	17
Color (tube type)	528	Hair Dryer	14
Space Heating Appliances		Heating Pad	10
Attic Fan	291	Sewing Machine	11
Dehumidifier	377	Vacuum Cleaner	46
Humidifier	163		
Window Fan	170		

Appendix C—Metric Decimal Prefixes

Multiplication factors	Prefix	Symbol
$1\ 000\ 000\ 000\ 000 = 10^{12}$	tera	T
$1\ 000\ 000\ 000 = 10^{9}$	giga	G
$1\ 000\ 000 = 10^{6}$	mega	M
$1\ 000 = 10^{3}$	kilo	k
$100 = 10^{2}$	hecto	h
$10 = 10^{1}$	deka	da
1	(Units)	
$0.1 = 10^{-1}$	deci	d
$0.01 = 10^{-2}$	centi	c
$0.001 = 10^{-3}$	milli	m
$0.000\ 001 = 10^{-6}$	micro	μ
$0.000\ 000\ 001 = 10^{-9}$	nano	n
$0.000\ 000\ 000\ 001 = 10^{-12}$	pico	p
$0.000\ 000\ 000\ 000\ 001 = 10^{-15}$	femto	f
$0.000\ 000\ 000\ 000\ 000\ 001 = 10^{-18}$	atto	a

Appendix D—Fuel and Energy Equivalents

1 bbl crude oil=	1 ton bituminous coal=	1000 ft³ natural gas=	1000 kwh electricity=
443 lb bituminous coal	4.52 bbl crude oil	79.01 lb bituminous coal	260.5 lb bituminous coal
5,604 ft³ natural gas	25,300 ft³ natural gas	74.95 gallons crude oil	3,397 ft³ natural gas
1,700 kwh electricity	7,679 kwh electricity	303.34 kwh electricity	0.588 bbl crude oil

Appendix E—Specific Heats and Heat Capacity of Materials

Material	Specific Heat (Btu/lb/°F)	Density (lb/ft³)	Heat Capacity (Btu/ft³/°F)	Material	Specific Heat (Btu/lb/°F)	Density (lb/ft³)	Heat Capacity (Btu/ft³/°F)
Air (at 1 atmosphere)	0.24 [75]	0.075	0.018	Gypsum	0.259	78	20.2
Aluminum (alloy 1100)	0.214	171	36.6	Hemp (fiber)	0.323	93	30.0
Asbestos fiber	0.25	150	37.5	Ice	0.487 [32]	57.5	28.0
Asbestos insulation	0.20	36	7.2	Iron, cast	0.12 [212]	450	54.0
Ashes, wood	0.20	40	8.0	Lead	0.031	707	21.8
Asphalt	0.22	132	29.0	Limestone	0.217	103	22.4
Bakelite	0.35	81	28.4	Magnesium	0.241	108	26.0
Brass, red				Marble	0.21	162	34.0
(85% Cu, 15% Zn)	0.09	548	49.3	Nickel	0.105	555	58.3
Brass, yellow				Octane	0.51	43.9	22.4
(65% Cu, 35% Zn)	0.09	519	46.7	Paper	0.32	58	18.6
Brick, building	0.2	123	24.6	Paraffin	0.69	56	38.6
Bronze	0.104	530	55.1	Porcelain	0.18	162	29.2
Cellulose	0.32	3.4	1.1	Rock salt	0.219	136	29.8
Cement (Portland clinker)	0.16	120	19.2	Salt water	0.75	72	54.0
Chalk	0.215	143	30.8	Sand	0.191	94.6	18.1
Charcoal (wood)	0.20	15	3.0	Silica	0.316	140	44.2
Clay	0.22	63	13.9	Silver	0.056	654	36.6
Coal	0.3	90	27.0	Steel (mild)	0.12	489	58.7
Concrete (stone)	0.22	144	31.7	Stone (quarried)	0.2	95	19.0
Copper (electrolytic)	0.092	556	51.2	Tin	0.056	455	25.5
Cork (granulated)	0.485	5.4	2.6	Tungsten	0.032	1210	38.7
Cotton (fiber)	0.319	95	30.3	Water	1.0 [39]	62.4	62.4
Ethyl alcohol	0.68	49.3	33.5	Wood, white fir	0.65	27	17.6
Fireclay brick	0.198 [212]	112	22.2	Wood, white oak	0.570	47	26.8
Glass, crown (soda-lime)	0.18	154	27.7	Wood, white pine	0.67	27	18.1
Glass, flint (lead)	0.117	267	31.2	Zinc	0.092	445	40.9
Glass, pyrex	0.20	139	27.8				
Glass, "wool"	0.157	3.25	0.5				

Source: Adapted from Anderson, pages 192–193.

Appendix F—Maximum Annual Collector Performance

Configuration	Latitude, degrees	Energy yield kBtu/ft^2 yr
Fully tracking	45	910
	30	985
Horizontal plate	45	520
	30	610
Fixed (+15°)	45	642
	30	713

Appendix G—Solar Energy Data

Description	Values
1. Total power radiated from the sun in all directions	3.8×10^{26} W
2. Solar constant—power per unit area at the top of the earth's atmosphere, for a surface directly facing the sun	1.353 kW/m^2 (approximate rounded value: 1.4 kW/m^2)
3. Amount incident at ground level per unit area, for a surface directly facing the sun (this amount varies with weather conditions and with the amount of atmosphere in the path; the value given here is typical for a time near noon on a clear and cloudless day)	1 kW/m^2
4. Energy delivered to a horizontal surface (approximate average rate for the 48 contiguous states, averaged over all hours of the day and night and averaged over a full year)	200 W/m^2

Appendix H—Solar Energy Conversions

The solar constant I_0=	$1 Btu/hr\text{-}ft^2$=	$1 Langley (Ly)$=	$1 Ly/min$=
429 Btu/hr-ft²	3.155 W/m²	1 cal/cm²	698 W/m²
1353 W/m²	7.54 x 10⁻⁴ kcal/sec-m²	11.63 W-hr/m²	221.2 Btu/hr-ft²
		3.69 Btu/ft²	

Appendix I—Radiation Exposure in the United States

Radiation source	Average dose rate per person (mrem/yr)
Natural sources	
Cosmic rays at ground level	44
Rocks, soil and building materials	40
Sources within the body (largely K^{40})	18
Subtotal	102
Artificial sources	
Fallout from nuclear weapons testing	4
Medical and dental diagnosis and treatment	73
Nuclear power installations	0.003
Occupational exposure	0.8
Miscellaneous (TVs, airplane travel, etc.)	2
Subtotal	80
Total	182

Appendix J—Exponential Growth

Multiplication factor in each unit of time	Percentage increase per unit time	Doubling time
1.0	0	Infinite
1.01	1	69.7
1.02	2	35.0
1.03	3	23.4
1.04	4	17.7
1.05	5	14.2
1.06	6	11.9
1.07	7	10.2
1.08	8	9.0
1.09	9	8.0
1.10	10	7.3
1.12	12	6.1
1.14	14	5.3
1.16	16	4.7
1.18	18	4.2
1.20	20	3.8

$$\text{DOUBLING TIME} = \frac{72}{\% \text{ OF CHANGE (INCREASE) PER UNIT OF TIME}}$$

Appendix K—Energy Requirements for Freight Transportation

Mode of transport	Energy consumption (Btu/ton-mile)	Mileage (ton-miles/gal)
Oil pipelines	450	275
Railroads	670	185
Waterways	680	182
Truck	2800	44
Airplane	42000	3

Appendix L—Energy Content of Fuels

Fuel	(Commonly used units)	Values (Btu/ton)
Coal (bituminous and anthracite)		25×10^6
Lignite		10×10^6
Peat		3.5×10^6
Crude oil	5.6×10^6 Btu/barrel	37×10^6
Gasoline	5.2×10^6 Btu/barrel	38×10^6
NGLs (Natural gas liquids)	4.2×10^6 Btu/barrel	37×10^6
Natural gas	1030 Btu/ft^3	47×10^6
Hydrogen gas	333 Btu/ft^3	107×10^6
Methanol (methyl alcohol)	6×10^4 Btu/gal	17×10^6
Charcoal		24×10^6
Wood	20×10^6 Btu/cord	12×10^6
Dung		15×10^6
Assorted garbage and trash		10×10^6
Fission	200 MeV/fission	7×10^{13}
		5×10^{11a}
D-D Fusion (deuterium)	7 MeV/deuteron	2.9×10^{14}
		8.6×10^{10}

These data are only intended for use in making estimates of available energy.

[a] per ton or kilogram of uranium metal, when only the U^{235} (abundance 0.72%) is used.

Appendix M—Energy Requirements for Passenger Transportation

Mode of transport	Maximum capacity (no. of passengers)	Vehicle mileage (miles/ gal)	Passenger mileage (passenger-miles/gal)	Energy consumption (Btu/ passenger-mile)
Bicycle	1	1560	1560	80
Walking	1	470	470	260
Intercity bus	45	5	225	550
Commuter train (10 cars)	800	0.2	160	775
Subway train (10 cars)	1000	0.15	150	825
Local bus	35	3	105	1180
Intercity train (4 coaches)	200	0.4	80	1550
Automobile	4	20	80	1550
Motorcycle	1	60	60	2060
747 jet plane	360	0.1	36	3440
Light plane (2 seat)	2	12	24	5160
Executive jet plane	8	2	16	7740
Concorde SST	110	0.12	13	9400
Snowmobile	1	15	15	10,300
Ocean liner	2000	0.005	10	12,400

Appendix N—Energy Units

1 electron-volt (eV) =	1 million electron-volts (MeV) =	1 calorie (cal) =
1 eV	10^6 eV	2.611×10^{19} eV
10^{-6} MeV	1 MeV	2.611×10^{13} MeV
3.829×10^{-20} cal	3.829×10^{-14} cal	1 cal
1.52×10^{-22} Btu	1.52×10^{-16} Btu	3.968×10^{-3} Btu
3.829×10^{-23} kcal	3.829×10^{-17} kcal	0.001 kcal
4.451×10^{-26} kWh	4.451×10^{-20} kWh	1.162×10^{-6} kWh
1.52×10^{-28} MBtu	1.52×10^{-22} MBtu	3.968×10^{-9} MBtu
1.854×10^{-30} Mw-day	1.854×10^{-24} Mw-day	4.843×10^{-11} Mw-day
5.077×10^{-33} Mw-yr	5.077×10^{-27} Mw-yr	1.326×10^{-13} Mw-yr

1 British thermal unit (Btu) =	1 kilocalorie (kcal or Cal)=	1 kilowatt-hour (kWh) =	1 million Btu (MBtu) =
6.581×10^{21} eV	2.611×10^{22} eV	2.247×10^{25} eV	6.581×10^{27} eV
6.581×10^{13} MeV	2.611×10^{16} MeV	2.247×10^{19} MeV	6.581×10^{21} MeV
252 cal	1000 cal	8.604×10^5 cal	2.52×10^8 cal
1 Btu	3.968 Btu	3413 Btu	10^6 Btu
0.252 kcal	1 kcal	860.4 kcal	2.52×10^5 kcal
2.929×10^{-4} kWh	1.162×10^{-3} kWh	1kWh	292.9 kWh
10^{-6} MBtu	3.968×10^{-6} MBtu	3.413×10^{-3} MBtu	1 MBtu
1.22×10^{-8} Mw-day	4.843×10^{-8} Mw-day	4.167×10^{-5} Mw-day	0.0122 Mw-day
3.341×10^{-11} Mw-yr	1.326×10^{-10} Mw-yr	1.141×10^{-7} Mw-yr	3.341×10^{-5} Mw-yr

1 megawatt-day (Mw-day) =	1 megawatt-year (Mw-yr) =
5.393×10^{29} eV	1.97×10^{32} eV
5.393×10^{23} MeV	1.97×10^{26} MeV
2.065×10^{10} cal	7.542×10^{12} cal
8.195×10^7 Btu	2.993×10^{10} Btu
2.065×10^7 kcal	7.542×10^9 kcal
2.4×10^4 kWh	8.766×10^6 kWh
81.95 MBtu	2.993×10^4 MBtu
1 Mw-day	365.2 Mw-day
2.738×10^{-3} Mw-yr	1 Mw-yr

Appendix O—Power Units

1 Btu per day =	1 kilowatt-hour per year (kWh/yr) =	1 watt (W) =	1 kilowatt (kW) =	1 megawatt (Mw) =
1 Btu/day	9.348 Btu/day	81.95 Btu/day	8.195×10^4 Btu/day	8.195×10^7 Btu/day
0.107 kWh/yr	1 kWh/yr	8.766 kWh/yr	8766 kWh/yr	8.766×10^6 kWh/yr
0.0122 W	0.1141 W	1W	1000 W	10^6 W
1.22×10^{-5} kW	1.141×10^{-4} kW	0.001 kW	1 kW	1000 kW
1.22×10^{-8} Mw	1.141×10^{-7} Mw	10^{-6} Mw	0.001 Mw	1 Mw

Other units of power:
  1 horsepower (hp) = 746 W.
  1 horsepower (hp) = 550 ft-lbs/second.
  1 Btu = 778 ft-lbs.

Appendix P—Annual Yields from a Barrel of Crude Oil

(1 Barrel of Oil Contains 42 Gallons)

Product	Gallons Per Barrel	% Yield
Leaded Gasoline	6.9	16.4
Unleaded Gasoline	12.3	29.4
Distillate Fuel Oil	9.1	21.6
Residual Fuel Oil	3.0	7.1
Jet Fuel	4.0	9.5
Petrochemical Feedstocks	1.2	2.8
Asphalt and Road Oil	1.3	3.2
Still Gas (Refinery Gas)	2.0	4.7
Coke	1.5	3.7
Liquefied Gases	1.3	3.1
Lubricants	0.5	1.2
Kerosine	0.3	0.8
Miscellaneous	0.2	0.5
Special Naphthas	0.2	0.4
Wax	0.1	0.1
Processing Gain	-1.9	-4.5
Totals	42.0	100.0

Notes:

(1) Leaded and unleaded gasoline includes both motor and aviation gasoline.
(2) Jet fuel includes both naphtha-type and kerosine-type fuel.
(3) Distillate fuel oil includes home-heating and diesel fuel, as well as No. 1 and No. 4 commercial fuel oils.
(4) Still gas (refinery gas) is that gas produced in refineries during the refining and cracking processes.
(5) Processing gain represents the amount by which total refinery output is greater than total input for a given period. The difference is due to the processing of crude oil into products which, in total, have less weight than the crude oil processed. Therefore, in terms of volume (barrels), the total output of products is greater than the input.

Source: American Petroleum Institute.

Appendix Q—Mathematical Summary

The following mathematical formulas and definitions will help you to solve various energy problems:

1. To calculate **percentage:**

Percentages can be calculated on any data if the total data is known and the part (or sought after) data is known.

$$\frac{\text{Part (Data)}}{\text{Total (Data)}} \times 100$$

For example, what is the percentage of foreign cars in a city that has 100,000 total cars and 28,000 foreign ones?

28,000 = part data
100,000 = total data

$$\frac{28,000}{100,000} \times 100 = 28\%$$

2. **Efficiency** is calculated by the formula:

$$\text{Efficiency} = \frac{\text{Output energy}}{\text{Input energy}} \times 100$$

3. **Total efficiency** of a system is calculated by:

Total efficiency = Efficiency of each converter multiplied together

4. **Doubling time** is calculated by the formula:

$$\text{Doubling Time} = \frac{72}{\text{Percentage of increase}}$$

5. **MB/DOE** is a unit to indicate how much energy in millions of barrels of oil per day are used.

6. Percentage of Growth or **Growth Rate** is calculated by:

$$\text{Growth Rate in \%} = \frac{\text{Change}}{\text{Original or starting point}} \times 100$$

Example: If one month an energy bill were \$22.00 and the next month the bill increased to \$24.00, the percentage of increase is calculated by:

$$\text{Growth Rate in \%} = \frac{(24 - 22) = 2}{22} \times 100 = 9.09\%$$

7. One **ccf** of natural gas = 100 cubic feet.
   One ccf is also equal to one **therm**.

8. To **convert ccf or therms to Btu's:**

$$1030 \text{ Btu's} \times \text{ccf or therm} = \text{Total Btu's}$$

9. **Cost Payback** is calculated by:

$$\text{Cost Payback} = \frac{\text{Initial capital cost}}{\text{Savings per year}}$$

10. To calculate the **amount of Btu's** necessary to raise water a certain temperature:

$$\text{Btu's necessary} = (\text{gallons of water}) \times (8 \text{ lbs}) \times$$
$$(\text{change in temperature in degrees})$$

For example, to raise a 25-gallon tank of water 45 degrees: 25 gal. $\times$ 8 lbs $\times$ 45° = 9000 Btu's

11. To calculate the amount of **energy** that can be taken from a stack of **wood** in Btu's:
    a. Multiply the dimensions of the wood pile—length $\times$ width $\times$ height—to get total cubic feet of wood.
    b. Divide the cubic feet of wood by 128 to get the number of cords.
    c. Multiply the number of cords $\times$ the energy content in the wood in Btu's.
    d. Multiply the energy content in the wood $\times$ the wood burning stove efficiency.

12. To **convert between R and U** values for insulation:

$$R = \frac{1}{U} \quad \text{and} \quad U = \frac{1}{R}$$

13. **Heating Degree Days** are calculated by:

$$\text{Degree Days} = 65 - \frac{\text{High + Low Temperature}}{2}$$

Appendices Bibliography

American Petroleum Institute. *Oil Barrel Brochure*. Washington, D.C., 1986.

Anderson/Riordan. *The New Solar Home Book*. Brickhouse Publishing Company. Andover, Massachusetts, 1987.

Glossary

Absorber: A material used in solar collectors that absorbs incoming solar radiation and converts it to thermal energy.

Absorptivity: The ability of a material or finish to absorb a maximum amount of solar energy. The term is used to define the performance characteristics of a solar absorber.

Accelerated Motion: A condition in which the velocity of a moving object increases as the motion continues.

Acceleration: The rate of change in velocity over time. The simplest type of acceleration is motion that occurs in a straight line over which the speed changes at a constant rate.

Acid Rain: The result of burning coal or other sulfur-containing fuels: the sulfur is introduced into the atmosphere from exhaust gases where it combines with airborne moisture to form this type of pollution.

Active Solar System: A solar energy system that uses pumps, fans, and an external energy source to transfer the heat collected.

Aerobic: Requiring oxygen to live. Aerobic digestion of organic material can be used to convert biomass to usable fuels.

Air-fuel Ratio: The ratio of air to fuel (in volume) entering a combustion chamber. Theoretically, the most efficient air-fuel ratio is 14.7 parts of air to 1 part of fuel.

Alpha Decay: The natural process in which a radioactive element emits two protons and two neutrons as a helium-4 nucleus (4He).

Alternating Current: (AC). The flow of an electrical charge through a conductor that reverses its direction of flow at regularly recurring intervals.

Ambient: Surrounding environment.

Amortization: Money set aside (installment payments) in a sinking fund for expenditures associated with a building project, equipment, or other capital projects.

Anaerobic: Able to live in the absence of free oxygen. Anaerobic digestion of organic material can be used to convert biomass to usable fuels.

Anthracite: The highest quality coal in terms of carbon content and heating value. These coals were formed over 300 million years ago and have been used as a fuel source for industrial, commercial, and residential heating applications.

Anticline: An upward fold or arch of strata with sloping downward rock formations on both sides of the crest.

Associated Gas: A term used to denote underground natural gas found in conjunction with crude oil.

Associated Gas Well: A natural gas well in which oil is found mixed with the natural gas.

Background Radiation: See *Natural Radiation*.

Barrel of Oil: A unit of volume used in oil-producing industries. One barrel of oil contains 42 gallons.

Beta Decay: The process in which a radioactive neutron spontaneously changes itself into a proton—or the reverse—and an electron is emitted.

Beta Particle: An electron emitted by a radioactive nucleus in beta decay.

Bioconversion: The process of producing usable energy from organic matter, such as plant and other organic wastes.

Biomass: Any organic material (alive or decaying), including various types of industrial and residential waste that contain chemical energy, used as a fuel source.

Bituminous Coal: A high quality coal, better than lignite and subbituminous in terms of heating value. These coals were formed about 300 million years ago, and their uses include burning for the production of electricity and making of coke for the steel industry.

Breeding Fuel: Radioactive elements that cannot be fissioned, which are transmuted to fissionable material.

British Thermal Unit (Btu): The amount of thermal energy needed to raise 1 pound of water 1 degree Fahrenheit.

Brush: A sliding electrical conductor that completes a circuit between a fixed and a moving part.

Bubble Cap: A device that aids in the condensing of vapors into a liquid in a fractionating column.

Camphene: A derivative of turpentine that was used as a fuel for lighting in the 1800s.

Carbon Content: The amount (percentage) of carbon in coal. Generally, the higher the carbon content, the more Btu's are available from the coal.

Catalyst: A substance that alters the rate of chemical reaction and is itself unchanged by the process.

CCF: A unit of measure used in the industries associated with natural gas to indicate 100 cubic feet of natural gas.

Change-of-state Storage: A solar-collector storage system in which the thermal energy is stored in a latent form. Change-of-state storage systems (like those that use eutectic salts, which change from a solid to a liquid when heated) allow for the storage of more thermal energy per pound than do other forms of storage.

Chimney Effect: Warm air rising in a structure, which causes cooler air to be drawn into the building at its lower level. The chimney effect results in increased infiltration.

Choke: A device that restricts the size of the opening through which a well emits oil or gas. The choke controls the flow rate of oil or gas extracted from a well.

Coal Gas: See *Town Gas.*

Coalification: The process whereby peat is changed into coal through accumulating sediments from other decaying matter in the absence of air. For this process to occur, heat and pressure are applied to the peat for millions of years.

Coal Slurry: Coal that is crushed into a fine powder, mixed with water, and sent through a pipe to a power plant for use.

Cogeneration: A process where the rejected or wasted heat from one electrical or mechanical generation system is used to operate a second generation system.

Coils: Insulated wire wound in a circle used to produce inductance into a circuit.

Collector Rings: Also called *slip rings*. A conducting contact between an electric circuit in a rotating member of a motor.

Combined-cycle Electrical Generation: A technology that uses gas turbines to convert the energy content of natural gas into electricity. Combined-cycle electrical systems have operating efficiencies as high as 50 percent.

Comfort Level: The temperature and humidity levels at which the occupants within a structure feel comfortable.

Commercial Sector: The societal sector that consumes energy for operating business establishments or governmental buildings and the enterprises contained in these buildings. The commercial sector includes hotels; restaurants; retail stores; religious and nonprofit organizations; health, social, and educational institutions; and federal, state, and local governments. This sector also encompasses public services such as street lights, pumps, bridges, and other services that require the use of electricity.

Conduction: The transfer of thermal energy involving objects at rest and touching each other. The transfer of this thermal energy occurs through molecular collisions of the materials in contact.

Control Rods: Materials used in nuclear reaction processes to absorb the excess neutrons, thus stopping or slowing down any chain reaction that exists. Control rods can be made from any moderator—cadmium, silver, indium alloys, and various boron mixtures.

Convection: The transfer of thermal energy by movement of a fluid from a warmer to a colder location.

Conventional Resources: Oil or natural gas energy resources that are able to be extracted by natural pressure, pumping, or by injection of water or gas into the well site.

Cracking: The processes in which crude oil is refined by molecular rearrangement.

Crude Oil: The natural (unrefined) state in which petroleum (a mixture of hydrocarbons) is found at the well site.

Curie: A unit of radioactivity. One curie is equal to 3.7×10^{10} radioactive disintegrations per second.

Deceleration: A term used to indicate reduction in speed. Deceleration may also be referred to as negative acceleration.

Decommissioned: To remove from service. All electric generating power plants must eventually be taken out of service or *decommissioned* due to their age.

Degree-day: A unit of measure indicating the severity of weather in a specific geographical location. The unit represents the amount of energy required to heat or cool a structure from a standard (as 65° F). Total degree-days for a given heating or cooling season are compared with records of other seasons to determine the relative severity and concurrent energy requirement for the given season.

Design Heat Loss: A mathematical computation used by building contractors to estimate the maximum hourly energy consumption of a structure. Building contractors used design heat loss calculations to determine a structure's heating and cooling requirements.

Deuterium: An isotope of hydrogen with an atomic weight of 2.0141. Deuterium (also called heavy hydrogen or heavy water) has a hydrogen isotope whose nucleus contains one proton and one neutron.

Development Well: A well drilled in an area where oil or natural gas has been previously located. Development wells are used in the extraction of crude oil or natural gas.

Devonian Shale Deposit: Natural gas found in (and obtained by fracturing) tight clay formations.

Direct Solar Energy: A term used to classify the solar energy reaching the earth. Energy conversion systems that immediately convert the sun's radiation into a usable form of energy are called direct solar energy systems. These systems can be subdivided into thermal and photovoltaic systems.

Discovery: A find of oil or natural gas not associated with a previous oil field.

Dry Gases: Natural gases that do not contain hydrocarbons heavier than butane. Dry gases include methane and ethane.

Dry Hole: An exploratory or development well that yields no oil or natural gas, or one that yields too little of these resources to allow it to be financially worthwhile to extract. Dry holes are sometimes called *dusters*.

Dry Steam Hydrothermal Reservoirs: A type of geothermal reservoir that is economically attractive since the steam produced is free of water particles and is of a higher temperature than other types of geothermal reserves.

Efficiency: A ratio of the output energy to the input energy required to accomplish a particular type of work. Efficiency is also a specification of the operating characteristics of a machine, engine, or energy converter.

Electrolysis: The decomposition of a substance by means of an electric current.

Electrolyte: A solution placed between two dissimilar materials to enable the flow of electricity in batteries.

Electromagnetic Spectrum: The entire range of radiation extending in wavelengths, from 10^{-13} centimeters to 10^4. The electromagnetic spectrum includes the following (in decreasing frequency): cosmic-ray photons, gamma rays, X rays, ultraviolet radiation, visible light, infrared radiation, microwaves, radio waves, heat, and electric currents.

Embargo: A legal withholding of a product (such as petroleum) from one country by another.

Emergency Core-cooling Systems: A series of safety mechanisms used in nuclear power plants. These systems are designed to extract the heat generated within the reactor should a loss of coolant occur.

Encasement: A term used in the decommissioning process of nuclear power plants. Encasement consists of sealing the radioactive part of the power plant to allow the radiation to diminish to safe levels for future disassembly.

Energy: The ability to do work or to exert a force over a distance.

Energy Consumption: The amount of energy that is utilized by a society in the commercial, industrial, residential, and transportation sectors of the economy. Total end-use energy consumption also includes electrical system energy losses in the generation, transmission, and distribution of electricity (including plant and unaccounted-for uses).

Energy Conversion: The transformation of one energy form into another. The direction of this process is always from a lower to a higher level of entropy.

Energy Demand: The amount of energy that is required in all sectors of society. Energy demand is what consumers, industries, and so on, need to maintain a certain standard of production or standard of living. Energy demand may be greater than energy consumption.

Energy Levels: An atom has different energy levels in each of its orbits. The inner orbit has a higher energy level; the outer orbit has a lower energy level.

Energy Management: The processes of reducing energy costs in the industrial sector. Energy management often involves methods to improve operating efficiencies, production operations, maintenance, scheduling, and use of newer technology. Energy management may also be referred to as *waste management*.

Energy Sector: That societal sector that consumes energy in the production of electrical energy.

Energy Storage: The process of placing energy into a state or form whereby it can be held for future use with minimum energy dissipated. The energy stored is called *static energy*.

Energy Utilization: The conversion of energy from one form to another to produce work.

Energy Winds: Winds that blow at a speed of 10 to 25 miles per hour for two to three days per week.

Enhanced Recovery: A method to increase the flow of crude oil from underground reservoirs. Enhanced recovery methods are sometimes called *secondary* or *improved recovery*.

Enrichment: The process of increasing the percentage of ^{235}U in a uranium sample.

Entombment: A term used in the decommissioning process of nuclear power plants. Entombment is the process of using concrete to seal all highly radioactive components inside the reactor building and the structure itself.

Entropy: A measure of the degree of disorder of a particular energy system. Entropy represents a means of quantifying the amount of disorder in an isolated system.

Envelope Design: A building design and construction practice that allows a building to operate as a solar collector. Control and distribution of thermal energy is accomplished from convection currents passing through the walls of the building.

Ethylene: An unsaturated hydrocarbon gas that is produced from ethane in a petrochemical industry; used to produce polyetheylene to make various plastics.

Existing Reserves: See *Proven Reserves*.

Fast Reactor: A nuclear reactor that does not use a moderator to slow down the neutrons that keep the chain reaction going.

Fault: A break in a rock formation that is caused by the shifting of the earth's crust. Faults result in the displacement of the adjacent surfaces of the strata parallel to the plane of the fracture.

Fire Flooding: See *In-situ Combustion*.

Fissile: Atoms that fission with the absorption of neutrons and, in the process, emit more neutrons to keep the fission process going.

Fission: Splitting an atomic nucleus into fragments of comparable mass with a resulting release of energy.

Flared Gas: Natural gas that is burned off as a waste by-product of petroleum production.

Fluidized Bed Combustion: The process of burning coal that has been mixed with limestone while suspended on a bed of heated air. This process helps to reduce both nitrogen oxide and sulfur dioxide emissions.

Fractional Distillation: The process of separating crude oil into its various component parts by their boiling temperatures.

Fractionation Process: A process that separates natural gas into its four major components: methane, ethane, propane, and butane.

Fuel Rods: Uranium dioxide pellets stacked into zirconium-alloy tubes for use in a nuclear reactor.

Fusion: A nuclear reaction in which atomic nuclei combine to form heavier nuclei with a resulting release of energy.

Future Reserves: Energy reserves that have not been located. Future reserves represent scientific estimates of energy reserves that may exist based upon exploration. Future reserves are also referred to as *potential reserves.*

Gamma Decay: The process in which a radioactive nucleus that has stored excess energy emits this energy in a sudden photon burst.

Gamma Rays: Electromagnetic radiation with greater energy than several hundred thousand electron volts.

Gaseous Diffusion: A process of uranium enrichment used to produce nuclear fuel or weapons-grade material.

Gasification: The process of producing the combustible gases of carbon monoxide, hydrogen, and methane by subjecting coal to react with steam and hydrogen in the presence of pressure and heat.

Gathering Lines: Pipes that carry petroleum from a storage tank to truck lines or to other pipelines.

Geothermal Energy: Energy derived or extracted from the internal heat of the earth. Geothermal energy includes residual heat, friction heat, or heat that results from radioactive decay.

Gravimeter: An instrument used to measure variations in the gravitational field of the earth.

Greek Fire: A petroleum-based incendiary weapon developed by Kallinikos in 673 B.C.

Gross Domestic Product (GDP): The total value of services and goods produced by labor and property within a country (as the United States). Like the Gross National Product, the GDP provides an indication of a nation's market value.

Gross National Product (GNP): The Gross National Product is a measure of a nation's market value. This measure is the total value of all goods and services produced by the residents of a nation in a year.

Growth Patterns: Methods of predicting how rapidly the use of certain energy forms will increase in the future.

Growth Rate: A number to show the percentage increase of any commodity over a certain period of time.

Half-life: The time required for half the nuclei in a radioactive substance to undergo radioactive decay.

Heat Gain: Thermal energy that enters a structure.

Heating Value: The thermal energy available in coal. This usually is represented by Btu's per pound (or kilogram) of coal.

Heat Loss: Thermal energy that passes out of a structure.

Heavy Hydrocarbons: A classification used in the refining industry that refers to natural gases with six or more carbon atoms. Hexane is an example of a heavy hydrocarbon.

Heliostats: A type of mirror that is automatically moved to focus the sun's energy onto a receiver. Heliostats are used in high-temperature solar collectors like central receivers.

High-level Waste: Nuclear waste in one or more of the following categories: spent fuel rods from nuclear reactors, radioactive defense waste from the federal government, artificially produced nuclear elements, or radioactive waste from mining and milling. High-level wastes have long half-lives—25,000 years or more.

High-temperature Collectors: See *Parabolic Dishes*.

Horsepower: A measure of work (force × distance) over a duration of time.

Hot-dry Rock: A type of geothermal reservoir in which rock has been heated by radioactive decay. To extract the energy contained in this type of reservoir, hydraulic fracturing is used and water is pumped into the reservoir to extract both hot water and steam.

Hydraulic Fracturing: The process of cracking hot rock and extracting the heat in a geopressured reservoir by pumping cold water down to the heated area. This cold water causes the rock to fracture. Once the rock is fractured, additional cold water is pumped into the area, heated by the rock, and allowed to return to the earth's surface for use.

Hydrocarbons: Organic chemical compounds consisting of only hydrogen and carbon.

Hydrocarbons can exist in gaseous, liquid, or solid phases. Their molecular structure varies from the simplest (i.e., methane) to very heavy and complex structures.

Hydro Energy: The kinetic energy of falling water.

Hydrologic Cycle: The continuous natural process where the sun evaporates water that is carried as a vapor into the atmosphere, then to land in the form of precipitation, then to the oceans by rivers and streams, and back again into the atmosphere as a vapor. The hydrologic cycle is responsible for the earth's weather systems and the development of rivers and streams.

Hydronic: Of, relating to, or being a heating or cooling system that uses water or a mixture of water and antifreeze as its working fluid.

Hydrothermal Reservoirs: A type of geothermal reservoir containing hot water or steam.

Impermeable Rock: Rock that allows no fluid (liquid or gas) to flow through it.

Improved Recovery: See *Enhanced Recovery*.

Incident Solar Radiation: See *Insolation*.

Indirect Solar Energy: A term used to classify the solar energy that has reached the earth and been stored in some form. Energy conversion systems that utilize the stored energy that has come from the sun over time are called indirect solar energy systems. Indirect solar energy sources must be converted to a usable form before they are able to be used.

Induction: The generation of electromotive force in a circuit by varying the magnetic flux through a circuit.

Industrial Sector: The societal sector that consumes energy for operating industry. This energy is used for operating machines, heating processes, and so forth. Manufacturing industries make up the largest part of the industrial sector, which also comprises construction, agriculture, fisheries, and forestry.

Inert: To exhibit no chemical or biological activity or to exhibit activity under special or extreme conditions.

Injection Well: A well used to pump water, gas, or chemicals into an underground oil reservoir of a producing field. Injection wells are used in enhanced recovery methods.

In situ Combustion: An enhanced oil recovery method where heat from a slow burning underground heat front causes the oil to thin out and increase the pressure in the reserve. In situ combustion is also known as *fire flooding*.

Insolation: Solar radiation received per unit area for a given unit of time on the surface of the earth. Insolation, also called incident solar radiation, is commonly measured in Btu/ft²/hr. Insolation received at the earth's surface can range from 0 to about 360 Btu's/ft²/hr.

Interstate Gas: Natural gas that is sold and consumed in a state other than the one in which it was produced.

Intrastate Gas: Natural gas that is sold and consumed in the state in which it was produced.

Isotope: An element that exists in one of two or more forms. Isotopes can be distinguished by small differences in atomic mass and physical and chemical properties but have the same atomic number.

Jack-up Drilling Rig: A self-contained platform used in offshore drilling operations. This type of drilling platform has legs that support the structure by extending down to the ocean floor; the legs can be raised for towing to and from a site.

Kerogen: An organic solid material found in oil shale. This insoluble material can be used to produce oil and natural gas when heated.

Licensing Basis: An evolving set of requirements (established by the United States Nuclear Regulatory Commission) that nuclear power plants must meet to maintain their operation.

Life-cycle Costing: The total cost of energy that a consumer would incur from installing and maintaining an energy conversion system. The life-cycle costs include the costs associated with fuel(s) used in the conversion system. Of all energy systems used, solar energy has the lowest life-cycle costs.

Light Hydrocarbons: A classification used in the refining industry that refers to natural gases with one to six carbon atoms. Methanc, cthane, propane, and butane are examples of light hydrocarbons.

Lignite: A low-grade, brownish-black, soft coal that was formed about 150 million years ago. Lignite is mostly used as an industrial fuel for the generation of electricity.

Liquefaction: Those processes designed to convert coal into synthetic oil or other liquid fuel products.

Liquefied Natural Gas: Natural gas in liquefied form. Liquefied natural gas is primarily methane that has been liquefied by reducing its temperature to $-2600°F$ at atmospheric pressure.

Low-level Waste: Nuclear waste that is **not** one of the following: spent fuel rods from nuclear reactors, radioactive defense waste from the federal government, artificially produced nuclear elements, or radioactive waste from mining and milling. Low-level wastes have relatively short half-lives—50 years or less.

LPG: Liquid petroleum gas. The chemical part of a natural gas sample (as propane or butane) that can easily be liquefied by pressurization.

Magnetometer: An instrument used to measure the intensity and direction of magnetic fields.

Manometer: An instrument used to measure pressure (of gases and vapors) and vacuum.

Meltdown: A severe nuclear reactor accident in which the radioactive fuel becomes overheated and melts the metal-encased fuel rods. A meltdown would likely cause the release of radioactive gases and particulates into the environment.

Millirem: One millirem is equal to 1/1000 rem, abbreviated as mrem. Radiation doses that are often small are measured in this unit.

Moderated: To be slowed down. A moderator is a substance (such as water, heavy water, or graphite) that is used to decrease the speed of fast neutrons in a nuclear reactor. Decreasing the speed of fast neutrons increases the likelihood of nuclear fission.

Monitored Retrievable Storage Facility: A temporary disposal site used to store high-level radioactive waste.

Mothballing: A term used in the decommissioning process of nuclear power plants. Mothballing refers to the practice of placing a plant in protective storage for future decontamination and disassembly.

Motor Octane: A laboratory measure of gasoline antiknock characteristics under severe engine operating conditions.

Multivane: A type of fan wind turbine often used for pumping water on farms.

Natural Radiation: Radiation exposure that a person receives from the sun, outerspace, other natural sources, and exposure to human-made sources. Each year the average person in the United States receives about 360 mrem of radiation exposure. This type of exposure is also called *background radiation*.

Neutral Wall: A type of wall that is highly insulated and allows very little heat gain or heat loss through a section.

Nonassociated Gas: Natural gas found in an underground reservoir that contains a minimum quantity of crude oil or no crude oil.

Nonassociated Gas Well: A well that is used to extract natural gas from an underground reservoir containing very little or no crude oil.

Nonenergy Sector: The societal sector that converts hydrocarbons into various products such as plastic, alcohol, and so on. The nonenergy sector may also be referred to as the *petrochemical sector* and is usually considered part of the industrial sector.

Normal Paraffins: Hydrocarbons found in natural gas. These hydrocarbons may also be called *saturated hydrocarbons* because each carbon atom is linked to a greater number of hydrogen atoms.

Normal Thermal Gradient: The degree of temperature rise as depth increases toward the center of the earth's core. The normal thermal gradient is estimated to be about 100°F for every mile toward the core of the earth.

Nuclear Fuel Cycle: The processes nuclear fuel undergoes before and after it is used within a nuclear power plant. These processes include mining, processing, enrichment, fuel fabrication, fuel reprocessing, and disposal of spent nuclear fuel.

Nucleons: A term used to reference both neutrons and protons.

Ocean Thermal Energy Conversion (OTEC): A process that utilizes the ocean temperature differential of 35°F to 50°F to vaporize and condense liquid refrigerants in order to operate a generator.

Oil Pools: Oil held in the pores of rock until released by movement of the earth, erosion, or oil drilling.

Oil Refining: The process in which crude oil is separated into a variety of useful hydrocarbon products.

Oil Shale: A fine-grained sedimentary rock (marlstone) that contains an organic material called kerogen. Kerogen, when heated to about 850°F, will yield oil and natural gas.

Older Petroleum: Crude oil that has hydrocarbons with low molecular weights. This type of crude oil is said to have lighter hydrocarbons.

Outer Continental Shelf: The shallow submerged portion of a continent that extends to an area of steep descent to the ocean floor.

Overburden: The material lying between the topsoil and the coal seam or stratum.

Oxygenated Fuels: Fuels that have additional oxygen added to them.

Parabolic Dish: A type of direct solar collector that concentrates the sun's parallel rays to a receiver or boiler, which is located at a focal point in front of the dish and is capable of

attaining temperatures up to 4,000°F.

Passive Solar System: A solar energy system that uses natural convection to move the thermal energy collected.

Peat: Partially carbonized vegetable matter that is often found in bogs. The decomposition of plant and animal remains form peat—the first geological step in the formation of coal. Peat has the lowest heating value of all coals.

Penstocks: Large pipes that direct water to a turbine for the generation of hydroelectric power.

Personal Energy Consumption: A term used to describe the amount of energy used by each person in a specific country.

Person-rem: A term used by scientists and health-care providers when measuring the exposure of a population to radiation. The person-rem is the added doses to each of the exposed individuals. For example, 50 persons exposed to 2 rem each would be 50 person-rem.

Petroleum: An inflammable liquid (yellow-to-black in color) found principally beneath the earth's surface. Petroleum can be processed by fractional distillation to yield natural gas, gasoline, naphtha, kerosene, lubricating oils, paraffin, asphalt, and a number of derivative products.

Phase Change: The temperature at which a material changes from a solid to a liquid. Eutectic salts change from a solid to a liquid at about 90°F.

Photoelectric Effect: The extraction of electrons from a substance by sunlight or incident electromagnetic radiation.

Photovoltaic Systems: A solar collector that is capable of producing a voltage when exposed to radiant energy (sunlight).

Pig: A 15,000-pound urethane ball used to clean natural gas pipelines.

Plutonium: An artificially produced radioactive substance produced by neutron bombardment of uranium. ^{239}Pu is a fuel used in nuclear reactors and in nuclear weapons.

Point Focus Collectors: See *Parabolic Dishes.*

Polyethylene: A petrochemical product used to make various types of plastic products.

Polyphase: Current in more than one phase; having or producing more than one phase.

Power: The measure of work done over time or rate of work.

Power Coefficient: A term used to identify the performance characteristics of a wind generator. The maximum efficiency of a wind generator is about 60 percent. The power coefficient is the percentage of power from a wind turbine system that can be extracted from the maximum speed of 60 percent.

Prevalent Winds: Winds that blow at a speed of 5 to 15 miles per hour for three to five days per week.

Price Controls: State or federal mandates to set the price structure of an energy resource. Generally price controls are used to stabilize inflation. However if kept on the commodity too long, controls decrease profits and reduce incentive for developing the product.

Proven Reserves: Estimated quantities of energy reserves that geological and engineering data show (with reasonable certainty) to be recoverable. Proven oil reserves are often termed *existing reserves.*

Rad: A unit of ionizing energy absorbed in or on an object. The rad represents radiation absorbed dose and is equal to radiation that imparts 100 ergs of energy to 1 gram of tissue.

Radiation: The transfer of thermal energy by electromagnetic waves.

Radioactive Decay: The disintegration of the nuclei of atoms. See *Radioactivity.*

Radioactivity: A natural process in which certain atoms give off various particles from their nuclei. Radioactive particles include alpha, beta, and gamma particles.

Reasonably Assured Resources: Energy resources that are considered recoverable with existing facilities, present technology, and at current costs and price levels.

Reformulated Gas (RFG): Gasolines that are blended with oxygen and other compounds to reduce air pollution.

Regenerative System: A mechanical energy storage system that utilizes the rotational forces of a flywheel.

Rem: The amount of ionizing radiation required to produce the same biological effect as one roentgen. The rem is an acronym for *roentgen equivalent man* and is equal to the amount of ionization produced in air by X rays or gamma rays.

Research Octane: A laboratory measure of gasoline antiknock characteristics under mild engine operating conditions of low speed and temperature.

Residential Sector: That societal sector that consumes energy in a dwelling. This energy is used for electrical appliances, heating, air-conditioning, and so on. The residential sector includes all private residences (both occupied and vacant), rented structures, mobile homes, and secondary homes. Structures used for institutional housing (school dormitories, hospitals, and military barracks) are not part of this sector; they are part of the commercial sector.

Retorting: A distillation process used to release oil and natural gas from oil shale.

Road Octane: A laboratory measure that represents gasoline characteristics under road driving conditions.

Rock Oil: A term used to distinguish petroleum from other vegetable oils and animal fats used prior to 1900.

Rotor: The rotating part of an electrical or mechanical device.

Saturated Hydrocarbons: See *Normal Paraffins.*

Scrubber: A device on a coal-fired electric generating power plant that chemically and mechanically processes the coal combustion gases with limestone, thus reducing the acidity of the exhaust.

Secondary Recovery: See *Enhanced Recovery.*

Sectors: Societal groupings that utilize or demand energy.

Seismograph: An instrument used for recording the intensity, direction, and duration of movement of the ground. Seismographs are used in oil exploration to determine various rock formations and their depth.

Semisubmersible Rig: A type of offshore drilling platform, which consists of a self-propelled drilling ship with anchors, cables, and winches to hold it above the site.

Slow Neutrons: Neutrons that have been moderated to allow nuclear fission to occur.

Solar Cells: See *Photovoltaic Systems.*

Solar Collector Efficiency: A measure of the rate of useful heat collected (output) to total solar radiation on the surface of the collector (input). Solar collector efficiency is useful to consumers and designers when comparing different types of collectors.

Solar Constant: The average intensity of incoming solar energy at the outer limits of the earth's atmosphere. The solar constant is equal to 1370 watts per square meter per second (measured on a plane perpendicular to its path).

Solar Orientation: The study of how a structure or a solar collector is positioned in relationship to the movement of the sun. Solar orientation is used to maximize the use of the sun's energy.

Sour Crudes: Crude oils that contain large amounts of sulfur and other mineral impurities.

Speculative Resources: Energy reserves that are believed to exist. With speculative resources, no consideration is given to the costs associated with extraction or availability of the reserves.

Speed: A unit of measure equal to the distance traveled divided by the time used to cover that distance.

Static Energy: Energy not in use, converted, or changed to another form. See *Energy Storage.*

Stator: The stationary part of a motor, turbine, or other working machine around which a rotor turns.

Steam-injected Gas Turbines: Turbines that use the steam from burning natural gas or other fossil fuels to produce electricity. In the production of electricity, these turbines inject some of the steam into the combustion chamber as a means of increasing the operating efficiencies to greater than 50 percent.

Steam Stimulation: A method of enhanced oil recovery by which steam is injected into an oil reservoir to thin out and pressurize the oil formation.

Strategic Petroleum Reserve: An oil reserve established by the federal government of the United States to be used in the event of supply disruptions from oil-exporting countries.

Stratigraphic Trap: A petroleum formation of permeable rock layers containing oil, which gradually taper off under impermeable layers of rock.

Strip- or Surface-mining: The process of mining and extracting coal from an open pit. This type of mining occurs where coal is found close to the surface of the earth. Generally, the soil is stripped away to expose and remove the coal seam.

Stripper Well: An oil well that is considered marginal in output (less than 10 percent of its original output per day). A stripper well can also be a natural gas well that produced less than 22.5 million cubic feet in its last 12 months of operation.

Subbituminous Coal: The coal having the second lowest heating value. This form of coal was formed about 200 million years ago and is used as a fuel for industrial processing, commercial and residential heating, and the generation of electricity.

Sweet Crudes: Crude oils having few sulfur and mineral impurities.

Syngas: A term commonly reserved for gaseous fuels that are obtained from a conversion process rather than by drilling. Synthetic gases can be obtained from coal, shale oils, tar sands, or biomass.

Synoil: A term commonly reserved for liquid fuels that are obtained from a conversion process rather than by drilling. Synthetic oils can be obtained from coal, shale oils, tar sands, biomass, or industrial wastes.

Tailings: The waste products (crushed rock) that remain after milling or mining.

TCF/yr: A unit of measure associated with natural gas; stands for trillion cubic feet per year.

Technically Recoverable Resources: A term used to denote petroleum reserves that should be able to produce crude oil in the future, but may not be economically profitable to extract.

Theory of Relativity: The physical theory of space and time advanced by Albert Einstein. The theory holds that mass and energy are interchangeable—mass can be transformed into energy and energy can be transformed into mass.

Therm: A unit used to measure a quantity of thermal energy from natural gas, typically measured as 100,000 Btu's.

Thermal Gradient: A term used to indicate the temperature increases in areas closer to the center or core of the earth. The degree of temperature increase is termed the normal thermal gradient. See *Normal Thermal Gradient.*

Thermal Pollution: A term used to refer to the discharge of waste heat into the environment.

Thermal Reactor: A nuclear reactor that requires the use of a moderator to slow down the neutrons to keep the chain reaction going.

Thermodynamics: An area of physics that analyzes the characteristics and properties of thermal energy.

Thermography: Infrared photography used to detect heat losses in buildings.

Tip Speed Ratio: The ratio of speed of the tip of a wind turbine blade to the speed of the wind.

Torque: A twisting force, applied to a shaft, that causes rotation.

Town Gas: A type of gaseous fuel produced by coal gasification.

Transmuted: The process by which one element is changed to another element by one or more nuclear reactions.

Transportation Sector: The societal sector that consumes energy for the movement of goods, services, people, or other species. This sector comprises the energy used for the propulsion systems in automobiles, conveyors, ships, and so forth.

Unconventional Resources: Oil or natural gas energy resources that are producible by enhanced oil recovery methods. Unconventional resources include natural gas produced from tar deposits, heavy oil deposits, oil shales, coal and peat formations, tight reservoirs, and gas in pressurized shales and brines.

Undiscovered Resources: A term used to denote petroleum resources found outside known gas and oil fields; these resources have been confirmed by exploratory drilling.

Universal Motor: A motor that can be operated on either AC or DC.

Uranium Oxide: A concentrated form of uranium or yellow cake. See *Yellow Cake.*

Velocity: The ratio of displacement (both distance and direction) to a time interval. Velocity is a vector quantity that describes a distance traveled in a specific direction.

Waste Management: See *Energy Management.*

Wet Gases: Natural gases that contains hydrocarbons heavier than butane. Wet gases have high heating values in Btu's per cubic foot of gas.

Wet Steam Reservoirs: A type of geothermal reservoir that produces steam with water. This type of reservoir requires filters to remove the water before the steam is used in a turbine.

Wildcatter: A term once used to refer to a person who drilled oil wells in areas not known to be oil fields.

Wildcat Well: An exploratory well drilled in a remote area where there has been no previous oil production.

Wind Farms: Clusters of wind turbine generators positioned in an area to produce electrical energy.

Yaw: To deviate from a straight course.

Yellow Cake: Uranium oxide concentrate resulting from the milling of uranium ore. Yellow cake contains 80 to 90 percent U_3O_8.

Young Petroleum: Crude oil that has hydrocarbons with high molecular weights.

Index